Science and Technology of Petroleum

Science and Technology of Petroleum

Edited by Michael Dedini

SYRAWOOD
PUBLISHING HOUSE

New York

Published by Syrawood Publishing House,
750 Third Avenue, 9ᵗʰ Floor,
New York, NY 10017, USA
www.syrawoodpublishinghouse.com

Science and Technology of Petroleum
Edited by Michael Dedini

International Standard Book Number: 978-1-68286-483-8 (Hardback)

Cataloging-in-Publication Data

Science and technology of petroleum / edited by Michael Dedini.
    p. cm.
Includes bibliographical references and index.
ISBN 978-1-68286-483-8
1. Petroleum. 2. Petroleum engineering. 3. Petroleum--Prospecting. 4. Petroleum--Geology. I. Dedini, Michael.
TN870 .S38 2017
665.5--dc23

Printed in the United States of America.

TABLE OF CONTENTS

PREFACE

Petroleum exploration and engineering is the study of methods and techniques that are used in petroleum extraction and refining. The petroleum industry relies on various processes that gather, process, transport, store and pipeline crude oil and petroleum. Liquefied petroleum gas is merely one of the many products of petroleum. Hydrocarbon distillation and purification are the main processes that take place in an oil refinery. The topics covered in this extensive book deal with the core subjects of petroleum engineering. The aim of this book is to present researches that have transformed this discipline and aided its advancement. It presents researches and studies performed by experts across the globe which will provide innovative insights to readers.

This book is a result of research of several months to collate the most relevant data in the field.

When I was approached with the idea of this book and the proposal to edit it, I was overwhelmed. It gave me an opportunity to reach out to all those who share a common interest with me in this field. I had 3 main parameters for editing this text:

1. Accuracy – The data and information provided in this book should be up-to-date and valuable to the readers.

2. Structure – The data must be presented in a structured format for easy understanding and better grasping of the readers.

3. Universal Approach – This book not only targets students but also experts and innovators in the field, thus my aim was to present topics which are of use to all.

Thus, it took me a couple of months to finish the editing of this book.

I would like to make a special mention of my publisher who considered me worthy of this opportunity and also supported me throughout the editing process. I would also like to thank the editing team at the back-end who extended their help whenever required.

Editor

Optimization of CO_2 separation technologies for Chinese refineries based on a fuzzy comprehensive evaluation model

Qian-Qian Song · Qing-Zhe Jiang · Zhao-Zheng Song

Abstract This study aims at determining the optimal CO_2 separation technology for Chinese refineries, based on current available technologies, by the method of comprehensive evaluation. Firstly, according to the characteristics of flue gas from Chinese refineries, three feasible CO_2 separation technologies are selected. These are pressure swing adsorption (PSA), chemical absorption (CA), and membrane absorption (MA). Secondly, an economic assessment of these three techniques is carried out in accordance with cash flow analysis. The results show that these three techniques all have economic feasibility and the PSA technique is the best. Finally, to further optimize the three techniques, a two-level fuzzy comprehensive evaluation model is established, including economic, technological, and environmental factors. Considering all the factors, PSA is optimal for Chinese refineries, followed by CA and MA. Therefore, to reduce Chinese refineries carbon emission, it is suggested that CO_2 should be captured from off-gases using PSA.

Keywords Chinese refineries · CO_2 emission · Separation technique · Economic evaluation · AHP-entropy method · Fuzzy comprehensive evaluation model

Q.-Q. Song · Q.-Z. Jiang (✉) · Z.-Z. Song (✉)
State Key Laboratory of Heavy Oil Processing, China University
of Petroleum, Beijing 102249, China
e-mail: jiangqz@cup.edu.cn

Z.-Z. Song
e-mail: song@cup.edu.cn

Edited by Xiu-Qin Zhu

1 Introduction

It is well accepted that carbon dioxide (CO_2), which is considered as the leading source of greenhouse gas (GHG) emissions, results in the climate change. Direct emissions of CO_2 from industry account for approximately 20 % of global CO_2 emissions (IEA 2010). Globally, the petroleum refining industry is one of the largest contributors to anthropogenic CO_2 emissions (IEA 2009; de Mello et al. 2009; Kuramochi et al. 2012). CO_2 emissions from refineries account for about 4 % of the global CO_2 emissions, close to 1 billion metric tons of CO_2 per year (Van Straelen et al. 2010).

China has surpassed U.S. as the world's largest emitter of CO_2. In 2012, China contributed 26.1 % of total world CO_2 emissions, equal to 8,254 million metric tons (Mt) (EIA 2013). On the other hand, the overall goal of the Chinese government is to reduce the CO_2 emissions per unit of GDP by 40–45 % by 2020 from 2005 levels. Nevertheless, with the ever-increasing enlargement of refining scale, the rising demand for high-quality products and the restricted supply of light and sweet crudes, CO_2 emissions of Chinese refineries have increased significantly. In 2012, Chinese crude oil refining reached 468 Mt. Consequently, it was estimated that CO_2 emissions would be around 133 Mt according to the CO_2 emission factor of 0.284 t/t crude oil (Ma et al. 2011). In this paper, a proper mitigation strategy for the Chinese oil refining sector is badly needed to enable China to appropriately implement its climate change policy with consideration of its socio-economic features.

At present, CO_2 emissions of oil refineries can be reduced in three ways. The first option requires energy conservation. The second option requires energy consumption switching to non-fossil fuels such as hydrogen, nuclear, biomass, and solar energy. The third option involves CO_2 capture and storage (CCS). However, the nature of oil refining implies that even if a refinery is highly energy efficient, it will consume considerable amounts of energy, and therefore produce considerable amounts of CO_2. The first two measures could reduce the current CO_2 emissions by 9–20 Mt/year, yet a 13–80 % reduction of CO_2 emissions could be achieved by implementing carbon capture (Johansson et al. 2012). Thus, to meet mid- to long-term CO_2 reduction, cost-effective CO_2 separation technologies are the key issue for oil refineries.

Most of the reported studies have focused on technical and economic performance of CO_2 capture technology for power generation, and only a very few studies are about CO_2 capture for the petroleum refining industry (IEA-GHG 2000; Van Straelen et al. 2010; de Mello et al. 2009; Kuramochi et al. 2012; Ho et al. 2011). Specifically, no studies have been reported on Chinese refineries.

The objective of this study is to optimize CO_2 separation technologies for Chinese refineries. Section 2 of this paper reviews CO_2 separation technologies and initially selects available technologies for Chinese refineries. Section 3 introduces the method of economic evaluation and fuzzy comprehensive evaluation. Section 4 presents the results. Section 5 gives the conclusions. The general frame of this study is shown in Fig. 1.

2 Preliminary selections of CO_2 separation technologies

Currently, the technologies for separating CO_2 from a flue gas include absorption, adsorption, use of membranes, cryogenics, and chemical-looping combustion (CLC) (Figueroa et al. 2008; Yang et al. 2008; Olajire 2010; Mondal et al. 2012). Choosing a suitable technology

depends on the characteristics of the flue gas stream and its separation requirements.

CO_2 may be emitted from a variety of sources at a refinery, even over 20 sources of emissions for a complex refinery. The main CO_2 emission sources include furnaces, boilers, catalyst regeneration, and hydrogen manufacturing. Figure 2 shows the CO_2 emissions of a typical Chinese refinery with 12 Mt/year of crude processing capacity (Jiang et al. 2013). The CO_2 concentration of flue gases changes in a wide range. For example, it is 10–20 vol% for catalyst regeneration in fluid catalytic cracking (FCC) (de Mello et al. 2009), and 50–55 vol% for hydrogen manufacturing units (Reddy and Vyas 2009). Table 1 shows the composition of some CO_2 emission sources for the typical Chinese refineries. Table 2 provides a summary of the pros and cons of CO_2 separation techniques. Based on the characteristics of the flue gases, three available CO_2 separation technologies are initially selected for Chinese refineries, as follows:

2.1 Pressure swing adsorption (PSA)

The PSA for CO_2 separation is based on the preferential or selective adsorption of CO_2 on a solid adsorbent at a relatively high pressure by contacting hot gas with solid adsorbent in a packed column. The adsorbed component (CO_2) is then desorbed from the solid by lowering the gas-phase partial pressures inside the column so that the adsorbent can be re-used. From the point of view of technology, PSA can capture CO_2 of almost all concentrations from flue gases. But some studies have proved that at the present commercial development state, PSA is not a suitable technology for bulk capture of CO_2 from postcombustion power plant flue gas (Reiner et al. 1994; Liu et al. 2012). However, due to relatively low energy consumption and easy operation, PSA is a promising technology

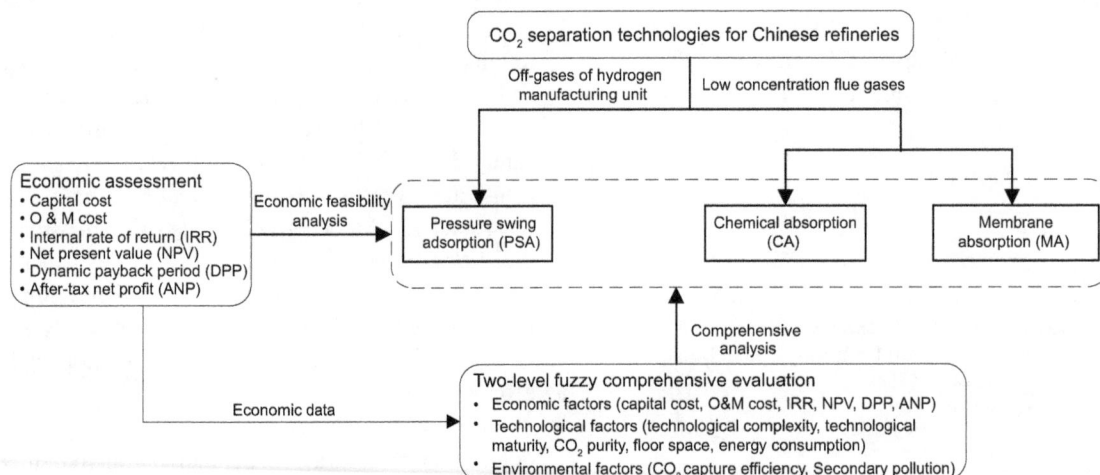

Fig. 1 The general frame diagram of this study

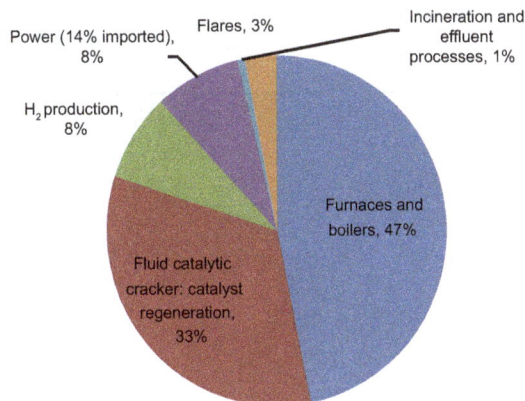

Fig. 2 CO_2 emissions of a Chinese refinery (Jiang et al. 2013)

Table 1 Compositions of some CO_2 emission sources for the typical Chinese refineries

Compositions, vol%	FCC	Hydrogen manufacturing
CO_2	16.6	50
N_2	82	–
O_2	0.8	–
H_2	–	30
CO	0.6	–
SO_x, ppmv	20	–
Other combustible gas (e.g., methane, light alkanes & alkenes)	–	20

for capturing CO_2 of high concentration (20–80 vol%) (Ho et al. 2008; Martunus et al. 2012). In oil refineries, most hydrogen manufacturing units use a PSA to recover H_2 from the syngas. PSA can only recover 75–90 % of the overall H_2 in the syngas. PSA off-gas containing CO_2, unrecovered H_2, and combustible gas is usually burned as fuel gas (shown in Table 1). The direct combustion of off-gases would lead to not only energy waste, but also the emission of large quantities of CO_2. Therefore, to maximize overall production and profitability, oil refineries should recover the CO_2 and remaining valuable H_2 in PSA off gas. Reddy and Vyas (2009) have investigated the recovery of CO_2 and H_2 from PSA off gas using the CO_2LDSep^{SM} plant. Beijing Yanshan Petrochemical Company (China) has constructed a PSA unit recovering hydrogen manufacturing off-gases, with which 0.2 Mt food-grade CO_2, 64.8 Mm^3 combustible gases and 36 Mm^3 H_2 can be produced annually. In the study, some relevant data of the PSA is determined based on this industrial application.

2.2 Chemical absorption (CA)

The chemical absorption process, using amine solutions, is a commercialized technology and has been used in natural gas

industry for 60 years. It is regarded as a mature process. The most commonly used solvent is monoethanolamine (MEA). Many studies have been reported about CO_2 capture from an existing coal-fired power plant using chemical absorption method based on MEA (Singh et al. 2003; Alie et al. 2005; Aroonwilas and Veawab 2007; Rubin et al. 2007; Abu-Zahra et al. 2007). They all have indicated that CO_2 capture using MEA is the only feasible option in the short term for flue gases with low concentration CO_2. Therefore, in this study, random packed columns and ordinary carbon steel are used to act as the absorber and stripper with anticorrosive fluororubbers and stainless steel internals. Furthermore, 20 wt% MEA is selected to absorb CO_2, and the CO_2 loadings of lean and rich solution are, respectively, set to 0.20 mol-CO_2/mol-MEA and 0.45 mol-CO_2/mol-MEA (Alie et al. 2005). An absorption temperature of 40 °C is used and the average pressure is around 0.01 MPa. The regeneration temperature is 110 °C and pressure is 0.025 MPa. Therefore, based on these operating conditions, the size of absorber and stripper needed can be determined using the model provided by Abu-Zahra et al. (2007).

2.3 Membrane absorption (MA)

MA, that is using a membrane in conjunction with chemical absorption, is considered as a suitable alternative to chemical absorption by many researchers, due to its unique advantages including large interfacial area, good device-modularity, good operational flexibility, and high mass transfer coefficient (Rangwala 1996; Falk-Pedersen and Dannström 1997; Feron and Jansen 2002; deMontigny et al. 2005; Yeon et al. 2005; Yan et al. 2007, 2008, 2011). In addition, membrane CO_2 absorption technology can also solve the operating problems, successfully including entrainment, flooding, and foaming. It means that solvent losses can be reduced substantially, thereby reducing the absorbent makeup cost. However, the barriers in coal-fired power plants are membrane wetting and plugging which degrade CO_2 separation performance. In order to maintain constant CO_2 removal efficiency, membrane modules need to be replaced again and again, and hence the cost of captured CO_2 will be increased (Mavroudi et al. 2003; Wang et al. 2005; Yan et al. 2008, 2011). Because there are no particles in the flue gas of refineries, membrane plugging will not occur. So, in the assessment of MA in this study, only the membrane wetting problem is considered. With higher specific surface area than packings, hydrophobic hollow fiber membrane contactor is used as the permeable barrier between gas and liquid phase, and the CO_2 can be captured by solvents (e.g., MEA) (Li and Chen 2005). Therefore, hydrophobic polypropylene membranes are selected owing to their relatively low price and commercial availability, with an assumption that the working life of the membranes is 5 years. In addition, the only difference between MA technique and CA technique is the

Table 2 Pros and cons of CO_2 separation technologies

Separation technologies	Applicable situations	Pros	Cons
Chemical absorption (CA)	Lower or medium CO_2 pressure	High selectivity	Large loss of absorbent
		Wide application	High energy consumption of regeneration
		Reproducible solvent	
		Quick reaction velocity	Strong corrosiveness
Membrane absorption (MA)	Lower CO_2 pressure	Large contact area	Porous membrane easily wetted or blocked
		No problems of bubbling, flooding and entrainment	Under research
		Better permeability and selectivity	
		Low energy consumption	
		Good modularity	
Pressure swing adsorption (PSA)	Higher CO_2 pressure	Easy process	Low recovery
		Low energy consumption	Big investment
		No pollution and corrosion	Large floor area
		High adaptability and product purity	
Physical absorption	Higher CO_2 and total pressure	Low energy consumption and corrosiveness	Low selectivity and separation efficiency
		Easy regeneration and lower dosage of adsorbent	Higher pressure
Cryogenics	High CO_2 pressure (concentration, ~ 60 vol%)	Higher separation efficiency and purity	High cost and energy consumption
Membrane technology	Higher CO_2 pressure	Simple device	High selectivity but not high permeability
		No pollution	Not high CO_2 purity
		Low energy consumption	Not resistant to high temperature
		High separation efficiency	Not easily cleaned
Chemical-looping combustion (CLC)	Under research	No NO_x	No large commercial application
		Low energy consumption	
		Low operation cost	

type of absorber. Others are designed as same as CA technique. Because of the constant liquid–gas contact area, the total membrane contact area can be easily scaled-up linearly based on some successful projects or even the experimental results. According to literature (Yeon et al. 2005; Yan et al. 2007, 2008, 2011), the total membrane contact area could be calculated.

3 Methods

3.1 Method for economic evaluation

Many studies on economic evaluation of CO_2 capture from industrial sources have been reported (Abu-Zahra et al. 2007; de Mello et al. 2009; Yan et al. 2011; Meerman et al. 2012). For Chinese refineries, the economic assessments are subject to two generally accepted guidelines: the Economic Assessment Method and Parameters for Capital Construction Projects (NDRC and MOC 2006), the

Economic Assessment Method and Parameters for Oil Construction Projects (MOHURD 2010). Although the economic indicators are not the same due to geographic differences, the methodology is similar.

In the study, the economic assessment using cash flow analysis are made up of the following six key economic categories: capital cost, operating and maintenance (O & M) cost, internal rate of return (IRR), net present value (NPV), dynamic payback period (DPP), and after-tax net profit (ANP). The cost model developed in the present study is based on the China-specific guidelines, and the methods for calculating the costs of each category and the data are drawn from the two guidelines (Table 3).

3.2 Method for fuzzy comprehensive evaluation

The evaluation of the optimal CO_2 separation technology for Chinese refineries is influenced by a number of factors, such as economic benefits, environmental policy, technological level, and social impact. However, because of the

Table 3 Cost estimation of CO$_2$ separation technologies

1	Capital cost	Formulas
1.1	Engineering	$C_1 + C_2 + C_3$
1.2	Miscellaneous	$1.1 \times C_4$
1.3	Budgetary reserves	$(1.1 + 1.2) \times C_5$
1.4	Annual interest	$(1.1 + 1.2 + 1.3)/C_6 \times (1 - C_7)/C_6 \times C_8$
1.5	Liquidity	
	Capital cost	$1.1 + 1.2 + 1.3 + 1.4 \times C_6 + 1.5$
2	**Operating and maintenance (O & M) cost**	**Formulas**
2.1	Materials	
2.2	Fuels & power	
2.3	Wage & Welfare	
2.4	Depreciation	Straight-line depreciation
2.5	Maintenance	
2.6	Miscellaneous	
2.7	Administration	
2.8	Financial	Straight-line amortization
2.9	Sales	
	Annual O & M cost	$\sum 2.i$ (where 2.i is 2.1–2.9)
3	**Annual cash flow**	**Formulas**
3.1	Income	
3.2	Cash inflow	3.1
3.3	Construction investment	$(1.1 + 1.2 + 1.3)/C_6$
3.4	Operating cost	$2.1 + 2.2 + 2.3 + 2.5 + 2.6 + 2.9$
3.5	Tax and extra charges	
3.6	Income tax	When $(3.1 - 3.7 - 3.6) > 0$, $(3.1 - 3.7 - 3.6) \times C_9$
3.7	Cash outflow	$3.3 + 3.4 + 3.5 + 3.6$
3.8	Net cash flow	$3.2 - 3.7$

C_1–C_9 are the parameters extracted from the China-specific Guidelines stated above. C_1 is the equipment procurement cost (10^4 Yuan); C_2 is the installation cost (10^4 Yuan); C_3 is the construction cost (10^4 Yuan); C_4 is the miscellaneous fixed asset cost in the proportion of engineering cost, 0.12; C_5 is the budgetary reserves ratio, 0.1; C_6 is the construction period (years); C_7 is the owner's capital–capital cost ratio; C_8 is the loan interest rate; and C_9 is the income tax rate

complexity and uncertainty involved in evaluation, a decision maker must consider these factors in a comprehensive evaluation to avoid one-sidedness. Fuzzy comprehensive evaluation is the process of evaluating an objective utilizing fuzzy set theory. When evaluating an objective, multiple related factors must be considered comprehensively in order to give an appropriate, non-contradicting, and logically consistent judgment (Jorge et al. 2000; Chen et al. 2002; Liang et al. 2006).

In this study, a two-level fuzzy comprehensive evaluation is used as follows.

- Assuming that the objective being evaluated contains n factors, i.e., the factor set is $U = \{U_1, U_2, U_3, \ldots, U_n\}$
- First-level fuzzy evaluation is carried out for each factor $U_k(k = 1, 2, \ldots, n)$ $A_k = (a_{k1}, a_{k2}, \ldots, a_{kp})$ is the fuzzy weight vector of each sub-factor in U_k, where a_{kl} is the relative importance of factor l, and $\sum_{l=1}^{p} a_{kl} = 1$.

The fuzzy appraisal matrix of all p sub-factors:

$$R_k = \begin{bmatrix} r_{k11} & r_{k12} & \cdots & r_{k1m} \\ r_{k21} & r_{k22} & \cdots & r_{k2m} \\ \cdots & \cdots & \cdots & \cdots \\ r_{kp1} & r_{kp2} & \cdots & r_{kpm} \end{bmatrix} \quad (1)$$

where r_{kij} is referred to the fuzzy membership degree of appraisal of factor i. In the study, based on the type of factor i, r_{kij} is defined as (Xie et al. 2012):
For benefit factors:

$$r_{ij} = \frac{x_{ij} - \min x_{ij}}{\max x_{ij} - \min x_{ij}} \quad (2)$$

For cost factors:

$$r_{ij} = \frac{\max x_{ij} - x_{ij}}{\max x_{ij} - \min x_{ij}}, \quad (3)$$

where x_{ij} is the eigenvalue of factor i, max x_{ij} and min x_{ij} is, respectively, the maximum, the minimum.
The first-level appraisal result, B_k, can be obtained:

$$B_k = (b_{k1}, b_{k2}, \ldots, b_{km}) = A_k \times R_k \quad (4)$$

- Second-level fuzzy evaluation is implemented for U According to the aforementioned results, the comprehensive appraisal result, B, is calculated as:

$$B = (b_1, b_2, \ldots, b_m) = A \times R$$
$$= (a_1, a_2, \ldots, a_n) \times \begin{bmatrix} b_{11} & b_{12} & \cdots & b_{1m} \\ b_{21} & b_{22} & \cdots & b_{2m} \\ \cdots & \cdots & \cdots & \cdots \\ b_{n1} & b_{n2} & \cdots & b_{nm} \end{bmatrix}$$
(5)

- The priority of evaluated objects can be obtained, and the optimal separation technology for Chinese refineries can finally be determined based on different scenarios.

4 Results

4.1 Economic evaluation

Three CO_2 separation technologies without considering CO_2 compression and transportation can recover 0.1 Mt/year liquid CO_2. Faced with the restricted downstream demand, CO_2 is usually consumed nearby. Hence, it is assumed for the boundary of the economic evaluation model that around oil refineries there are enough market demands and CO_2 can be completely sold as a commodity. Additionally, Chinese government has established 7 CO_2 pilot emission trading markets in Beijing, Shanghai, Tianjin, Shenzhen etc., and will establish a China national CO_2 emission trading market in the 13th Five-Year Plan. Therefore, some mandatory measures, such as CO_2 emissions taxation, must be taken in the near future in China for refineries as large emission sources. Considering the carbon emissions taxation, CO_2 capture can obtain additional economic benefit. Due to many uncertainties in the pilot carbon market, the CO_2 price may fluctuate greatly, for example, the lowest CO_2 price is 28 Yuan/t and the highest CO_2 price can reach 140 Yuan/t in Shenzhen carbon market. So, in the economic evaluation in this work, a CO_2 price of 140 Yuan/t is used. However, it is worth noting that with the PSA technique, 18 Mm^3/year of H_2 can be recovered simultaneously, and can be sold at a price of 2.0 Yuan/m^3. The common basis for the three techniques is shown in Table 4.

The cost of major equipment for the PSA technique is estimated in terms of investment per unit capacity. Based on the PSA unit of Beijing Yanshan Petrochemical Company (China), the major equipment investment for the PSA technique is estimated to be 243.6 Yuan/t, accounting for

Table 4 Assumptions for economic evaluation with PSA, MA and CA techniques

Items	Value
CO_2 capture efficiency, %	90
CO_2 purity, %	>99
Project working life, years	15
Construction period, years	1
Plant operating time, h/year	8000
Discount rate, %	12
CO_2 sale price, Yuan/t	260
CO_2 transaction price, Yuan/t	140
Owner's capital/capital cost	0.3
Loan interest rate, %	6.55
Income tax rate, %	25

Table 5 Economic evaluation results for the three techniques

Items	CA	MA	PSA
Capital cost, 10^4 Yuan	3453.3	3613.9	3671.3
O & M cost, 10^4 Yuan/year	3407.1	3397.0	3348.3
IRR, %	17.55	17.75	86.05

approximately 72 % of total construction investment. For the CA and MA techniques, the cost of the absorber and stripper is firstly calculated from the above designs of CA and MA. Then, other costs are estimated in terms of certain proportions. The economic evaluation results for the three CO_2 separation techniques are shown in Table 5.

Table 5 shows that the capital cost of the PSA technique is the highest, but its other economic indicators are optimal. The reason is that it can simultaneously recover CO_2 and by-product H_2. The capital cost of the MA technique is higher than that of CA. That is because the membrane price is high (about 13 Yuan/m^2), and the membrane needs to be replaced about every 5 years. However, the O & M cost of CA is higher than that of MA, because of lower CO_2 loading capacity, higher solvent losses, and regeneration heat consumption, meaning that using membrane gas absorption process can save the O & M cost in capture of CO_2.

4.2 Fuzzy comprehensive evaluation

4.2.1 Establishment of the factor set

A factor set consists of various factors affecting the evaluation objective. The evaluation objective of this study is to optimize CO_2 separation technologies. Combining experts' knowledge and experience with the actual production condition of Chinese refineries, the factor set

Table 6 Factors for CO_2 separation technologies

Items	Types	Attributes
Economy (C_1)		
IRR (I_1)	Benefit	Quantitative
NPV (I_2)	Benefit	Quantitative
DPP (I_3)	Cost	Quantitative
Capital cost (I_4)	Cost	Quantitative
O&M cost (I_5)	Cost	Quantitative
ANP (I_6)	Benefit	Quantitative
Technology (C_2)		
Technological complexity (I_7)	Cost	Qualitative
Technological maturity (I_8)	Benefit	Qualitative
CO_2 purity (I_9)	Benefit	Quantitative
Floor space (I_{10})	Cost	Qualitative
Energy consumption (I_{11})	Cost	Qualitative
Environment (C_3)		
CO_2 capture efficiency (I_{12})	Benefit	Quantitative
Secondary pollution (I_{13})	Cost	Qualitative

Table 7 Results of the weight vector by AHP

C	C_1 0.4000	C_2 0.2000	C_3 0.4000	W_i'
I_1	0.4149	0	0	0.1660
I_2	0.2592	0	0	0.1037
I_3	0.0501	0	0	0.0200
I_4	0.0848	0	0	0.0339
I_5	0.1609	0	0	0.0644
I_6	0.0300	0	0	0.0120
I_7	0	0.0732	0	0.0146
I_8	0	0.4269	0	0.0854
I_9	0	0.1824	0	0.0365
I_{10}	0	0.0413	0	0.0083
I_{11}	0	0.2762	0	0.0552
I_{12}	0	0	0.5000	0.2000
I_{13}	0	0	0.5000	0.2000

C is the optimization of CO_2 separation technologies for Chinese refineries; W_i' is the weight of the factor set

Table 8 Results of the weight vector by entropy method

Items		E_i	d_i	W_i''
Benefit				
I_1	IRR	0.0181	0.9819	0.1212
I_2	NPV	0.0337	0.9663	0.1192
I_5	ANP	0.0022	0.9978	0.1231
I_8	Technological maturity	0.5119	0.4881	0.0602
I_9	CO_2 purity	0.6219	0.3781	0.0467
I_{12}	CO_2 capture efficiency	0.6072	0.3928	0.0485
Cost				
I_3	DPP	0.0960	0.9040	0.1116
I_4	Capital cost	0.4659	0.5341	0.0659
I_6	O & M cost	0.3793	0.6207	0.0766
I_7	Technological complexity	0.5119	0.4881	0.0602
I_{10}	Floor space	0.4555	0.5445	0.0672
I_{11}	Energy consumption	0.5794	0.4206	0.0519
I_{13}	Secondary pollution	0.6126	0.3874	0.0478

E_i is the entropy; d_i is the diversity; W_i'' is the entropy weight

involving economic, technological, and environmental factors has been established (Table 6).

4.2.2 Determination of the weight vector

There are dozens of methods to determine a weight, including Delphi (Ma et al. 2003), AHP (Saaty and Shang 2011; Jiang et al. 2012) and entropy (Lai et al. 2012; Khan and Bhuiyan 2014) methods. Each method has its background, significance, merits, shortcomings, and specific areas of application. Combination of different methods can offset the shortcomings of each, integrate their various advantages, and result in an advanced evaluation method (Refat et al. 2011). In this study, the weight of factors is determined by a combination of subjective (AHP) and objective (entropy) methods, that is, the AHP-entropy method takes into account the data and the subjective preferences of decision-makers to achieve a synthesis of subjective and objective and to make the results more realistic and reliable.

4.2.2.1 Determination of the weight vector by AHP method
AHP, originally developed by Saaty (1977, 1990, 1994), is a mathematical method for analyzing complex decision problems with multiple criteria and through combination of quantitative analysis with qualitative ones. In this paper, the model is based on the combination of qualitative opinions from experts and references and quantitative analysis from aforementioned economic evaluation, and the results are listed in Table 7.

4.2.2.2 Determination of the weight vector by entropy method
Entropy, first proposed by Shannon, is a concept originating from information theory (Chen et al. 2012). It can calculate the amount of information in the message, which consists of a countable length of character combination. By computing the total probabilities of all the characters in the message, the information entropy of the message is obtained. Generally, the smaller the entropy of a certain factor is acquired, the more weight the factor is given, and vice versa. Table 8 presents the results of the weight vector by the entropy method.

Table 9 Results of the weight vector by AHP-entropy method

Items	AHP	Entropy	AHP-entropy
Economy (C_1)			
I_1	0.4149	0.1962	0.4392
I_2	0.2592	0.1931	0.2700
I_3	0.0501	0.1806	0.0488
I_4	0.0848	0.1067	0.0488
I_5	0.1609	0.1994	0.1731
I_6	0.0300	0.1240	0.0201
Technology (C_2)			
I_7	0.0732	0.2105	0.0791
I_8	0.4269	0.2105	0.4614
I_9	0.1824	0.1630	0.1526
I_{10}	0.0413	0.2348	0.0498
I_{11}	0.2762	0.1813	0.2571
Environment (C_3)			
I_{12}	0.5000	0.5035	0.5035
I_{13}	0.5000	0.4965	0.4965

4.2.2.3 Determination of the weight vector by AHP-entropy method   The combined weights are determined by Eq. (6), and the results are shown in Table 9.

$$a_i = \frac{\left(\omega_i' \times \omega_i''\right)}{\left(\sum_{i=1}^{m} \omega_i' \times \omega''\right)} \qquad (6)$$

4.2.3 Determination of the fuzzy appraisal matrix

Based on Eqs. (1), (2), and (3), the fuzzy appraisal matrix for the economic, technological, and environmental factors is obtained as follows:

$$R_{C1} = \begin{bmatrix} 0.0000 & 0.0029 & 1.0000 \\ 0.0000 & 0.0061 & 1.0000 \\ 0.0000 & 0.0224 & 1.0000 \\ 1.0000 & 0.2633 & 0.0000 \\ 0.0003 & 0.0000 & 1.0000 \\ 0.0000 & 0.1718 & 1.0000 \end{bmatrix},$$

$$R_{C2} = \begin{bmatrix} 1.0000 & 0.0000 & 0.3333 \\ 1.0000 & 0.0000 & 0.3333 \\ 1.0000 & 0.0000 & 0.7538 \\ 0.0000 & 1.0000 & 0.2500 \\ 0.5000 & 1.0000 & 0.0000 \end{bmatrix},$$

$$R_{C3} = \begin{bmatrix} 1.0000 & 0.6296 & 0.0000 \\ 0.0000 & 0.6667 & 1.0000 \end{bmatrix}$$

4.2.4 Results and analysis

(1)   Results of first-level fuzzy comprehensive evaluation
According to Eq. (4), the following results of the single-level fuzzy evaluation are obtained:

- For the economic factors, $B_1 = [0.0489\ \ 0.0203\ \ 0.9512]$. The order of the economic benefit of the three techniques is PSA>CA>MA.
- For the technological factors, $B_2 = [0.8217\ \ 0.3069\ \ 0.3076]$. The order of the technological level is CA>PSA>MA.
- For the environmental factors, $B_3 = [0.5035\ \ 0.6480\ \ 0.4965]$. The order of the environmental benefit is MA>CA>PSA.

(2)   Results of two-level fuzzy comprehensive evaluation
The B_1, B_2, and B_3 constitute the fuzzy appraisal matrix of second-level factor, that is,

$$R_C = \begin{bmatrix} 0.0489 & 0.0203 & 0.9512 \\ 0.8217 & 0.3069 & 0.3076 \\ 0.5035 & 0.6480 & 0.4965 \end{bmatrix}.$$

The economic, technological, and environmental factors all belong to benefit attribute, and based on the AHP-entropy method, the weigh vector is $A = [0.3247\ \ 0.3664\ \ 0.3089]$. The overall fuzzy comprehensive evaluation is $B = A \times R_C = [0.3942\ \ 0.3220\ \ 0.6349]$. The order of the three separation technologies is PSA>CA>MA.

5 Conclusions and implications

Chinese refineries should prefer PSA, due to its unique advantage of recovering both CO_2 and H_2. With the increasing amounts of heavy and high-sulfur crude oils to be processed, demand for H_2 in Chinese refineries is rising. The recovery of hydrogen from manufacturing off-gases with PSA can not only reduce total emissions from refineries, but also expand their production and profitability. On the other hand, MA, if the obstacle of the membrane wetting is overcome, will be the most promising alternative to chemical absorption to capture CO_2 from oil refinery flue gas in the future. In addition, the methodology presented might be beneficial to decision-making for CO_2 capture projects in oil refineries in other countries. The result is important for evaluating the deployment of CCS in Chinese refining sector.

Utilization of technological options for separation and/or capture of CO_2 from flue gases will make Chinese refining industry realize their target of emission reduction in the future. However, without mandatory cuts, incentives, and a major technological break-through, the high costs impose restrictions on the implementation of carbon capture at Chinese refineries. There are a lot of uncertainties about which technologies could lead to real improvements and which have no real prospects for reducing the cost of capture. On the other hand, the use of the captured CO_2 as a secondary product appears to be attractive. The utilization includes producing chemical substances through the application of organic chemistry, biofuels, etc., or injecting pure captured CO_2 underground for enhanced oil recovery (EOR), enhanced gas recovery (EGR), and enhanced coalbed methane (ECBM), although this is not mentioned in the paper.

Acknowledgments The authors gratefully acknowledge the general support from the China University of Petroleum Foundation, and the Research Institute of Safety and Environment Technology, China National Petroleum Corporation.

References

Abu-Zahra MRM, Niederer JPM, Feron PHM, et al. CO_2 capture from power plants: part II. A parametric study of the economical performance based on mono-ethanolamine. Int J Greenh Gas Control. 2007;1(2):135–47.

Alie C, Backham L, Croiset E, et al. Simulation of CO_2 capture using MEA scrubbing: a flowsheet decomposition method. Energy Convers Manag. 2005;46(3):475–87.

Aroonwilas A, Veawab A. Integration of CO_2 capture unit using single- and blended-amines into supercritical coal-fired power plants: implications for emission and energy management. Int J Greenh Gas Control. 2007;1(2):143–50.

Chen JH, Sheng DR, Li W, et al. A model of multi-objective comprehensive evaluation for power plant projects. Proc CSEE. 2002;22(12):152–5 (in Chinese).

Chen XL, Wang RM, Cao YF, et al. A novel evaluation method based on entropy for image segmentation. Procedia Eng. 2012;29:3959–65.

de Mello LF, Pimenta RDM, Moure GT, et al. A technical and economical evaluation of CO_2 capture from FCC units. Energy Procedia. 2009;1(1):117–24.

deMontigny D, Tontiwachwuthikul P, Chakma A. Comparing the absorption performance of packed columns and membrane contactors. Ind Eng Chem Res. 2005;44(15):5726–32.

Falk-Pedersen O, Dannström H. Separation of carbon dioxide from offshore gas turbine exhaust. Energy Convers Manag. 1997;38(S):81–9.

Feron PHM, Jansen AE. CO_2 separation with polyolefin membrane contactors and dedicated absorption liquids: performances and prospects. Sep Purif Technol. 2002;27(3):231–42.

Figueroa JD, Fout T, Plasynski S, et al. Advances in CO_2 capture technology-The U.S. Department of Energy's Carbon Sequestration Program. Int J Greenh Gas Control. 2008;2(1):9–20.

Ho MT, Allinson GW, Wiley DE. Comparison of MEA capture cost for low CO_2 emissions sources in Australia. Int J Greenh Gas Control. 2011;5(1):49–60.

Ho MT, Allinson GW, Wiley DE. Reducing the cost of CO_2 capture from flue gases using pressure swing adsorption. Ind Eng Chem Res. 2008;47(14):4883–90.

IEA Greenhouse Gas R&D Programme (IEA GHG). CO_2 abatement in oil refineries: fired heaters. Report no. IEA/CON/99/61. Cheltenham: U.K. 2000.

IEA. Energy technology transitions for industry—strategies for the next industrial revolution. Paris. 2009.

International Energy Agency (IEA). Energy Technology Perspective. Paris. 2010.

Jiang QZ, Ma JK, Chen GS, et al. Estimation and analysis of carbon dioxide emissions in refineries. Mod Chem Ind. 2013;33(4):1–6 (in Chinese).

Jiang QZ, Xu YM, Xin WJ, et al. SWOT-AHP hybrid model for vehicle lubricants from CNPCLC, China. Pet Sci. 2012;9(4):558–64.

Johansson D, Rootzén J, Berntsson T, et al. Assessment of strategies for CO_2 abatement in the European petroleum refining industry. Energy. 2012;42(1):375–86.

Jorge H, Antunes CH, Martins AG. A multiple objective decision support model for the selection of remote load control strategies. IEEE Trans Power Syst. 2000;15(2):865–72.

Khan JF, Bhuiyan SM. Weighted entropy for segmentation evaluation. Opt Laser Technol. 2014;57:236–42.

Kuramochi T, Ramírez A, Turkenburg W, et al. Comparative assessment of CO_2 capture technologies for carbon-intensive industrial processes. Prog Energy Combust Sci. 2012;38(1):87–112.

Lai WK, Khan IM, Poh GS. Weighted entropy-based measure for image segmentation. Procedia Eng. 2012;41:1261–7.

Li JL, Chen BH. Review of CO_2 absorption using chemical solvents in hollow fiber membrane contactors. Sep Purif Technol. 2005;41(2):109–22.

Liang ZH, Yang K, Sun YW, et al. Decision support for choice optimal power generation projects: fuzzy comprehensive evaluation model based on the electricity market. Energy Policy. 2006;34(17):3359–64.

Liu Z, Wang L, Kong XM, et al. Onsite CO_2 capture from flue gas by an adsorption process in a coal-fired power plant. Ind Eng Chem Res. 2012;51(21):7355–63.

Ma JK, Jiang QZ, Song ZZ, et al. Construction of refinery carbon industry chain in low carbon economy perspective. Mod Chem Ind. 2011;31(6):1–6 (in Chinese).

Ma YT, Wang ZG, Yang Z, et al. Fuzzy comprehensive method for gas turbine evaluation. Proc CSEE. 2003;23(9):218–20 (in Chinese).

Martunus, Helwani Z, Wiheeb AD, et al. In situ carbon dioxide capture and fixation from a hot flue gas. Int J Greenh Gas Control. 2012;6:179–88.

Mavroudi M, Kaldis SP, Sakellaropoulos GP. Reduction of CO_2 emission by a membrane contacting process. Fuel. 2003;82(15–17):2153–9.

Meerman JC, Hamborg ES, van Keulen T, et al. Techno-economic assessment of CO_2 capture at steam methane reforming facilities using commercially available technology. Int J Greenh Gas Control. 2012;9:160–71.

Ministry of Housing and Urban-Rural Development (MOHURD). The economic assessment method and parameters for oil construction projects. 1st ed. Beijing: China Planning Press; 2010 (in Chinese).

Mondal MK, Balsora HK, Varshney P. Progress and trends in CO_2 capture/separation technologies: a review. Energy. 2012;46(1): 431–41.

National Development and Reform Commission (NDRC) and Ministry of Construction (MOC). The economic assessment method and parameters for capital construction projects. 3rd ed. Beijing: China Planning Press; 2006 (in Chinese).

Olajire AA. CO_2 capture and separation technologies for end-of-pipe applications-a review. Energy. 2010;35(6):2610–28.

Rangwala HA. Absorption of carbon dioxide into aqueous solutions using hollow fiber membrane contactors. J Membr Sci. 1996;112(2):229–40.

Reddy S, Vyas S. Recovery of carbon dioxide and hydrogen from PSA tail gas. Energy Procedia. 2009;1(1):149–54.

Refat AG, Muhammad H, Shesha J. Transformer insulation risk assessment under smart grid environment due to enhanced aging effects. Electrical Insulation Conference. Annapolis MD. 5–8 June 2011; 276–279.

Reiner P, Audus H, Smith AR. Carbon dioxide capture from fossil fuel power plants, Report SR2, IEA Greenhouse Gas R&D Programme. Cheltenham. 1994.

Rubin ES, Yeh S, Antes M, et al. Use of experience curves to estimate the future cost of power plants with CO_2 capture. Int J Greenh Gas Control. 2007;1(2):188–97.

Saaty TL, Shang JS. An innovative orders-of-magnitude approach to AHP-based multi-criteria decision making: Prioritizing divergent intangible humane acts. Eur J Oper Res. 2011;214(3):703–15.

Saaty TL. A scaling method for priorities in hierarchical structures. J Math Psychol. 1977;15(3):234–81.

Saaty TL. Highlights and critical points in the theory and application of the analytic hierarchy process. Eur J Oper Res. 1994;74(3):426–47.

Saaty TL. How to make a decision: the analytic hierarchy process. Eur J Oper Res. 1990;48(1):9–26.

Singh D, Croiset E, Douglas PL, et al. Techno-economic study of CO_2 capture from an existing coal fired power plant: MEA scrubbing vs. O_2/CO_2 recycle combustion. Energy Convers Manag. 2003;44(19): 3073–91.

U.S. Energy Information Administration (EIA). International Energy Outlook 2013. Washington D.C. 2013.

Van Straelen J, Geuzebroek F, Goodchild N, et al. CO_2 capture for refineries: a practical approach. Int J Greenh Gas Control. 2010;4(2):316–20.

Wang R, Zhang HY, Feron PHM, et al. Influence of membrane wetting on CO_2 capture in microporous hollow fiber membrane contactors. Sep Purif Technol. 2005;46(1–2):33–40.

Xie CS, Dong DP, Hua SP, et al. Safety evaluation of smart grid based on AHP-entropy method. Syst Eng Procedia. 2012;4:203–9.

Yan SP, Fang MX, Wang Z, et al. Economic analysis of CO_2 separation from coal-fired flue gas by chemical absorption and membrane absorption technologies in China. Energy Procedia. 2011;4:1878–85.

Yan SP, Fang MX, Zhang WF, et al. Comparative analysis of CO_2 separation from flue gas by membrane gas absorption technology and chemical absorption technology in China. Energy Convers Manag. 2008;49(11):3188–97.

Yan SP, Fang MX, Zhang WF, et al. Experimental study on the separation of CO_2 from flue gas using hollow fiber membrane contactors without wetting. Fuel Process Technol. 2007;88(5):501–11.

Yang HQ, Xu ZH, Fan MH, et al. Progress in carbon dioxide separation and capture: a review. J Environ Sci. 2008;20(1):14–27.

Yeon SH, Lee KS, Sea B, et al. Application of pilot-scale membrane contactor hybrid system for removal of carbon dioxide from flue gas. J Membr Sci. 2005;257(1–2):156–60.

2

Experimental study and a proposed new approach for thermodynamic modeling of wax precipitation in crude oil using a PC-SAFT model

Taraneh Jafari Behbahani[1]

Abstract A powerful method is necessary for thermodynamic modeling of wax phase behavior in crude oils, such as the perturbed-chain statistical associating fluid theory (PC-SAFT). In this work, a new approach based on the wax appearance temperature of crude oil was proposed to estimate PC-SAFT parameters in thermodynamic modeling of wax precipitation from crude oil. The proposed approach was verified using experimental data obtained in this work and also with those reported in the literature. In order to compare the performance of the PC-SAFT model with previous models, the wax precipitation experimental data were correlated using previous models such as the solid solution model and multi-solid phase model. The results showed that the PC-SAFT model can correlate more accurately the wax precipitation experimental data of crude oil than the previous models, with an absolute average deviation less than 0.4 %. Also, a series of dynamic experiments were carried out to determine the rheological behavior of waxy crude oil in the absence and presence of a flow improver such as ethylene–vinyl acetate copolymer. It was found that the apparent viscosity of waxy crude oil decreased with increasing shear rate. Also, the results showed that the performance of flow improver was dependent on its molecular weight.

Keywords Wax precipitation · PC-SAFT model · Solid solution model · Multi-solid phase model · Crude oil

✉ Taraneh Jafari Behbahani
  jafarit@ripi.ir

[1] Research Institute of Petroleum Industry (RIPI),
   P.O.Box 14665-1998, Tehran, Iran

Edited by Xiu-Qin Zhu

1 Introduction

Crude oil is a complex mixture of hydrocarbons, consisting of waxes, asphaltenes, resins, aromatics, and naphthenics. Wax precipitation has a substantial effect on oil production and transportation. Wax precipitation can result many problems, such as decreased production rates of crude oil, increased power requirements, and failure of facilities. Wax is the high molecular weight paraffin fraction of crude oil and can be separated by reduction in oil temperature below the pour point of crude oil. The solubility of high molecular weight waxes in crude oil decreases with decreasing temperature. In the transportation of waxy crude oil in a cold environment at temperatures below the oil pour point, the temperature gradient of the oil creates a concentration gradient of the dissolved waxes due to their difference in solubility. The driving force, created by the concentration gradient, transfers the waxes from the oil toward the pipe wall where they precipitate and form a solid phase. The solid phase reduces the available area for oil flow, and in turn causes a drop in the pipe flow capacity. In order to predict the wax precipitation conditions, a reliable thermodynamic model is necessary. Several thermodynamic models for estimation of wax precipitation have been reported in the literature (Barker and Henderson 1969; Chen et al. 2009). A literature review (Dalirsefat and Feyzi 2007) indicates that the models for wax precipitation have been developed by two different approaches. The first important approach for modeling of wax precipitation uses a cubic equation of state (EOS) for vapor–liquid equilibrium and an activity coefficient model for solid–liquid equilibrium. These models are based on solid solution (SS) theory which assumes that all the components in the solid phase are miscible in all proportions (Won 1968, 1989; Hansen et al. 1988; Pedersen et al. 1991; Zuo et al. 2001; Ji et al. 2004). Chen et al. (2009)

proposed new correlations for the melting points and solid–solid transition temperatures of treated paraffin based on the experimental results from differential scanning calorimetry (DSC). They first estimated the required thermodynamic properties of pure n-paraffin and then a new approach based on the UNIQUAC equation was described. Finally, the impact of pressure on wax phase equilibrium was studied. The second approach based on a multi-solid (MS) phase model uses only an EOS for all phases in equilibrium. In fact, an EOS is used directly for vapor–liquid equilibrium, and solid phase is described indirectly from the EOS by fugacity ratio. The multi-solid (MS) phase model assumes that each pure or pseudo-component precipitate constitutes a separate solid phase which is not miscible with other solid phases (Lira-Galeana et al. 1996; Pan and Firoozabadi 1997; Nichita et al. 2001; Escobar-Remolina 2006; Dalirsefat and Feyzi 2007). The thermodynamic models are developed based on complex properties, such as interaction coefficient, critical properties, acentric factor, solubility parameter, and molecular weight. They are not specified for long chain waxes in crude oil. In order to develop a thermodynamic model for modeling wax phase behavior in crude oils, a powerful method is necessary. The perturbed-chain statistical associating fluid theoretical (PC-SAFT) equation of state is especially useful for modeling phase behavior of complex structure, such as wax in crude oil. Recently, the asphaltene precipitation in Iranian crude oils was investigated by using the PC-SAFT equation of state (Sedghi and Goual 2014; Punnapala and Vargas 2013; María 2014; Panuganti et al. 2012; Leekumjorn and Krejbjerg 2013; Jafari Behbahani et al. 2011a, b, c, 2012, 2013a, b, c, 2014a, b, c, 2015). The effect of wax inhibitors on pour point and rheological properties of Iranian waxy crude oils were investigated (Jafari Ansaroudi et al. 2013; Jafari Behbahani 2008, 2014a, b). In this work, a PC-SAFT model has been proposed to predict the wax appearance temperature and the amount of precipitated wax. A new approach for estimation of PC-SAFT parameters of wax precipitation in crude oil has been proposed in this work. In order to compare the performance of the PC-SAFT model, the wax precipitation experimental data obtained in this work and reported in the literature were correlated using previous models such as solid solution and multi-solid phase models. Also, the pour points and viscosity of the studied crude oil were measured at different concentrations of flow improvers.

2 Materials and methods

2.1 Material

A waxy crude oil was used for investigation of rheological behavior in the absence/presence of flow improver such as ethylene–vinyl acetate copolymer. The physical characteristic of crude oil is shown in Tables 1 and 2.

Two types of polymers with different properties were selected as flow improver (denoted as EVA#1 and EVA#2). The characteristics of the flow improvers are provided in Table 3.

2.2 Experimental apparatus

Apparent viscosity as a function of temperature was measured with a Haake RV12 concentric cylinder viscometer equipped with double gap geometry. Molecular weights of polymers were determined by Waters gel permeation chromatography (GPC) in a Shimadzu LC10AD system equipped with a refractive index detector and ultrastyragel columns of 106, 105, 104, and 500 A connected in series. Tetrahydrofuran was used as the mobile phase, at a flow rate of 1 mL/min. Composition of the polymer was measured by an elemental analyzer (PerkinElmer 2400). This elemental analyzer is a proven instrument for rapid determination of the carbon, hydrogen, nitrogen, sulfur, or oxygen content in organic and other types of materials. It has the capability of handling a wide variety of sample types in the fields of pharmaceuticals, polymers, chemicals, environment, and energy, including solids, liquids, volatile, and viscous samples.

Based on the classical Pregl-Dumas method, samples are combusted in a pure oxygen environment, with the

Table 1 The characterization of studied waxy crude oil

Specifications	Test method	Value
Specific gravity @15.56/15.56 °C	ASTM D-4052	0.8240
API	ASTM D-287	40.22
Sulfur content, wt%	ASTM D-2622	0.1
H_2S content, ppm	FIP	1.0
Base sediment and water, vol%	ASTM D-96	1.0
Salt content, P.T.B.	ASTM D-3230	1
Kinematic viscosity @10 °C cSt	ASTM D-445	0
Pour point, °C	ASTM D-97	21
Reid vapor pressure, psi	ASTM D-323	3.5
Asphaltenes, wt%	IP-143	0.3
Wax content, wt%	BP-327	15.7
Drop melting point of wax, °C	IP-31	60
Carbon residue, wt%	IP-13	4.0
Ash content, wt%	ASTM D-482	0.05
Acidity, total mg-KOH/g	ASTM D-664	0.01
Nickel, ppm	ASTM D-5708	9.1
Vanadium, ppm	ASTM D-5708	<3
Iron, ppm	ASTM D-5708	9.9
Lead, ppm	ASTM D-5708	1.1
Sodium, ppm	ASTM D-5708	193.0

Table 2 The composition and molecular weight of the studied crude oil

Composition	Mole, %	Molecular weight
C_1	0.005	
C_2	0.132	
C_3	2.124	
C_4	1.002	
C_5	3.008	
C_6	3.256	
C_7	2.958	93.8
C_8	8.236	105.0
C_9	13.247	121.0
C_{10}	10.214	135.0
C_{11}	8.215	147.0
C_{12}	6.258	161.0
C_{13}	4.879	175.0
C_{14}	3.147	189.0
C_{15}	3.492	202.0
C_{16}	2.956	215.0
C_{17}	2.149	231.0
C_{18}	2.136	245.0
C_{19}	2.458	258.0
C_{20+}	20.128	412.0

Table 3 Characteristics of used polymers

Issue	Test method	EVA#1	EVA#2
N content	ASTM D-5291	<0.5	<0.5
C content	ASTM D-5291	88.5	83.5
H content	ASTM D-5291	12.1	13.1
Molecular weight	GPC	816,896	725,981

resultant combustion gases measured in an automated fashion. Wax and asphaltene contents were determined according to BP-327, IP-143, respectively. Pour points were measured by ASTM D-97 method.

2.3 Experimental procedure

An appropriate quantity of flow improvers were dissolved in cyclohexane (improver to cyclohexane molar ratio of 1:2), and added to crude oil and then heated in a thermostatic bath maintained at 50 °C. The viscosity and shear stress of the studied crude oil was measured at different shear rates in the range of 10–100 s^{-1}. The rheological data cover the temperature range of 0–27 °C. Also, the pour points and viscosity of the studied crude oil were measured at different concentrations of flow improvers.

3 Theoretical section

The existing thermodynamic models (solid solution (SS) theory and multi-solid (MS) phase model) are based on complex properties, such as interaction coefficient, critical properties, acentric factor, solubility parameter, and molecular weight which are not specified for long chain waxes. A reliable model is necessary to predict the wax precipitation condition in crude oil. In this work, in order to model wax precipitation, the PC-SAFT EOS was used for prediction of the wax appearance temperature and the mass of precipitated wax. Also, a new approach for investigating the PC-SAFT EOS parameters of wax in the crude oil was proposed in this work. In order to examine the performance of the PC-SAFT model, the wax precipitation experimental data reported in the literature were correlated using the above-mentioned models such as the Ji model (Ji et al. 2004) from the solid solution (SS) category and the Lira-Galeana model (Lira-Galeana et al. 1996) from the multi-solid (MS) phase category.

3.1 PC-SAFT model

To calculate the equilibrium between oil and wax phases, it is obviously necessary to use an accurate thermodynamic model. The PC-SAFT equation of state properly predicts the phase behavior of mixtures containing long chain fluids similar to the large wax molecules. The thermodynamic phase behavior of fluid mixtures can be described by perturbation theory. In this approach, the properties of a fluid are obtained by expanding about the same properties of a reference fluid. The statistical associating fluid theory (SAFT) equation of state was developed by Chapman et al. (1989) by applying and extending Wertheim's first-order perturbation theory (Wertheim 1984, 1986) to chain molecules, and Gonzalez et al. (2007) used the PC-SAFT EOS for asphaltene phase behavior modeling. In this theory, molecules are modeled as chains of bonded spherical segments and the properties of a fluid are obtained by expanding about the same properties of a reference fluid. Gross and Sadowski (2001) proposed the perturbed-chain modification (PC-SAFT) to account for the effects of chain length on the segment dispersion energy, by extending the perturbation theory of Barker and Henderson (1969) to a hard chain reference (Koyuncu et al. 2002). This version of SAFT properly predicts the phase behavior of mixtures containing high molecular weight fluids such as the large wax molecules.

The PC-SAFT model describes the residual Helmholtz free energy (A^{res}) of a mixture of nonassociating fluid as follows:

$$\frac{A^{\text{res}}}{RT} = \frac{A^{\text{seg}}}{RT} + \frac{A^{\text{chain}}}{RT} = m \times \left(\frac{A_0^{\text{hs}}}{RT} + \frac{A_0^{\text{disp}}}{RT} \right) + \frac{A^{\text{chain}}}{RT}, \quad (1)$$

where A^{seg} is segment contribution to the mixture Helmholtz free energy, A^{chain} is the chain contribution to the mixture Helmholtz free Energy, A_0^{hs} is the hard-sphere contribution to the mixture Helmholtz free energy, and A_0^{disp} is the dispersion contribution to the mixture Helmholtz free energy.

The PC-SAFT equations are described in Appendix 1.

3.1.1 Proposed approach for estimation of the PC-SAFT parameters in modeling of wax precipitation

When describing a component using an equation of state, values of critical temperature, critical pressure, and acentric factor are required. However, it is almost impossible to measure directly critical properties for longer-chain wax due to thermal decomposition at high temperatures. As a result, different correlations have been suggested in the literature for estimation of critical properties and acentric factors for these components. However, estimated values using these correlations can differ considerably, potentially having a negative effect on the reliability of wax precipitation prediction. In this work, a new approach based on the wax appearance temperature of crude oil was proposed to estimate PC-SAFT parameters in thermodynamic modeling of wax precipitation from crude oil as follows:

(1) Three pseudo-components describe the liquid phase: wax, aromatics/resins, and asphaltenes. The characterization of this phase is conducted based on the liquid fluid compositional information and SARA (waxes, aromatics/resins, and asphaltene) analysis.

(2) The wax PC-SAFT EOS parameters in the crude oil are tuned to meet the wax appearance temperature of the crude oil. In order to characterize wax cut, as shown in Fig. 1, saturates were divided into two sections: one precipitated and another not precipitated, and the perturbed-chain form of the statistical associating fluid theory (PC-SAFT) parameters were tuned to match the wax appearance temperature of crude oil.

(3) The PC-SAFT EOS parameters for asphaltene are fitted to precipitation onset measurements based on ambient titrations and/or depressurization measurements.

(4) The PC-SAFT parameters for aromatics/resins pseudo-components are calculated from their average molecular weight. The aromatics/resins pseudo-component is linearly weighted by the aromaticity parameter between poly-nuclear-aromatic (PNA)

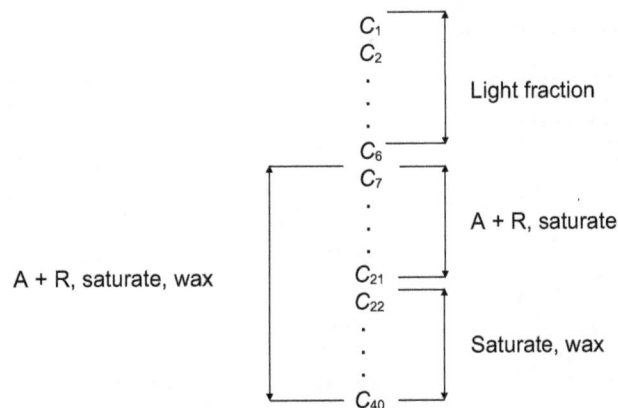

Fig. 1 The proposed characterization of wax cut in crude oil (A is asphaltenes and R is resins)

and benzene-derivative components, characterized by following equations:

$$m = 0.0139\text{MW} + 1.2988 \quad (2)$$

$$\frac{\varepsilon}{k} = 119.4 \ln(\text{MW}) - 230.21 \quad (3)$$

$$\sigma = (0.0597\text{MW} + 4.2015) \times 10^{-10}/m. \quad (4)$$

(5) The wax phase behavior modeling procedure for an oil starts with the definition of four pseudo-components that represent the gas phase: nitrogen (N_2), carbon dioxide (CO_2), methane (CH_4), and light pseudo-components (hydrocarbons C_2 and heavier). The average molecular weight of the light pseudo-component is used to estimate the PC-SAFT EOS parameters. Gross and Sadowski (2001) identified the three pure component parameters required for nonassociating molecules of n-alkanes by correlating their vapor pressures and liquid volumes.

$$m = 0.0253\text{MW} + 0.9263 \quad (5)$$

$$\sigma = (0.1037\text{MW} + 2.7985) \times 10^{-10}/m \quad (6)$$

$$\frac{\varepsilon}{k} = 32.8 \ln(\text{MW}) + 80.398, \quad (7)$$

where MW is the molecular weight, m is the average of pure species segment number, σ is the segment diameter, and ε is the depth of the pair potential.

One advantage of the SAFT-based equations of state is their ability to predict chain length of pure component parameters. In this model, the parameters for the pseudo-components in oil can be determined on the basis of their average molar mass correlations for aromatic and n-alkane fractions.

3.2 Multi-solid phase model

In multi-solid (MS) wax models developed by Lira-Galeana et al. (1996), each solid phase is considered as a pure component which does not mix with other solid phases and can exist as a pure solid (solid assumption). The number and the identity of precipitating components are obtained from Michelsen's phase stability analysis (Michelsen 1982) which states that component i may exist as a pure solid:

$$f_i(P,T,z) - f^s_{\text{pure},i}(P,T) \geq 0.0 \quad i = 1,2,\ldots,n, \tag{8}$$

where $f_i(P,T,z)$ is the fugacity of component i with feed composition z and SSS is the fugacity of pure component i in solid phase. This model is based on the precipitation of certain heavy components of crude with average properties and performs calculations for the liquid/multi-solid phase. The criterion of vapor–liquid–solid equilibrium is that the fugacities for every component i, must satisfy the following equations:

$$f^V_i = f^L_i = f^s_{\text{pure},i}(P,T) \quad i = 1,2,\ldots,n \tag{9}$$

$$f^V_i = f^L_i \quad i = 1,2,\ldots,n, \tag{10}$$

where f is the fugacity, N is the total number of components, and $f^s_{\text{pure},i}$ is the fugacity of the pure solid phase. The fugacities of each component in the vapor and liquid phases are calculated by the equation of state. This model is described in Appendix 2.

3.3 Solid solution (SS) theory

These models are based on solid solution (SS) theory which assumes that all the components in the solid phase are miscible in all proportions. In solid solution theory, the solid phase is a solid solution, stable or not, of all the components that crystallize. To solve the equilibrium problems, an equation of state (EOS) plus activity coefficient are used. Produced reservoir hydrocarbon fluids under pipeline conditions commonly consist of liquid and vapor phases. This model uses a cubic EOS for vapor–liquid equilibrium and an activity coefficient model for solid–liquid equilibrium. In the solid solution model, the selection of suitable activity coefficient models for description of solid and liquid phase behavior is the main part of the modeling. This model is described in Appendix 3.

3.4 Model validation

To verify the performance of the proposed model for calculating wax precipitation from crude oil, numerical simulation runs were conducted for experiments performed in this work and given in the literature (Pauly et al. 2004;

Pedersen et al. 1991; Chen et al. 2009). In this work, solutions (Tables 6, 10) (Chen et al. 2009) and mixtures 1–5 (Fig. 5) (Pauly et al. 2004) were used to verify the performance of the proposed model for calculating wax precipitation from crude oil. Wax precipitation experimental data in crude oil were correlated by adjusting the parameters of the PC-SAFT model to achieve the best match with the experimental data using MATLAB software. For optimization and determination of the model parameters [i.e., the segment diameter (σ), depth of pair potential (ε/k), and the average of the pure species segment number (m)], history matching was used. In this study, the square root of the sum of differences between measured and calculated wax appearance temperature data was defined as the objective function:

$$\text{Objective Function} = \sqrt{\sum_{i=1}^{n} \left((T)_{\text{measured}} - (T)_{\text{calculated}} \right)^2}. \tag{11}$$

As a result, the model parameters obtained by optimization are σ, ε/k, and m. A genetic optimization algorithm was used in combination with the numerical scheme to obtain parameters of the PC-SAFT model (σ, ε/k, and m) by fitting the experimental data. The performance of the proposed model for correlating the wax precipitation of the experimental data reported in the literature was compared to those obtained using the Ji model (Ji et al. 2004) from solid solution (SS) category and Lira-Galeana model (Lira-Galeana et al. 1996) from multi-solid (MS) phase category. The absolute deviation of the correlated WAT values obtained for the studied models from their experimental values was calculated by the following relation:

$$AD(\%) = \frac{|T_{\text{EXP}} - T_{\text{Cal}}|}{T_{\text{EXP}}} \times 100. \tag{12}$$

The amount of the precipitated wax for one mole of oil feed is calculated as follows:

$$\text{Precipitated wax weight \%} = \frac{\text{Weight of precipitated wax}}{\text{Weight of oil feed}} \times 100. \tag{13}$$

Also, the aromaticity parameter (γ) determines the aromatics/resins trend to behave as PNA or benzene derivatives. To quantify the degree of aromaticity, γ will take a value between one, for PNA, and zero, for benzene derivatives. The aromaticity parameter is tuned for the fluid to meet the experimental values of stock-tank-oil (STO) density, STO refractive index, and bubble point for crude oil (Jafari Behbahani et al. 2011a, 2012, 2013a). The aromaticity value for this crude oil was determined to be 0.98 which indicates a close behavior to PNA components.

4 Results and discussion

The main objective of this section is to investigate the performance of the PC-SAFT model for calculating WAT and the weight percent of wax precipitation using a proposed approach to estimation the PC-SAFT parameters based on the wax appearance temperature of crude oil. Also, the performance of the PC-SAFT model for calculating WAT and the weight percent of wax precipitation has been compared with those obtained using the Ji model from the solid solution (SS) category and the Lira-Galeana model from the multi-solid (MS) phase category for the experimental data given in this work and given in the literature. A series of experiments was carried out to investigate the rheological behavior of crude oil using the waxy crude oil sample in the absence/presence of flow improver such as ethylene–vinyl acetate copolymer. The rheological data of the studied crude oil are shown in Fig. 2 at temperature range of 0–27 °C and shear rate range of 10–100 s^{-1}.

Figure 2 shows that with increasing shear rate, the apparent viscosity decreases dramatically. Results show that the shear rate has a considerable effect on viscosity particularly at temperatures below the pour point. At high temperatures above the pour point, the waxy crude oil behaves like a typical homogeneous isotropic liquid with Newtonian characteristics. At temperatures below the pour point, the amount of dissolved wax starts to reach its saturation limit, forming a solid solution in the crude, which leads to a sharp increase of viscosity. Under these circumstances, the viscosity is influenced by two parameters: the effect of temperature reduction that causes viscosity increase against the shear rate that tends to lower it. Further cooling causes formation of a gel network, leading to a progressive rise in viscosity at relatively small dynamic gel strength. On the other hand, the energy exerted by shearing and dissipated in the crude leads to disruption of these bonds and accumulated deformation of the crude gel structure occurs.

Figure 3 shows the influence of the flow improver on the pour points. Results demonstrate a significant reduction in the pour point of the studied crude oil for different concentrations of flow improver, in particular for EVA#1.

It can be found that the higher molecular weight flow improver displays better efficiency on pour point of the studied crude oil. The experimental data for the amount and composition of wax precipitated from treated and untreated oil at different temperatures are shown in Tables 4, 5.

Table 6 shows the values of correlation parameters in the PC-SAFT model for crude oil used in this work and for crude oil taken from the literature (Pedersen et al. 1991) (oil 10, 12, 15). It should be noted that the parameters of saturates, aromatics, and asphaltenes were calculated from crude oil in this work and other parameters taken from the literature (Pedersen et al. 1991) (oil 10, 12, 15).

The data points which are used in the tuning procedure were taken from the literature (Pedersen et al. 1991) (oil 10, 12, 15) and the data points which are predicted by the PC-SAFT model and new approach in this work were taken from the literature (Pauly et al. 2004) (oil 1–5).

Figure 4 shows the performance of the PC-SAFT model and other thermodynamic models in correlating the wax precipitation weight percent in crude oil used in this work.

Tables 7, 8, and 9 show the average absolute deviations (AADs) of wax precipitation from the experimental data of crude oil in this work and given in the literature (Pauly et al. 2004) using the PC-SAFT model and other thermodynamic models.

Figure 5 shows the deviation of the predicted wax precipitation weight percent from their experimental results by using studied models.

It should be noted that the solution 1, solution 2, and mixture 5 are waxy solutions.

Also, Table 10 shows the values of parameters for the MS model.

The results show that PC-SAFT model can predict more accurately than the Ji model from the solid solution (SS) category and the Lira-Galeana model from the multi-solid (MS) phase category, with deviation between 2.3 % and 5.5 %.

Wax molecules contain highly nonspherical and associating molecules. Also, waxes containing long chains are mixtures with large-size asymmetry. In such cases, a more appropriate model is the one that incorporates both the chain length (molecular shape) and molecular association. The PC-SAFT model provides a method for describing the thermodynamics of the complex molecules such as wax, because the PC-SAFT model is based on statistical mechanics and can accurately predict the phase behavior of long chain of wax in crude oil.

The chain term in SAFT is successful in reproducing the equilibrium properties of wax with long chains. Also, the obtained results confirm that the Lira-Galeana model from the multi-solid (MS) phase category is capable of correlating the wax precipitation experimental data with the AADs of 11.5 %–16.8 %, whereas the Ji model from the solid solution (SS) category predicts the wax precipitation experimental data with the AADs of 19.2 %–25.3 %. It should be noted that the Ji model based on solid solution (SS) theory used two types of thermodynamic models to describe the nonideality of liquid phase, which makes this

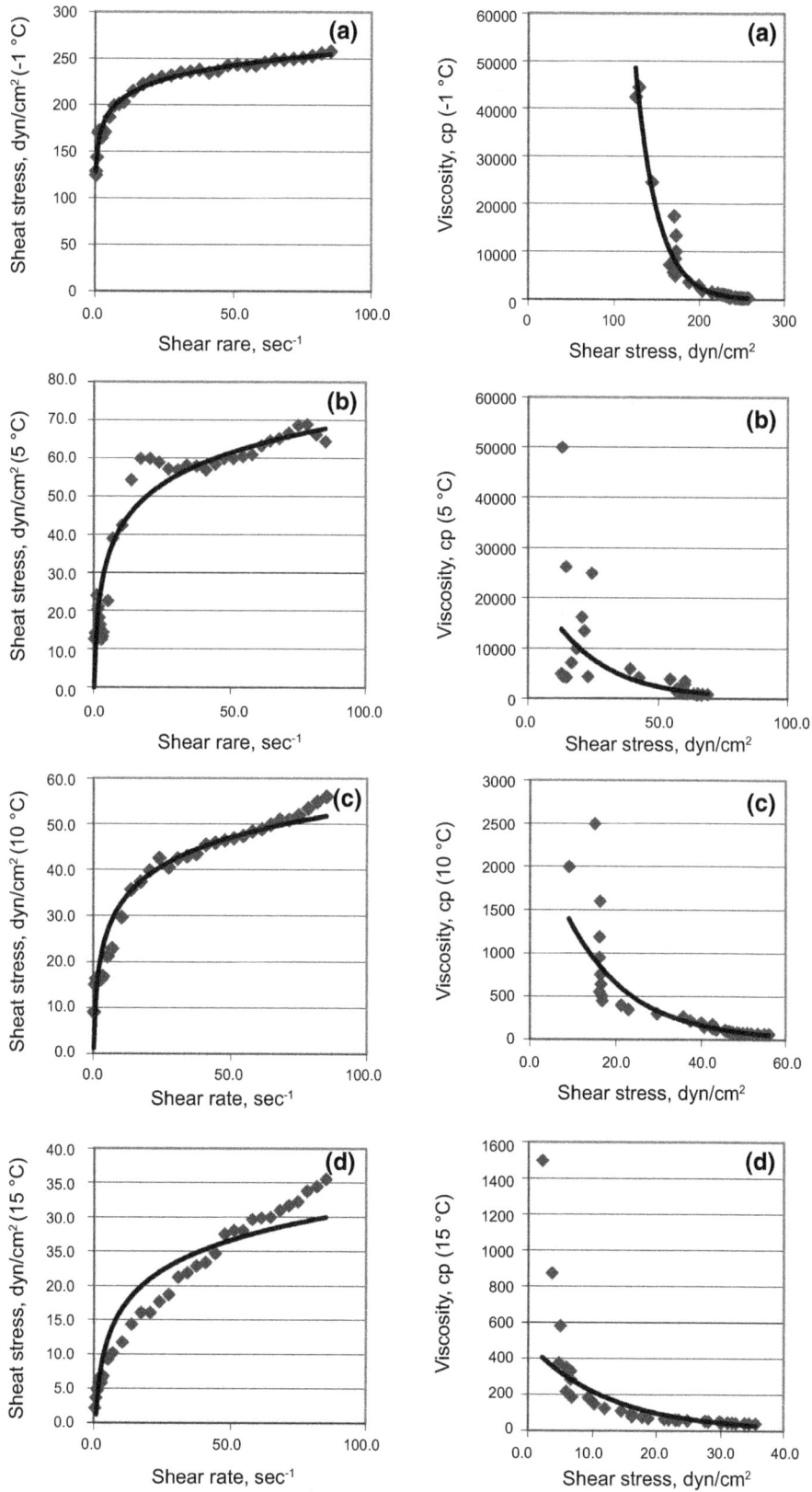

Fig. 2 The rheological behavior of the studied crude oil at different temperatures and shear rates, **a** at −1 °C, **b** 5 °C, **c** 10 °C, **d** 15 °C

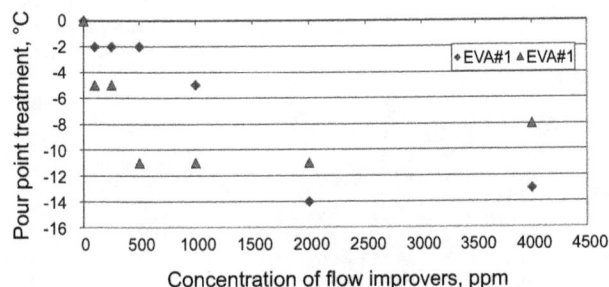

Fig. 3 The pour point behavior of the studied crude oil at different concentrations of flow improvers

model thermodynamically inconsistent. It is observed from the curves that the Lira-Galeana model and Ji model overestimate the amount of precipitated wax. It should be noted that the amount of flow improvers in wax solution is so little that its contribution to the molar composition of the original paraffin solution can be neglected. In other words, the molar composition of solution with and without flow improvers is the same. The effect of flow improvers on the thermodynamic behavior of waxy oil is modeled as a reduction in the WAT and wax precipitation. In this case, the approaches which are used to describe wax precipitation in waxy oil are suitable for the wax-flow improvers system. Also, it can be observed from sensitivity analysis

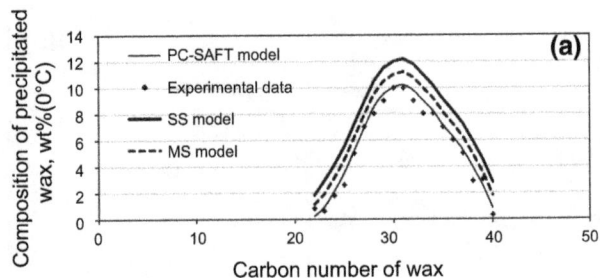

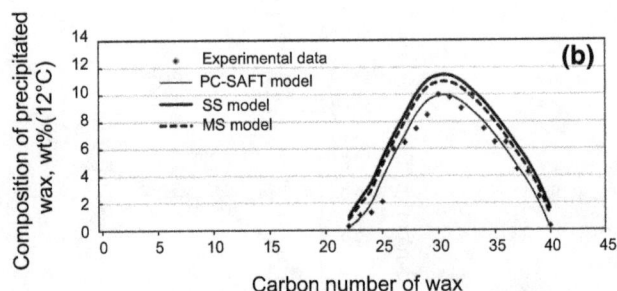

Fig. 4 Comparison of the performance of the PC-SAFT model with other thermodynamic models in correlating the wax precipitation weight percent in crude oil used in this work at 0 °C (**a**) and 12 °C (**b**)

that the model is also sensitive to the techniques for estimation of critical properties (T_c, P_c) and acentric factor pseudo-components.

Table 4 The weight percent of wax precipitated from pure crude oil and crude oil treated with EVA#1

Temperature, °C	Crude oil	Crude oil with different contents of EVA#1					
		100 ppm	250 ppm	500 ppm	1000 ppm	2000 ppm	4000 ppm
27	2.68	2.23	1.85	1.46	1.12	0.89	0.56
22	3.58	2.98	2.38	1.95	1.68	1.58	1.44
17	5.68	4.64	3.24	3.08	2.65	2.37	2.18
12	6.36	5.95	5.38	4.84	4.67	4.05	3.88
0	8.15	7.57	7.52	6.74	6.62	6.36	6.21

Table 5 The weight percent of wax precipitated from crude oil treated with EVA#2

Temperature, °C	100 ppm	250 ppm	500 ppm	1000 ppm	2000 ppm	4000 ppm
27	2.41	2.35	1.59	1.32	1.02	0.85
22	3.14	2.94	2.12	1.94	1.79	1.52
17	5.11	4.87	4.28	2.95	2.54	2.38
12	6.12	5.91	5.11	4.8	4.68	4.21
0	7.87	7.42	7.14	6.81	6.64	6.51

Table 6 The values of approach parameters for PC-SAFT model

SAFT Parameter	Saturates	Aromatics + resins	Asphaltenes	C_1	C_2	C_3	C_4
m	6.314	6.753	29.8	1	3.5206	2.0020	2.3316
σ, °A	3.982	3.754	4.5	3.7039	1.6069	3.6184	3.7086
ε/k, K	256.8	321.7	410	150.03	191.42	208.11	222.88
Molecular weight	213.5	295.4	2300	16	30	44	58

Table 7 Deviation (%) of the correlated wax precipitation weight percent from the experimental results by the PC-SAFT model and the studied thermodynamic models (for crude oil used in this work)

Temperature, °C	PC-SAFT model	SS model	MS model
0	4.2	21.3	14.8
12	2.6	24.3	15.7
17	3.2	22.5	13.2
22	2.3	23.6	16.3
27	4.1	20.8	14.7

Table 8 Deviation (%) of the correlated wax precipitation weight percent from the experimental results by the PC-SAFT model and the studied thermodynamic models (for crude oil treated with EVA#1)

Temperature, °C	PC-SAFT model	SS model	MS model
0	5.1	19.2	13.1
12	4.2	23.1	16.2
17	3.9	24.1	11.6
22	4.6	20.3	15.8
27	3.5	19.8	13.2

Table 9 Deviation (%) of the correlated wax precipitation weight percent from experimental results by the PC-SAFT model and the studied thermodynamic models (crude oil treated with EVA#2)

Temperature, °C	PC-SAFT model	SS model	MS model
0	2.9	23.3	12.4
12	5.2	20.1	14.7
17	4.8	25.3	11.5
22	3.6	22.5	16.8
27	5.1	23.2	12.9

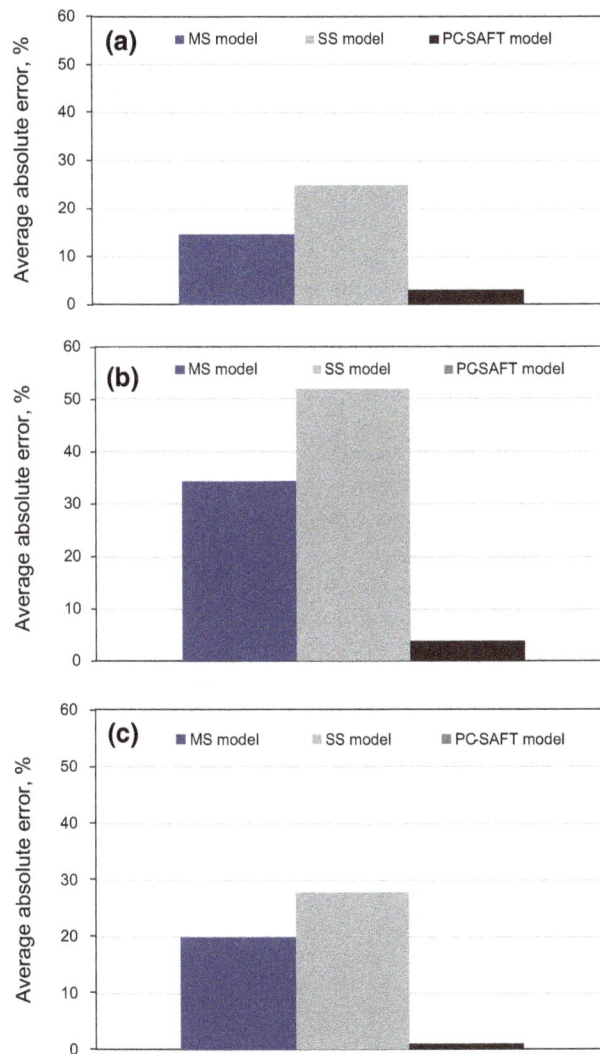

Fig. 5 Deviation of the correlated wax precipitation weight percent from the experimental results of crude oil in this work by the studied thermodynamic models—**a** solution 1, **b** solution 2, **c** mixture 5

Table 10 The values of parameters for MS model

	Molecular weight	T_c, °C	P_c, bar	Ω
Light	69	187	35	0.24
Asphaltene +Resin	162	423	24	0.49
Saturate	210	432	15	0.69
Wax	398	567	8	1.22

T_C critical temperature, P_c critical pressure and ω is acentric factor

5 Conclusions

In this work, the PC-SAFT model was used to estimate precipitated wax weight percent in crude oil and the model was verified using the experimental data obtained in this work and also given in the literature. The performance of the PC-SAFT model in prediction of precipitated wax weight percent was compared with those obtained using the Ji model based on solid solution (SS) theory and Lira-Galeana model from the multi-solid (MS) phase category as follows:

(1) In this work, a new approach based on the wax appearance temperature of crude oil has been proposed to estimate the PC-SAFT parameters in thermodynamic modeling of wax precipitation from crude oil.

(2) It can be concluded that the PC-SAFT model can predict more accurately the experimental data when compared with the Ji model from the solid solution (SS) category and the Lira-Galeana model from the multi-solid (MS) phase category with a deviation of

between 2.3 % and 5.5 %. One advantage of the SAFT-based equations of state is their ability to predict the chain length dependence of the pure component parameters. In this model, the parameters for the pseudo-components in the oil can be determined on the basis of their average molar mass correlations for aromatic and n-alkane.

(3) The obtained results indicate that the Lira-Galeana model from the multi-solid (MS) phase category can predict more accurately the experimental data than the Ji model from the solid solution (SS) category.

(4) It is obvious that by increasing the shear rate, the apparent viscosity decreases dramatically. Results show that the shear rate has a considerable effect on decreasing viscosity particularly at temperatures below the pour point.

(5) Results demonstrate a significant reduction in the pour point of the studied crude oil for different concentrations of flow improver, in particular for EVA#1. It was found that the higher molecular weight flow improver displays a greater effect on the pour point of the studied crude oil.

(6) It is obvious that increasing the shear rate decreases the apparent viscosity dramatically. Results show that the shear rate has a considerable effect on decreasing viscosity particularly at temperatures below the pour point.

Appendix 1

The average segment number of the mixture, m, is an average of the pure species' segment number, m_i, weighted by the species' compositions:

$$m = \sum_i x_i \cdot m_i. \tag{14}$$

The Mansoori–Carnahan–Starling–Leland (1971) equation of state provides the free energy contribution of the hard-sphere mixtures:

$$\frac{A_0^{hs}}{RT} = 6\left[\frac{(\xi_2^3 + 3\xi_1\xi_2\xi_3 - 3\xi_1\xi_2\xi_3^2)}{\xi_3(1-\xi_3)^2}\right. \left. -(\xi_0 - \xi_2^3/\xi_3^2)ln(1-\xi_3)\right]/\pi\rho \tag{15}$$

$$\xi_k = \left(\frac{\pi\rho}{6}\right)\sum_i x_i \cdot m_i \cdot d_{ii}^k \tag{16}$$

$$d_{ii} = \sigma_i\left\{1 - 0.12\exp\left(\frac{\varepsilon}{k}\right)\right\}. \tag{17}$$

The contribution to A^{res} due to chain formation is given by

$$\frac{A^{chain}}{RT} = \sum_i x_i(1 - m_i)\ln g_{hs}(d_{ii}), \tag{18}$$

where $g_{hs}(d_{ii})$ is the hard-sphere pair correlation function at contact given by

$$g_{hs}(d_{ii}) = \frac{1}{1-\xi_3} + 3d_{ii}/2\left(\frac{\xi_2}{1-\xi_3^2}\right) + 2\left(\frac{d_{ii}}{2}\right)^2\left(\xi_2^2/(1-\xi_3)^3\right). \tag{19}$$

The PC-SAFT model incorporates the effects of chain length on the segment dispersion energy. The perturbed-chain dispersion contribution is given by

$$\frac{A_0^{disp}}{RT} = \frac{A_1}{RT} + \frac{A_2}{RT} \tag{20}$$

$$\frac{A_1}{RT} = -2\pi\rho I_1(\eta,m)\sum_i\sum_j x_i \cdot x_j \cdot m_i \cdot m_j \cdot \sigma_{ij}^3\left(\frac{\varepsilon_{ij}}{kT}\right) \tag{21}$$

$$\frac{A_2}{RT} = -\pi\rho I_2(\eta,m) \cdot 1/\sum_i\sum_j x_i \cdot x_j \cdot m_i \cdot m_j \cdot \sigma_{ij}^3\left(\frac{\varepsilon_{ij}}{kT}\right) \tag{22}$$

$$\sigma_{ij} = 0.5(\sigma_i + \sigma_j) \tag{23}$$

$$\varepsilon_{ij} = (1 - k_{ij})\sqrt{\varepsilon_{ij}\varepsilon_{jj}}. \tag{24}$$

I_1 and I_2 are functions of the system packing fraction and average segment number.

Appendix 2

The solid phase fugacities of the pure components, $f_{pure,i}^s$, can be calculated from the fugacity ratio expressed as follows:

$$\ln\left(\frac{f^s}{f^L}\right)Pure, i = \frac{\Delta h_i^t}{RT}\left(\frac{T}{T_i^f} - 1\right)\frac{\Delta C_{Pi}}{R}\left[1 - \frac{T}{T_i^f} + \ln\left(\frac{T_i^f}{T}\right)\right], \tag{25}$$

where T_i^f is the fusion (melting) temperature.

$$\Delta C_{Pi} = C_{pi}^L - C_{pi}^s, \tag{26}$$

where C_{pi}^L and C_{pi}^s are the heat capacity of pure component i at constant pressure corresponding to liquid and solid phases, respectively.

$$\Delta h_i^t = \Delta h_i^f - \Delta h_i^{tr}, \tag{27}$$

where Δh_i^f and Δh_i^{tr} are the enthalpy of fusion and the enthalpy of first solid-state transition, respectively. By using the above equation and an EOS, fugacity in solid and liquid phases, and the numbers of the precipitated solid phases can be calculated. Solid–liquid equilibrium calculations have been performed by using equilibrium and material balance equations. The fugacity coefficient of component i is calculated by an EOS model. Among the EOS models available, the modified PR equation of state is used.

$$P = \frac{RT}{V-b} - \frac{a}{V(V+b)+b(V-b)}. \tag{28}$$

The parameters a and b of a pure component are described by the conventional critical parameters approach. The critical properties and acentric factor required in the evaluation of equation of state parameters are obtained from the Gasem's correlations (Ghanaei et al. 2007). For mixtures, the conventional linear mixing rule is kept for the parameter b:

$$b = \sum_i x_i b_i, \tag{29}$$

whereas for the parameter a, the LCV mixing rule is used.

$$\frac{a}{bRT} = \left(\frac{\lambda}{A_v} + \frac{1-\lambda}{A_m}\right)\left(\frac{G^E}{RT}\right) + \frac{1-\lambda}{A_m}\sum_i \ln\left(\frac{b}{b_i}\right) + \sum_i x_i \frac{a_i}{b_i RT}, \tag{30}$$

where A_m, A_v are constant, then the fugacity coefficient of component i in a mixture, for the PR EOS, is given by the following equation:

$$\ln \varphi_i^L = \frac{b_i}{b}\left(\frac{PV}{RT}-1\right) - \ln\frac{P(V-b)}{RT} - \frac{\alpha_i}{2\sqrt{2}}\ln\left[\frac{V+(1+\sqrt{2})b}{V+(1-\sqrt{2})b}\right] \tag{31}$$

$$\alpha_i = \left(\frac{\lambda}{A_v} + \frac{1-\lambda}{A_m}\right)\ln\gamma_i + \frac{1-\lambda}{A_m}\ln\left(\left(\frac{b}{b_i}\right)+\frac{b}{b_i}-1\right) + \frac{a_i}{b_i RT}. \tag{32}$$

Appendix 3

The criterion of vapor–liquid–solid equilibrium is that the fugacities for every component i must satisfy the following equations:

$$f_i^V = f_i^L = f_{pure,i}^s(P,T) \quad i=1,2,\dots,n \tag{33}$$

$$f_i^V = f_i^L \quad i=1,2,\dots,n, \tag{34}$$

where f is the fugacity, n is the total number of components, and N_s is the number of solid phases determined. The fugacity of each component in the vapor and liquid phases is calculated by the equation of state. The fugacity coefficient of component i in the liquid phase is calculated by an EOS/G^E model. The modified PR equation of state is used. The activity coefficient of component i is calculated using the UNIFAC method. It is assumed that for mixtures containing alkanes only, the residual term in the UNIFAC model is zero. Thus, mixtures containing different alkanes are described by the combinatorial term. The Staverman–Guggenheim combinatorial term, which is used in UNIFAC, is (Gao et al. 2001)

$$\ln\gamma_i = \ln\frac{\phi_i}{x_i} + 1 - \frac{\phi_i}{x_i} - \frac{Z}{2}q_l\left(\ln\left(\frac{\phi_i}{\theta_i}\right) + 1 - \frac{\phi_i}{\theta_i}\right), \tag{35}$$

where Z is the coordination number. In this work, ri and qi have been obtained from the following relations by data fitting presented in the literature (Ji et al. 2004):

$$r_i = 0.6744 C_{ni} + 0.4534 \tag{36}$$

$$q_i = 0.54 C_{ni} + 0.616. \tag{37}$$

References

Barker JA, Henderson D. Perturbation theory and equation of state for fluids. J Chem Phys. 1969;47:4714–21.

Chapman WG, Gubbins KE, Jackson G, Radosz M. SAFT: equation-of-state solution model for associating fluids. Fluid Phase Equilib. 1989;52:31–8.

Chen WH, Zhang XD, Zhao ZC, Yin CY. UNIQUAC model for wax solution with pour point depressant. Fluid Phase Equilib. 2009;280:9–15.

Dalirsefat R, Feyzi F. A thermodynamic model for wax deposition phenomena. Fuel. 2007;86:1402–8.

Escobar-Remolina JCM. Prediction of characteristics of wax precipitation in synthetic mixtures and fluids of petroleum. Fluid Phase Equilib. 2006;240:197–203.

Gao W, Robinson RL Jr, Gasem KAM. Improved correlations for heavy n-paraffin physical properties. Fluid Phase Equilib. 2001;179:207–16.

Ghanaei E, Esmaeilzadeh F, Fathi Kaljahi J. A new predictive thermodynamic model in the wax formation phenomena at high pressure condition. Fluid Phase Equilib. 2007;254:126–37.

Gonzalez DL, Hirasaki GJ, Chapman WG. Modeling of asphaltene precipitation due to changes in composition using the perturbed chain statistical associating fluid theory equation of state. Energy Fuels. 2007;21(3):1231–42.

Gross J, Sadowski G. Perturbed-chain SAFT: an equation of state based on a perturbation theory for chain molecules. Ind Eng Chem Res. 2001;40:1244–60.

Hansen JH, Fredenslund A, Pedersen KS, Ronningsen HP. A thermodynamic model for predicting wax formation in crude oils. AIChE J. 1988;34:1937–42.

Jafari Ansaroudi HR, Vfaei-Safti M, Msoudi SH, Jafari Behbahani T, Jafari H. Study of the morphology of wax crystals in the presence of ethylene-co-vinyl acetate copolymer. Pet Sci Technol. 2013;31:643–51.

Jafari Behbahani T. A new investigation on wax precipitation in petroleum fluids: Influence of activity coefficient models. Pet Coal. 2014a;56(2):157–64.

Jafari Behbahani T. Experimental Investigation of the Polymeric flow improver on Waxy oils. Pet Coal. 2014b;56(2):139–42.

Jafari Behbahani T, Golpasha R, Akbarnia H, Dahaghin A. Effect of wax inhibitors on pour point and rheological properties of Iranian waxy crude oil. Fuel Process Technol. 2008;89:973–7.

Jafari Behbahani T, Dahaghin A, Kashefi K. Effect of solvent on rheological behavior of iranian waxy crude oil. Pet Sci Technol. 2011a;29:933–41.

Jafari Behbahani T, Ghotbi C, Taghikhani V, Shahrabadi A. Experimental investigation and thermodynamic modeling of asphaltene precipitation. J Sci Iran C. 2011b;18(6):1384–90.

Jafari Behbahani T, Ghotbi C, Taghikhani V, Shahrabadi A. Investigation of asphaltene deposition mechanisms during primary depletion and CO_2 injection, paper 143374, European formation damage conference, Noordwijk; 2011c, June 7–10.

Jafari Behbahani T, Ghotbi C, Taghikhani V, Shahrabadi A. Investigation on asphaltene deposition mechanisms during CO_2 flooding processes in porous media: A novel experimental study and a modified model based on multilayer theory for asphaltene adsorption. J Energy Fuels. 2012;26:5080–91.

Jafari Behbahani T, Ghotbi C, Taghikhani V, Shahrabadi A. A modified scaling equation based on properties of bottom hole live oil for asphaltene precipitation estimation under pressure depletion and gas injection conditions. Fluid Phase Equilib. 2013a;358:212–9.

Jafari Behbahani T, Ghotbi C, Taghikhani V, Shahrabadi A. Asphaltene deposition under dynamic conditions in porous media: theoretical and experimental investigation. Energy Fuels. 2013b;27:622–39.

Jafari Behbahani T, Ghotbi C, Taghikhani V, Shahrabadi A. Experimental study and mathematical modeling of asphaltene deposition mechanism in core samples. J. Oil Gas Sci Technol Rev. IFp 2013c. doi: 10.2516/ogst/2013128.

Jafari Behbahani T, Dahaghin A, JafariBehbahani Z. Experimental investigation and thermodynamic modeling of phase behavior of reservoir fluids. Energy Sources Part A. 2014a;36:1256–65.

Jafari Behbahani T, Ghotbi C, Taghikhani V, Shahrabadi A. A new model based on multilayer kinetic adsorption mechanism for asphaltenes adsorption in porous media during dynamic condition. Fluid Phase Equilib. 2014b;375:236–45.

Jafari Behbahani T, Ghotbi C, Taghikhani V, Shahrabadi A. Investigation of asphaltene adsorption in sandstone core sample during CO_2 injection: experimental and modified modeling. Fuel. 2014c;133:63–72.

Jafari Behbahani T, Dahaghin A, JafariBehbahani Z. Modeling of flow of crude oil in a circular pipe driven by periodic pressure variations. Energy Sources Part A. 2015;37:1406–14.

Ji HY, Tohidi B, Danesh A, Todd AC. Wax phase equilibria: developing a thermodynamic model using a systematic approach. Fluid Phase Equilib. 2004;216:211–7.

Koyuncu M, Demirtas A, Ogul R. Excess properties of liquid mixtures from the perturbation theory of Barker-Henderson. Fluid Phase Equilib. 2002;193:87–95.

Leekumjorn S, Krejbjerg K. Phase behavior of reservoir fluids: comparisons of PC-SAFT and cubic EOS simulations. Fluid Phase Equilib. 2013;359:17–23.

Lira-Galeana C, Firoozabadi A, Prauznits JM. Thermodynamics of wax precipitation in petroleum mixtures. AIChE J. 1996;42:239–48.

Mansoori GA, Carranhan N, Starling K, Leland T. Equilibrium thermodynamic properties of the mixture of hard spheres. J Chem Phys. 1971;54:1523–31.

Michelsen ML. The isothermal flash problem: 1 stability. Fluid Phase Equilib. 1982;9:1–19.

Nichita DV, Goual L, Firoozabadi A. Wax precipitation in gas condensate mixtures. SPE Prod Facil. 2001;16:250–9.

Pan H, Firoozabadi A. Pressure and composition effect on wax precipitation: experimental data and model results. SPE Prod Facil. 1997;12:250–9.

Panuganti SR, Vargas FM, Gonzalez DL, Kurup AS, Chapman WG. PC-SAFT characterization of crude oils and modeling of asphaltene phase behavior. Fuel. 2012;93:658–69.

Pauly J, Daridon JL, Coutinho JAP. Solid deposition as a function of temperature in the nC10 + (nC24–nC25–nC26) system. Fluid Phase Equilib. 2004;224:237–44.

Pedersen KS, Skovborg P, Ronningsen HP. Wax precipitation from North Sea crude oils. 4. Thermodynamic modeling. Energy Fuel. 1991;5:924–32.

Punnapala S, Vargas FM. Revisiting the PC-SAFT characterization procedure for an improved asphaltene precipitation prediction. Fuel. 2013;108:417–29.

Sedghi M, Goual L. PC-SAFT modeling of asphaltene phase behavior in the presence of nonionic dispersants. Fluid Phase Equilib. 2014;369:86–94.

Wertheim MS. Fluids with highly directional attractive forces. III. Multiple attraction site. J Stat Phys. 1986;42:459–76.

Wertheim MS. Fluids with highly directional attractive forces. II. Thermodynamic perturbation theory and integral equations. J Stat Phys. 1984;35:35–40.

Won KW. Thermodynamics for solid solution–liquid–vapor equilibria: wax phase formation from heavy hydrocarbon mixtures. Fluid Phase Equilib. 1968;30:265–79.

Won KW. Thermodynamic calculation of cloud point temperatures and wax phase compositions of refined hydrocarbon mixtures. Fluid Phase Equilib. 1989;53:377–96.

Zúñiga-Hinojosa MA, Justo-García DN, Aquino-Olivos MA, Román-Ramírez LA, García-Sánchez F. Modeling of asphaltene precipitation from n-alkane diluted heavy oils and bitumens using the PC-SAFT equation of state. Fluid Phase Equilib. 2014;376:210–24.

Zuo JY, Zhang DD, Ng HJ. An improved thermodynamic model for wax precipitation from petroleum fluids. Chem Eng Sci. 2001;56:6941–7.

3

Study of the properties of non-gas dielectric capacitors in porous media

Hong-Qi Liu · Yan Jun · You-Ming Deng

Abstract The size of pores and throats is at the nano-meter scale in tight oil and shale gas zones, and the resistivity of these reservoirs is very high, so the reservoirs show more dielectric properties than conductivity proper-ties. The conductive and dielectric characteristics of a parallel plate capacitor full of fresh water, NaCl solutions, and solid dielectrics, for example, sands are investigated in this paper, and the capacitance data of the non-gas capac-itor are measured at different salinities and frequencies by a spectrum analyzer. The experimental results illustrate that the capacitance of this kind of capacitor is directly pro-portional to the salinity of the solutions and inversely proportional to the measuring frequency, the same as a vacuum parallel plate capacitor. The remarkable phenom-enon, however, is that the capacitance is inversely pro-portional to the square of the distance between two plates. The specific characteristic of this capacitor is different from the conventional parallel plate capacitor. In order to explain this phenomenon, the paper proposed a new con-cept, named "single micro ion capacitor", and established a novel model to describe the characteristics of this par-ticular capacitor. Based on this new model, the theoretical capacitance value of the single micro ion capacitor is calculated, and its polarization and relaxation mechanisms are analyzed.

Keywords NaCl solution · Debye model · Single micro ion capacitor · Dielectrics · Micro capacitivity

1 Introduction

As we know, rocks have both conductive and dielectric properties. The conductivity is completely determined by various anions and cations, such as Na^+, Mg^{2+}, Ca^{2+}, K^+, Cl^-, OH^-, HCO_3^-, SO_4^{2-}, and CO_3^{2-} in water solution in the intergranular pores of rocks. Which property of the rocks will play a predominant role depends on the salinity, porosity, permeability, and the geometric structure of the pores. In fact, almost all substances in nature are dielec-trics, and only a few have conductivity (σ) (Havriliak and Negami 1966). The conductive paths of these charges are usually continuous in relatively high porosity and high permeability reservoirs, but in most occasions, they are discontinuous in tight oil and shale gas zones, or in the low porosity and low permeability reservoirs (Gasparrini et al. 2014; Ghanizadeh et al. 2014). Therefore, the electric properties of rocks include two major aspects: one is the conductive capability of free positive and negative charges in the water solution through the paths formed by pores and throats which connected with each other; the other is the dielectric capability of the bound charges of the non-con-ductive substances and particles, including rock matrix particles, oil or gas molecules, and pure water molecules (Freedman and Vogiatzis 1979; Jonscher 1983; Endres and Bertrand 2006). Nevertheless, whether conductive or dielectric, the current paths in the formation are influenced by the geometric pore structure of rocks. The relationship

H.-Q. Liu (✉) · Y.-M. Deng
State Key Laboratory of Oil and Gas Reservoir Geology and Exploitation, Southwest Petroleum University, Chengdu 610500, Sichuan, China
e-mail: lhqjp1@126.com

Y. Jun
Centrica Energy (E&P) Upstream Kings Close, 62 Huntly Street, Aberdeen AB 101 RS, UK

Edited by Jie Hao

between the conductive and dielectric properties of rocks and the pore structure has been widely discussed in detail in the literature (Toumelin and Torres-Verdin 2009), so it is not described here. Prior to 1940, studies on the electrical properties of rocks mainly focused on conductivity. In 1941, K. S. Cole and R. H. Cole established the Cole–Cole model for dielectric constants (Cole and Cole 1941); later, many scholars studied ionic conduction and rock polarization processes, analyzing the dielectric constant property of non-homogeneous multi-pore materials (Hilfer et al. 1995, 1999; Nover et al. 2000; Ruffett et al. 1991; Davidson and Cole 1950, 1951).

With regard to the electrical logging technology in the petroleum industry, almost all research has focused on the detection and interpretation of formation resistivity. However, more and more complex and unconventional reservoirs have been discovered, and the traditional, simple geophysical conductive model cannot solve the problems of strongly heterogeneous reservoirs, such as low porosity and low permeability oil reservoirs, tight oil zones, and shale gas zones. Although scholars have proposed various solutions and models for some important questions to describe the electrical conductivity mechanism in rocks (Chelidze and Gueguen 1999; Chelidze et al. 1999; Asami 2002), it is increasingly difficult to accurately determine fluid types and calculate the saturation of hydrocarbon. In 1988, Clark et al. proposed an electromagnetic propagation tool (EPT) to measure the dielectric constant of the formation at frequencies of 2 MHz–1.1 GHz (Glover et al. 1994a, b, 1996; Clark et al. 1990). This method can identify fluid types and calculate the fluid saturation in high porosity and high permeability reservoirs. Dong and Wang (2009) studied the dielectric constant of several common minerals, including quartz, calcite, and dolomite, within a frequency range from Hz to GHz level, and they identified pore structures and the distribution of formation water using the dielectric constant. In 1985, Lockner and Byerlee (1985) studied the complex resistivity of rocks; later, other scholars also studied the complex resistivity of rocks. However, due to the extraordinarily complex heterogeneity of porous rocks, there are still many disputes over the conductive and dielectric mechanisms in rocks (Clavier et al. 1984; Zemanek 1989; Hamada and Al-Awad 1998). Scholars have proposed many conductive models for rocks, of which the most representative is the dual-water argillaceous sandstone conductive model proposed by Waxman and Smits in 1968 (Waxman and Smits 1968; Knight and Nur 1987; Hassoun et al. 1997).

2 Micro ion capacitor model

Obviously, the conductive and dielectric capabilities of rocks depend on the amount of different positive and negative ions dissolved in water solution. Generally, these different ions can be equivalent to Na^+ and Cl^- in view of their conductive property. Therefore, it is crucial to study the conductive and dielectric properties of NaCl solutions (Jonscher 1999; Lesmes and Morgan 2001). The pure water and hydrocarbon molecules cannot conduct current, and the conductive ions are separated from each other by water and hydrocarbon molecules, as shown in Fig. 1. So the solution in the pores can be regarded as a mixture of non-conductive molecules and conductive free ions (Fig. 2). Considering the water and hydrocarbon molecules as non-conductive land, the Na^+ and Cl^- ions are an isolated conductive river separated by this land, and the ion river moves through this land in a tortuous path. Then a model with "ion flux" and "molecule land" can be established to explain the conductive and dielectric mechanisms. In Fig. 1, the yellow or gray blocks denote rock particles of different minerals, the blue parts represent the formation water, the red belts denote hydrocarbons dispersed in the water, which are non-conductive parts; and the green solid ball denotes free Cl^-, the red solid ball denotes free Na^+, which are conductive parts. Of course, the rock matrix is also non-conductive.

Figure 2 illustrates the distribution of the non-conductive molecules of water and hydrocarbon, and the conductive ions of Na^+ and Cl^- in an external electromagnetic field (EMF). If an alternating EMF, such as $I = I_0 \sin(\omega t + \theta_0)$, is applied on both sides of rocks, the free ions will be rearranged quickly according to the external EMF, resulting in a regular arrangement of ions. The rearranged ions move directionally under the effect of the external EMF as shown in Fig. 3. These ions gradually move to the ends of mineral grains through pores. Generally the Cl^- ions gather at the anode, and Na^+ ions gather at the cathode. An internal electric field is created in the rock, whose field direction is opposite to the external field. The time of this process depends on the conductive properties of rocks. But after the established relatively stable EMF within a remarkably short

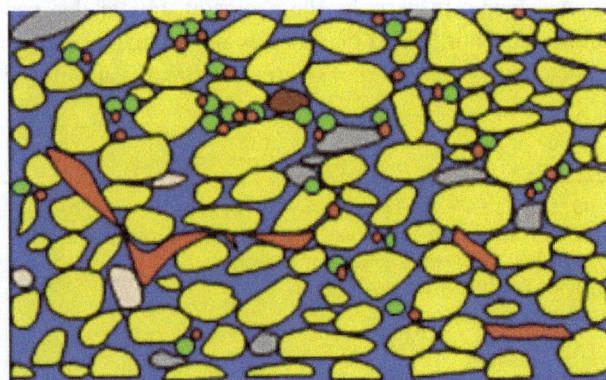

Fig. 1 Schematic diagram of a rock with pure water (*blue*), hydrocarbon (*red belt*), and Na^+ (*small red ball*) and Cl^- (*large green ball*) in the rock

Fig. 2 Distribution of the non-conductive water and hydrocarbon molecules and ion flux without external EMF

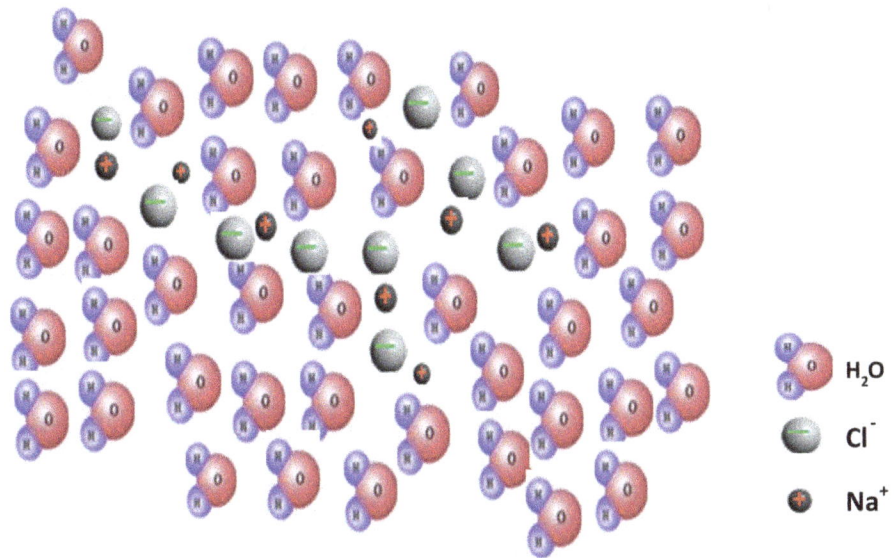

H₂O

Cl⁻

Na⁺

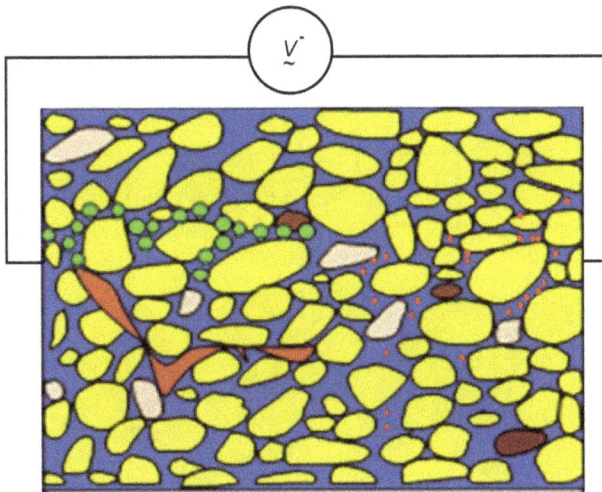

Fig. 3 Schematic diagram of a rock with pure water (*blue*), hydrocarbon (*red belt*), and ions traversing through the rock under the influence of an external electromagnetic field (EMF)

time, the amount of free ions in rock pores becomes less and less, and the ions and molecules form a new distribution, as shown in Fig. 4.

The movements of the positive and negative ions result in the establishment of a capacitor. The positive pole is composed of cations, such as Na^+, Ca^{2+}, Mg^{2+}, and K^+, the negative pole is composed of anions, such as Cl^-, OH^-, and SO_4^{2-}, and the dielectric is water or hydrocarbon molecules. From this moment on, the rocks will show dielectric properties rather than conductivity at the macro level.

The polarization process above, perhaps will not exist in high porosity and especially high permeability reservoirs. As for tight formations, the pore throats are very narrow. For example, when the pores are at the nanometer scale

with poor connectivity, the occurring phenomenon is shown in Fig. 5. In tight oil and shale gas zones, several anions and cations respectively gather at the opposite ends of the particle as polar plates like a usual parallel plate capacitor, whereas the non-conductive molecules in the middle are dielectrics, thus a microscopic capacitor, called a "micro ion capacitor", is established. Numerous such micro ion capacitors can be created almost at the same time in the formation, and they may be connected in series or parallel.

3 Theoretical value of single ion capacitor

The radii of Na^+, Cl^-, and H^+ ions are 0.95, 1.81, and 2.08 Å, respectively. The size of the pore throats in rocks ranges from several to more than 10 microns, and the thickness of the water membrane is less than 1 micron. The space for ion transportation is tens of thousands of times larger than the ion diameter. As shown in Fig. 5, there may be at least tens of thousands of positive and negative ions gathering at the ends of rock particles in the water-wet phase, and a microscopic capacitor is formed with Na^+ and Cl^- ions as two polar plates, with the rock particles wrapped by water membrane with a certain thickness as a dielectric. Although the capacitance of this micro ion capacitor cannot be accurately calculated at present, a theoretical value can be estimated.

If the pores are very small and are not connected, ions are likely to form isolated capacitors with water molecules as dielectric. In case of extreme conditions, a single ion and a single water molecule constitute a micro ion capacitor, as shown in Fig. 6. According to the definition of a capacitor, assuming the voltage of an external EMF is 1 V, the

Fig. 4 The non-conductive
water and hydrocarbon
molecules and ions form a
micro ion capacitor with an
external EMF

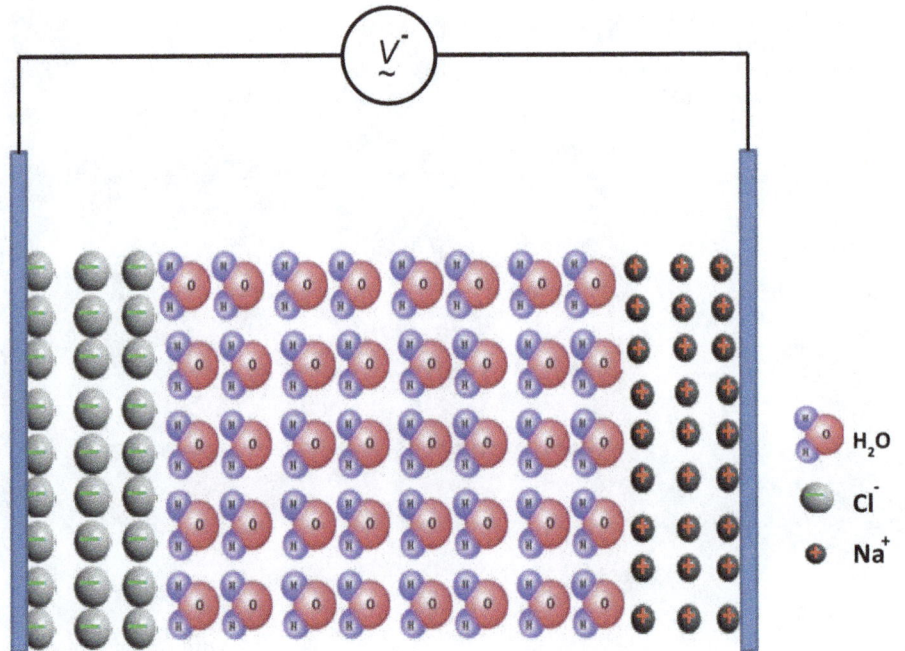

H$_2$O

Cl$^-$

Na$^+$

Fig. 5 A single water-wet mineral particle with ions at two sides
forming a micro ion capacitor

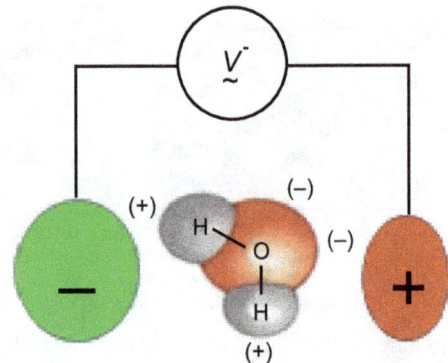

Fig. 6 A single ion capacitor, one sodium ion and one chloride ion
respectively as positive and negative plates with a non-conductive
H$_2$O molecule as media

quantity of the charge for example Na$^+$ is 1.6×10^{-19} C,
then the capacitance of a single ion capacitor is:

$$C_0 = \frac{Q}{V} = \frac{1.6 \times 10^{-19} \text{ C}}{1 \text{ V}} = 1.6 \times 10^{-19} \text{ F} \qquad (1)$$

4 Experiments

4.1 Experimental instrument

Figure 7 illustrates a measuring device using a PVC pipe
with a diameter of 2.5 cm and a length of 150 cm. Two
plates were placed, one was fixed on the bottom, the other
was movable on the top, and the pipe was full of NaCl
solution. So the resistivity and capacitance of NaCl solution can be detected at different distances, salinities, and
frequencies. One important aspect must be noted that the
material of plates should be gold, rather than iron, alumi-
num, copper, and others, to avoid corrosion by the NaCl
solution.

4.2 Experimental process

First, fresh water with different volumes was injected into
the PVC pipe, and the upper plate was put at different
distances corresponding to the fluid volume. Then the
resistance (R_p) and capacitance (C_p) were tested with a
frequency spectrum analyzer. In this experiment, 12 groups
of data were recorded at 12 different distances. At each test
point, R_p and C_p were tested at frequencies of 100 Hz and
1 kHz, where the subscript "p" means parallel connection.

Second, NaCl solution was injected into the same pipe.
Then, we repeat the experimental process like that of fresh
water, 12 groups of data of R_p and C_p of NaCl solution with
salinities of 0.2 and 1.2 g/L were recorded. The capacitance
data are listed in Table 1 and resistance data in Table 2.

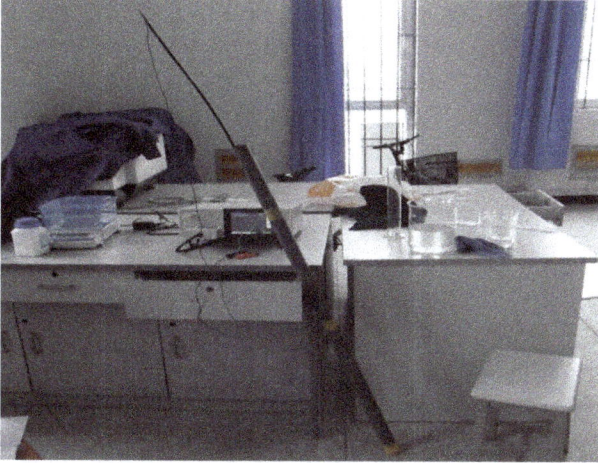

Fig. 7 A PVC pipe with copper wireline and full of NaCl solution, the dimension is length × diameter = 150 cm × 2.5 cm

Figure 8 illustrates a plastic container, with a dimension of 17 cm × 5.0 cm × 4.7 cm, filled with sand which is saturated with NaCl solution. The sand was collected from a river, and was washed many times. Moreover, the sand was filtered by a sorting sieve in order to keep sand almost at the same size. Its mineral composition is mainly quartz. During the experiments, the conditions were changed as follows:

(1) The distance between the polar plates varied from 2 to 17 cm, with eight test points;
(2) The frequency changed from 4 Hz to 5 MHz, with 100 test points, only seven points listed in the tables;
(3) The salinity varied from 1.5 to 200 g/L, with eight test points.

Table 1 Capacitance (C_p, μF) at different salinities (S), frequencies (f), and distances (d) in the PVC pipe

d (m)	S (g/L)				
	Water f (Hz)	0.2	1.2	0.2	1.2
	100			1 k	
0.1	970	23,500	286,000	642.0	11,200
0.2	252	6,170	83,800	160.0	3,010
0.3	112	2,730	39,900	71.0	1,340
0.4	61	1,550	22,500	39.0	763
0.5	41	976	15,100	24.0	491
0.6	25	694	10,300	18.1	345
0.7	21	510	7,730	13.0	248
0.8	15	371	5,760	10.0	192
0.9	12	303	4,820	8.0	153
1.0	9	241	3,740	6.0	121
1.1	7	201	3,150	5.0	98
1.2	6	165	2,550	4.5	82

Table 2 Resistance (R_p, Ω) at different salinities (S), frequencies (f), and distances (d) in the PVC pipe

d (m)	S (g/L)				
	Water f (Hz)	0.2	1.2	0.2	1.2
	100			1 k	
0.1	23.1	3.19	0.56	3.08	0.51
0.2	45.1	6.24	1.01	6.14	0.96
0.3	67.45	9.35	1.47	9.25	1.43
0.4	90.44	12.42	1.92	12.32	1.88
0.5	113	15.56	2.38	15.46	2.34
0.6	136	18.61	2.85	18.5	2.81
0.7	159	21.77	3.3	21.67	3.26
0.8	183	24.9	3.78	24.79	3.74
0.9	205	27.92	4.22	27.82	4.18
1.0	228	31.1	4.69	30.96	4.65
1.1	253	34.21	5.15	34.1	5.12
1.2	277	37.3	5.61	37.2	5.57

The measurement results of C_p are listed in Table 3 and R_p in Table 4, the distance and frequency changed with a fixed salinity of 1.5 g/L.

Table 5 lists the capacitance data of sand saturated with NaCl solution with different salinities and a fixed distance of 17 cm, and Table 6 lists the resistance data of sand saturated with NaCl solution with different salinities and a fixed distance of 17 cm.

5 Analysis of experimental results

5.1 Relationship between conductive and dielectric properties and the distance between plates

Based on the experimental results in the PVC pipe, and the data in Table 1, Fig. 9 demonstrates the characteristics of this capacitor with fresh water and 0.2 and 1.2 g/L NaCl solutions at 100 Hz and 1 kHz. The capacitance changes with different salinities of the solution and different distances between the two plates. Generally, we can conclude as follows:

(1) Under the same conditions, the higher the salinity, the larger the capacitance;
(2) Under the same conditions, the higher the frequency, the lower the capacitance;
(3) Under the same conditions, the capacitance is approximately inversely proportional to the square of the distance.

As shown in Fig. 9, there are five different curves, corresponding to 1.2 g/L salinity at 100 Hz, 0.2 g/L

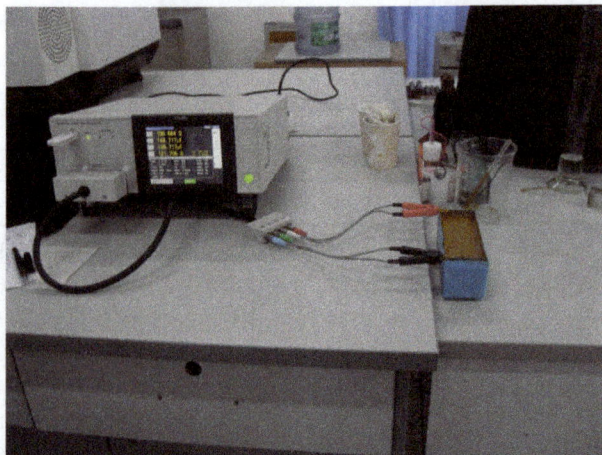

Fig. 8 A plastic container with copper plate and sand saturated with NaCl solution, the dimension is length × width × height = 17 cm × 5.0 cm × 4.7 cm

salinity at 100 Hz, 1.2 g/L salinity at 1 kHz, fresh water at 100 Hz, 0.2 g/L salinity at 1 kHz, respectively. It can be seen that with either the fresh water or the NaCl solution, either at 100 Hz or 1 kHz, the capacitance is a function of the square of the distance. By the least square method, we know that the distance's exponent of each curve is very close to two. We test it at many different frequencies from

4 Hz to 5 MHz and with different salinities, and all of the results prove the same rule, that is, the capacitance is approximately inversely proportional to the square of the distance. Why does this phenomenon occur?

It is well known that the capacitance of a parallel plate capacitor is proportional to the area of the plate (A), and is inversely proportional to the distance between the plates (d), denoted by Eq. (2)

$$C = \varepsilon_0 \frac{A}{d} \tag{2}$$

But for the PVC pipe capacitor, according to the definition of capacitance and the rule mentioned above, the equation of C_p can be expressed by

$$C_p = \epsilon \frac{A}{d^2}, \tag{3}$$

where ϵ is a coefficient. The key point should be noted that ϵ is not the dielectric constant ε_0 because the units of these two parameters are different, and then the physical definition is also different. Based on SI unit, the dimension of ϵ is F, which is the same as capacitance. Here, we call ϵ as micro capacitivity.

According to Eq. (3), ϵ can be expressed as follows:

$$\epsilon = C_p \frac{d^2}{A} \tag{4}$$

Table 3 Capacitance (C_p, μF) of sands saturated with NaCl solution at different frequencies and distances in a plastic container

f, Hz	d, cm							
	17	15	13	10	8	5	3	2
4	1.40E+02	1.79E+02	2.39E+02	4.22E+02	6.37E+02	1.60E+03	4.41E+03	9.88E+03
10.2	7.06E+01	8.93E+01	1.19E+02	2.11E+02	3.14E+02	8.02E+02	2.23E+03	5.01E+03
70.1	6.95E+00	8.92E+00	1.29E+01	2.01E+01	3.13E+01	8.02E+01	2.29E+02	5.00E+02
286.8	1.40E+00	1.79E+00	2.38E+00	4.01E+00	6.04E+00	1.58E+01	4.36E+01	9.74E+01
1019.1	1.87E−01	2.42E−01	3.15E−01	5.38E−01	8.31E−01	2.14E+00	5.41E+00	1.32E+01
5029.1	1.40E−02	1.70E−02	2.30E−02	3.80E−02	6.10E−02	1.57E−01	4.34E−01	9.81E−01
10,171	3.00E−03	4.00E−03	6.00E−03	1.00E−02	1.60E−02	4.00E−02	1.20E−01	2.23E−01

Table 4 Resistance (R_p, Ω) of sands saturated with NaCl solution at different frequencies and distances in a plastic container

f, Hz	d, cm							
	17	15	13	10	8	5	3	2
4	1.67E+02	1.64E+02	1.60E+02	1.54E+02	1.51E+02	1.47E+02	1.42E+02	1.35E+02
10.2	1.17E+02	1.13E+02	1.08E+02	1.01E+02	9.65E+01	9.17E+01	8.86E+01	8.66E+01
70.1	7.42E+01	6.86E+01	6.24E+01	5.38E+01	4.81E+01	4.07E+01	3.66E+01	3.48E+01
286.8	6.19E+01	5.60E+01	4.94E+01	4.01E+01	3.37E+01	2.49E+01	1.98E+01	1.75E+01
1019.1	5.77E+01	5.17E+01	4.50E+01	3.54E+01	2.88E+01	1.92E+01	1.33E+01	1.03E+01
5029.1	5.60E+01	4.99E+01	4.32E+01	3.35E+01	2.68E+01	1.70E+01	1.07E+01	7.43E+00
10,171	5.57E+01	4.96E+01	4.28E+01	3.32E+01	2.65E+01	1.67E+01	1.03E+01	6.99E+00

Table 5 Capacitance (C_p, pF) varies with frequency (f) and salinity (S)

f, Hz	S, g/L							
	1.5625	3.125	6.25	12.5	25	50	100	200
4	8.31E+03	1.95E+04	4.32E+04	8.23E+04	1.36E+05	2.13E+05	3.03E+05	3.43E+05
6.1	4.75E+03	1.16E+04	2.70E+04	5.52E+04	9.69E+04	1.57E+05	2.28E+05	2.60E+05
9.8	2.43E+03	6.24E+03	1.53E+04	3.35E+04	6.39E+04	1.08E+05	1.65E+05	1.87E+05
15.6	1.20E+03	3.21E+03	8.28E+03	1.93E+04	3.98E+04	7.30E+04	1.17E+05	1.34E+05
25	5.73E+02	1.60E+03	4.29E+03	1.05E+04	2.35E+04	4.62E+04	8.07E+04	9.23E+04
39.9	2.68E+02	7.66E+02	2.12E+03	5.42E+03	1.28E+04	2.75E+04	5.27E+04	6.14E+04
63.9	1.23E+02	3.61E+02	1.03E+03	2.67E+03	6.66E+03	1.53E+04	3.18E+04	3.83E+04
102.1	5.53E+01	1.66E+02	4.86E+02	1.29E+03	3.30E+03	7.97E+03	1.78E+04	2.25E+04
163.3	2.46E+01	7.53E+01	2.26E+02	6.12E+02	1.60E+03	3.96E+03	9.35E+03	1.25E+04
261.1	1.08E+01	3.36E+01	1.03E+02	2.85E+02	7.58E+02	1.92E+03	4.68E+03	6.86E+03
417.6	4.80E+00	1.48E+01	4.64E+01	1.31E+02	3.55E+02	9.08E+02	2.28E+03	3.80E+03
1,708	4.50E−01	1.26E+00	3.99E+00	1.17E+01	3.35E+01	8.97E+01	2.34E+02	7.68E+02
4368.4	1.20E−01	2.70E−01	7.80E−01	2.26E+00	6.57E+00	1.81E+01	4.83E+01	2.72E+02
17,867	4.00E−02	5.00E−02	8.00E−02	1.70E−01	4.50E−01	1.16E+00	3.11E+00	4.99E+01
32,895	4.00E−02	4.00E−02	4.00E−02	4.00E−02	7.00E−02	8.00E−02	1.20E−01	2.05E+01
45,695	3.00E−02	3.00E−02	3.00E−02	1.00E−02	−2.00E−02	−1.60E−01	−5.80E−01	1.16E+01
186,900	3.00E−02	2.00E−02	1.00E−02	−3.00E−02	−1.20E−01	−4.40E−01	−1.35E+00	−1.81E+00
764,420	3.00E−02	2.00E−02	1.00E−02	−3.00E−02	−1.20E−01	−4.30E−01	−1.25E+00	−2.94E+00
1,955,000	2.00E−02	2.00E−02	1.00E−02	−3.00E−02	−1.20E−01	−4.00E−01	−1.09E+00	−2.44E+00
3,126,500	2.00E−02	2.00E−02	1.00E−02	−3.00E−02	−1.20E−01	−3.90E−01	−9.40E−01	−1.92E+00
5,000,000	2.00E−02	2.00E−02	0.00E+00	−3.00E−02	−1.30E−01	−3.70E−01	−7.40E−01	−1.32E+00

The experiments with the capacitor filled with sand saturated with NaCl solution in a plastic container also show the same rule. From Fig. 10, it can be seen that the capacitance is also approximately inversely proportional to the square of the distance between plates.

The six curves in Fig. 10 were tested at six different frequencies. In fact, from 4 Hz to 5 MHz, we recorded 100 groups of capacitance data, and each group shows the inversely proportional relationship between the capacitance and the square of the distance.

In the above experiments, the dielectrics are fresh water, saline solution, or sand mixed with saline solution, which have different salinities and different saturations. We can see that whether the measurement was carried out in a PVC pipe or in a plastic container, and whether the measured dielectric is liquid or solid, all of the results indicate that the capacitance is approximately inversely proportional to the square of the distance between plates. Such phenomenon may be related to the polarization and relaxation processes. As early as 1929, Debye assumed that the dipole relaxation of a dielectric is a purely viscous process without elastic forces, and established an equation to describe the relationship of the dielectric constant and the angular frequency as shown in Eq. (5) (Debye 1929). In 1941, K. S.

Cole and R. H. Cole found that the relaxation of most solid dielectrics does not satisfy the Debye model. They corrected the Debye model by taking into account the electrical conductance, and they proposed the so-called Cole–Cole model, as shown in Eq. (6)

$$\varepsilon^*(\omega) = \varepsilon_\infty + \frac{\varepsilon_s - \varepsilon_\infty}{1 + i\omega\tau} \qquad (5)$$

$$\varepsilon^*(\omega) = \varepsilon_\infty + \frac{\varepsilon_s - \varepsilon_\infty}{1 + (i\omega\tau)^{1-\alpha}}, \qquad (6)$$

where, ε_∞ is the optical frequency dielectric constant, ε_s is the static dielectric constant, $i = \sqrt{-1}$, τ is the time constant, and α is the empirical coefficient, $0 < \alpha < 1$.

The above two and other models, such as Lorentz–Lorenz, Maxwell–Wagner, and Onsager models, describe the dielectric constant of isolated material. In our experiment, the medium is a mixture of conductive materials, such as Na^+, Cl^-, and other cations or anions, and insulating matter, such as water and hydrocarbon molecules. In the pores of rocks, the conductive and non-conductive materials coexist, and in most cases, they can form numerous micro capacitors, that is the micro ion capacitor mentioned before. Obviously, the inversely proportional relationship between ionic capacitance and the square of

Table 6 Resistance (R_p, Ω) varies with frequency (f) and salinity (S)

f, Hz	S, g/L							
	1.5625	3.125	6.25	12.5	25	50	100	200
4	7.94E+02	4.66E+02	2.83E+02	1.88E+02	1.27E+02	8.73E+01	6.52E+01	6.10E+01
6.1	7.64E+02	4.42E+02	2.62E+02	1.70E+02	1.14E+02	7.87E+01	5.86E+01	5.46E+01
9.8	7.40E+02	4.20E+02	2.44E+02	1.54E+02	1.01E+02	6.93E+01	5.15E+01	4.78E+01
15.6	7.21E+02	4.03E+02	2.30E+02	1.40E+02	8.98E+01	6.05E+01	4.46E+01	4.12E+01
25	7.09E+02	3.92E+02	2.19E+02	1.30E+02	8.05E+01	5.29E+01	3.81E+01	3.54E+01
39.9	7.00E+02	3.83E+02	2.12E+02	1.23E+02	7.38E+01	4.68E+01	3.25E+01	3.02E+01
63.9	6.94E+02	3.78E+02	2.07E+02	1.18E+02	6.92E+01	4.24E+01	2.83E+01	2.60E+01
102.1	6.91E+02	3.74E+02	2.03E+02	1.15E+02	6.61E+01	3.94E+01	2.53E+01	2.31E+01
163.3	6.88E+02	3.71E+02	2.01E+02	1.13E+02	6.40E+01	3.75E+01	2.33E+01	2.10E+01
261.1	6.86E+02	3.70E+02	1.99E+02	1.11E+02	6.27E+01	3.63E+01	2.20E+01	1.96E+01
417.6	6.85E+02	3.69E+02	1.98E+02	1.10E+02	6.17E+01	3.54E+01	2.12E+01	1.86E+01
1,708	6.84E+02	3.67E+02	1.97E+02	1.09E+02	6.03E+01	3.42E+01	2.00E+01	1.64E+01
4368.4	6.83E+02	3.67E+02	1.96E+02	1.08E+02	6.00E+01	3.39E+01	1.97E+01	1.52E+01
17,867	6.83E+02	3.67E+02	1.96E+02	1.08E+02	5.98E+01	3.37E+01	1.95E+01	1.38E+01
32,895	6.82E+02	3.67E+02	1.96E+02	1.08E+02	5.98E+01	3.37E+01	1.95E+01	1.34E+01
45,695	6.82E+02	3.66E+02	1.96E+02	1.08E+02	5.98E+01	3.36E+01	1.95E+01	1.32E+01
186,900	6.80E+02	3.66E+02	1.96E+02	1.08E+02	5.98E+01	3.37E+01	1.95E+01	1.29E+01
764,420	6.74E+02	3.64E+02	1.95E+02	1.08E+02	5.99E+01	3.39E+01	1.99E+01	1.33E+01
1,955,000	6.56E+02	3.57E+02	1.93E+02	1.07E+02	5.98E+01	3.46E+01	2.15E+01	1.58E+01
3,126,500	6.31E+02	3.47E+02	1.89E+02	1.05E+02	5.96E+01	3.55E+01	2.38E+01	1.98E+01
5,000,000	5.74E+02	3.22E+02	1.78E+02	1.01E+02	5.86E+01	3.75E+01	2.85E+01	2.92E+01

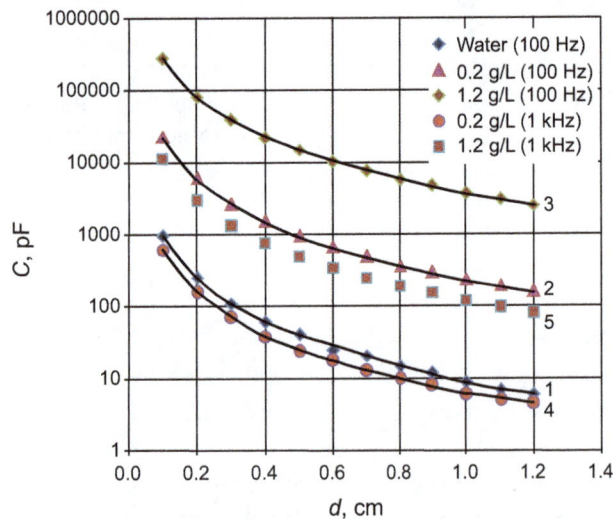

Fig. 9 The relationship of NaCl solution's capacitance and the distance between plates at different salinities and frequencies. The capacitance increases with increasing salinity and decreasing frequency, and decreases with the increase of the square of the distance. 1. $C_p = 9.2428d^{-2.0048}$, $R^2 = 0.9987$; 2. $C_p = 243.92d^{-1.9980}$, $R^2 = 0.9998$; 3. $C_p = 3841.50d^{-1.9038}$, $R^2 = 0.9992$; 4. $C_p = 6.2724d^{-2.009}$, $R^2 = 0.9996$; 5. $C_p = 122.11d^{-1.981}$, $R^2 = 0.9997$

the distance cannot be explained by either the Debye model or the Cole–Cole model. We should create a novel model to explain this rule.

5.2 Preliminary explanation

The main reason of this remarkable phenomenon is that in the pores and fractures, especially in tight oil and shale gas zones, various ions, water molecules, and hydrocarbon molecules can form numerous micro ion capacitors, which will parallel or series connect with each other. Compared with conventional capacitors, the outstanding characteristic of this micro capacitor is that the length of the plates nearly equals the distance between the plates. The parameter A is the surface area of the ion, and the distance d is the diameter of water or hydrocarbon molecules.

For the micro ion capacitor mentioned above, if one or several water molecules are dielectric, and Na^+ and Cl^- or other cations and anions are plates, the distance between plates almost equals the size of ionic plates. Then the capacitance of micro ion capacitor also accords with the rule as shown in Eq. (3). The diameter of a water molecular is about 4 $\overset{\circ}{A}$, and we can estimate the capacitance of such a micro ion capacitor as follows:

$$C_{Na^+} = \epsilon \frac{A}{d^2} = \epsilon \frac{\pi r^2}{d^2} = \epsilon \frac{\pi(1.81/2)^2}{(4 + 0.95/2 + 1.81/2)^2}$$
$$= 0.088851\epsilon \tag{7}$$

Fig. 10 The relationship between capacitance and the distance between two plates at different frequencies. The *figure* illustrates the increase of the capacitance with decreasing frequency, and the increase with a decrease in the square of the distance, which is different from conventional capacitors

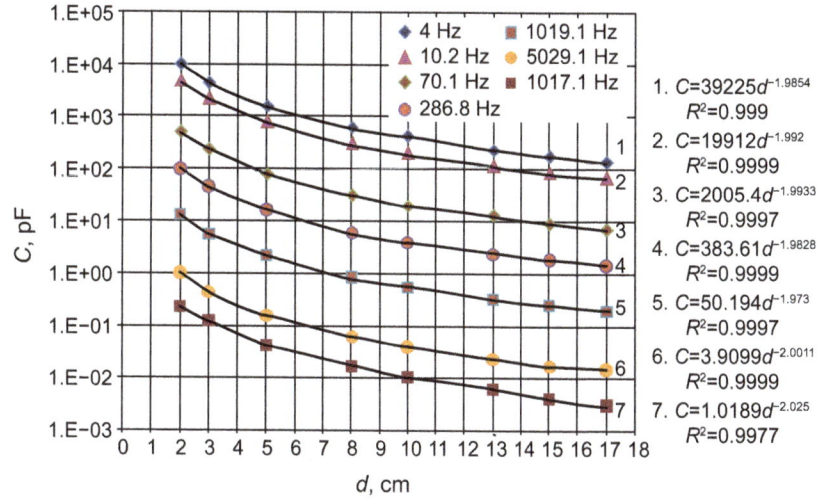

$$C_{Cl^-} = \epsilon \frac{A}{d^2} = \epsilon \frac{\pi r^2}{d^2} = \epsilon \frac{\pi(0.95/2)^2}{(4 + 0.95/2 + 1.81/2)^2}$$
$$= 0.024477\epsilon \tag{8}$$

According to the theoretical value estimated in Eq. (1), the value of micro capacitivity ϵ can be calculated:

$$\epsilon_{Na^+} = 1.80 \times 10^{-6} \text{ pF}, \tag{9}$$

$$\epsilon_{Cl^-} = 6.54 \times 10^{-6} \text{ pF}, \tag{10}$$

Because the different diameters of Na^+ and Cl^-, the micro capacitivity ϵ varies from 1.8×10^{-6} to 6.54×10^{-6} pF.

For capacitors with liquid and solid dielectrics, many aspects remain unclear, especially the mechanisms of polarization and relaxation. The reasons may be related to the single ionic capacitors formed between free Na^+ and Cl^- ions in the non-conductive liquid or solid molecules, and the single ionic capacitors are connected in series or parallel. However, the distance between molecules in a gas dielectric or vacuum dielectric capacitor is too long to form a microscopic capacitor. This may be the important reason for the remarkable difference between these two types of capacitors. Another reason may be the complexity of channels for ion transportation (Ma et al. 2014).

novel concept and a model, single micro ion capacitor, are first proposed in this paper. Based on the experimental results, we found that:

(1) The ionic capacitance is inversely proportional to the square of the distance, which remarkably differs from that of parallel plate capacitors with air dielectric;

(2) Compared with conventional capacitors, the outstanding characteristic of a micro ion capacitor is that the length of the plate nearly equals the distance between the plates;

(3) Based on the micro ion capacitor model, the micro capacitivity ϵ varies from 1.8×10^{-6} to 6.54×10^{-6} pF.

Such phenomenon may be also related to the relatively complex polarization and relaxation mechanisms of the numerous single micro ion capacitors, and another important reason may be the tortuosity of the ion conductive path in porous solid media, which will be discussed later.

Acknowledgments The authors are grateful for the financial support from Basic Science Program of Advanced Well Logging Technology of CNPC (2014A-2319) and support from the Science and Technology Program (G12-3) of State Key Laboratory of Oil and Gas Reservoir Geology and Exploitation of SWPU (Southwest Petroleum University).

6 Conclusions

We measured the capacitance and resistance of non-gas dielectrics in a PVC pipe and a plastic container, and the experimental results illustrated an unusual rule in the ionic capacitance of fluid or solid dielectrics. Although many models have been established, there are still many problems about the polarization and relaxation of the ions or molecules under the external EMF that are not solved. A

References

Asami K. Characterization of heterogeneous systems by dielectric spectroscopy. Prog Polym Sci. 2002;27(8):1617–59.

Chelidze TL, Gueguen Y. Electrical spectroscopy of porous rocks: a review—I. Theoretical model. Geophys J Int. 1999;137(1):1–15.

Chelidze TL, Gueguen Y, Ruffet C. Electrical spectroscopy of porous rocks: a review—II. Experimental results and interpretation. Geophys J Int. 1999;137(1):16–34.

Clark B, Allen DF, Best DL, et al. Electromagnetic propagation logging while drilling: theory and experiment. 1990. SPE 18117-PA.

Clavier C, Coates G, Dumanoir J. Theoretical and experimental bases for the dual-water model for interpretation of shaly sands. Soc Petrol Eng J. 1984;24:153–68.

Cole KS, Cole RH. Dispersion and absorption in dielectrics. J Chem Phys. 1941;9(4):341–51.

Davidson DW, Cole RH. Dielectric relaxation in glycerine. J Chem Phys. 1950;18(10):1417.

Davidson DW, Cole RH. Dielectric relaxation in glycerol, propylene glycol and n-propanol. J Chem Phys. 1951;19(12):1484–90.

Debye PW. Polar molecules. New York: Chemical Catalog Co.; 1929.

Dong XB, Wang YH. A broadband dielectric measurement technique: theory, experimental verification, and application. J Environ Eng Geophys. 2009;14(1):25–38.

Endres AL, Bertrand EA. A pore-size scale model for the dielectric properties of water-saturated clean rocks and soils. Geophysics. 2006;71:F185–93.

Freedman R, Vogiatzis JP. Theory of microwave dielectric constant logging using the electromagnetic wave propagation method. Geophysics. 1979;44(5):969–86.

Gasparrini M, Sassi W, Gale JFW. Natural sealed fractures in mudrocks: a case study tied to burial history from the Barnett Shale, Fort Worth Basin, Texas, USA. Mar Petrol Geol. 2014;55:122–41.

Ghanizadeh A, Gasparik M, Amann-Hildenbrand A, et al. Experimental study of fluid transport processes in the matrix system of the European organic-rich shales: I. Scandinavian Alum Shale. Mar Pet Geol. 2014;51:79–99.

Glover PWJ, Gomez JB, Meredith PG, et al. Modelling the stress–strain behavior of saturated rocks undergoing triaxial deformation using complex electrical conductivity measurements. Surv Geophys. 1996;17(3):307–30.

Glover PWJ, Meredith PG, Sammonds PR, et al. Ionic surface electrical conductivity in sandstone. J Geophys Res. 1994;99(B11):21635–50.

Glover PWJ, Meredith PG, Sammonds PR, et al. Measurements of complex electrical conductivity and fluid permeabilities in porous rocks at raised confining pressures. Rock Mechanics in Petroleum Engineering. Delft, Netherlands. 29–31 August 1994b. pp. 29–36.

Hamada GM, Al-Awad MNJ. Petrophysical evaluation of low resistivity sandstone reservoirs. International Symposium of Core Analysts. The Hague. 14–16 Sept 1998.

Hassoun TH, Zainalabedin K, Minh CC. Hydrocarbon detection in low contrast resistivity pay zones, capillary pressure and ROS determination with NMR logging in Saudi Arabia. SPE Paper 37770. 10th MEOS, Bahrain. 15–18 March 1997.

Havriliak S, Negami S. A complex plane analysis of α-dispersions in some polymer systems. J Polym Sci Part C. 1966;14(1):99–117.

Hilfer R, Prigogine I, Rice SA. Transport and relaxation phenomena in porous media. Adv Chem Phys. 1995;92:299.

Hilfer R, Widjajakusuma J, Biswal B. Macroscopic dielectric constant for microstructures of sedimentary rocks. Granular Matter 2. Berlin: Springer-Verlag; 1999. pp. 137–41.

Jonscher AK. Dielectric relaxation in solids. London: Chelsea Press; 1983. pp. 219–26.

Jonscher AK. Dielectric relaxation in solids. J Phys D. 1999;32:57–70.

Knight RJ, Nur A. The dielectric constant of sandstones, 60 kHz to 4MHz. Geophysics. 1987;52(5):644–54.

Lesmes DP, Morgan FD. Dielectric spectroscopy of sedimentary rocks. J Geophys Res. 2001;106(B7):13329–46.

Lockner DA, Byerlee JD. Complex resistivity measurements of confined rock. J Geophys Res. 1985;90:7837–47.

Ma JS, Sanchez JP, Wu KJ, et al. A pore network model for simulating non-ideal gas flow in micro- and nano-porous materials. Fuel. 2014;116(15):498–508.

Nover G, Heikamp S, Freund D. Electrical impedance spectroscopy used as a tool for the detection of fractures in rock samples exposed to either hydrostatic or triaxial pressure conditions. Nat Hazards. 2000;21(2–3):317–30.

Ruffett C, Gueguen Y, Darot M. Complex conductivity measurements and fractal nature of porosity. Geophysics. 1991;56(6):758–68.

Toumelin E, Torres-Verdin C. Pore-scale simulation of kHz-GHz electromagnetic dispersion of rocks: effects of rock morphology, pore connectivity, and electrical double layers. In: SPLWA 50th Annual Logging Symposium. 21–24 Jun 2009.

Waxman MH, Smits LJM. Electrical conductivities in oil-bearing shaly sands. Soc Petrol Eng J. 1968;8(2):107–22.

Zemanek J. Low-resistivity hydrocarbon-bearing sand reservoirs. SPE Form Eval. SPE-15713-PA. 1989;4(4):515–21.

4

Quantification and timing of porosity evolution in tight sand gas reservoirs: an example from the Middle Jurassic Shaximiao Formation, western Sichuan, China

Zheng-Xiang Lü[1] · Su-Juan Ye[2] · Xiang Yang[1] · Rong Li[3] · Yuan-Hua Qing[1,4]

Abstract The diagenesis and porosity evolution of the Middle Jurassic Shaximiao sandstones were analyzed based on petrographic observations, X-ray diffractometry, scanning electron microscopy observations, carbon and oxygen stable isotope geochemistry, fluid inclusion microthermometry, and thermal and burial history modeling results. The point count data show that secondary pores (av. 5.5 %) are more abundant than primary pores (av. 3.7 %) and are thus the dominant pore type in the Shaximiao sandstones. Analysis of porosity evolution indicates that alteration of sandstones mainly occurred during two paragenetic stages. Mechanical compaction and cementation by early chlorite, calcite, and quartz typically decrease the depositional porosity (40.9 %) by an average of 37.2 %, leaving porosity of 3.7 % after stage I (<85 °C, 175–145 Ma). The original intergranular porosity loss due to compaction is calculated to be 29.3 %, suggesting that mechanical compaction is the most significant diagenetic process in primary porosity destruction. Stage II can be further divided into two sub-stages (Stage II$_a$ and Stage II$_b$). Stage II$_a$ (85–120 °C, 145–125 Ma) is characterized by late dissolution, which enhanced porosity by 8.8 %, and the porosity increased from 3.7 % to 12.5 %. During stage II$_b$ (>120 °C, 125–0 Ma), the precipitation of late chlorite, calcite, quartz, and kaolinite destroyed 3.3 % porosity, leaving porosity of 9.2 % in the rock today.

Keywords Diagenesis · Porosity evolution · Tight gas sandstones · Jurassic · Western Sichuan

1 Introduction

Reservoir quality is one of the key controls on prospectivity during petroleum exploration and exploitation. With the development of experimental methods, petroleum scientists have become more concerned about quantification and timing of porosity evolution, water–rock–hydrocarbon interaction, diagenetic kinetics, multi-scale evaluation of diagenesis, retention diagenesis, and reservoir quality prediction (Wood and Byres 1994; Bjørlykke and Jahren 2012; Taylor et al. 2004, 2010; Ajdukiewicz and Lander 2010). Due to the increasing market demand for gas and also to technology advances, tight gas plays have received considerable attention in recent years, and it is now important from both a scientific and a practical standpoint to investigate diagenesis in these low permeability sandstones (Stroker et al. 2013). However, the diagenesis and reservoir quality evolution pathways in tight gas sand reservoirs are more complex than those in conventional reservoirs, and so qualitative analysis and inference-based quantitative study cannot provide satisfactory results. In order to quantitatively predict reservoir quality, numerical simulations of diagenetic processes have been carried out by some researchers (Prodanović et al. 2013; Dai et al. 2003; Liu et al. 2014), and many critical concerns, including the selection

✉ Su-Juan Ye
sujuan_ye@hotmail.com

[1] State Key Laboratory of Oil and Gas Reservoir Geology & Exploitation, Chengdu University of Technology, Chengdu 610059, Sichuan, China

[2] Exploration and Production Research Institute, Sinopec Southwest, Chengdu 610041, Sichuan, China

[3] Chengdu Institute of Geology and Mineral Resources, Chengdu 610081, Sichuan, China

[4] PetroChina Tarim Oilfield Company, Korla 841000, Xinjiang, China

Edited by Jie Hao

of diagenetic environment parameters, the quantitative calibration of various diagenetic processes, and the delimitation of diagenetic time, present great challenges to diagenesis and reservoir quality research. Studies in tectonics, stratigraphy/sedimentology, geochemistry, and burial–thermal history should be combined together to achieve integrated research (Zhang et al. 2013). In other words, what are the temporal and spatial relationships, both relative and absolute, among different diagenetic reactions and what controls reservoir quality?

The Middle Jurassic Shaximiao Formation in western Sichuan, China is such a reservoir, which contains sandstones with 50 % of samples having porosity lower than 8 % and permeability less than 0.1 mD and was previously regarded as tight sandstones with no gas storage and flow capacity. However, commercial gas flows have been obtained from the Shaximiao Formation with high-yield gas flows produced from many wells, demonstrating its significant gas reserve and production potential. The Shaximiao sandstones have been deeply buried with maximum depths >3.5 km and have undergone complex diagenetic alternations, which thus cause the sandstones to be tight and highly heterogeneous. Previously, studies of the Shaximiao sandstones have mainly focused on investigating the overall reservoir characteristics, including depositional systems (Tan et al. 2008; Wang and Wang 2010; Ye and Lü 2010; Yang et al. 2012), petrology, diagenetic processes, and reservoir classification and evaluation (Lü et al. 2000; Ye and Lü 2010; Yang et al. 2012). However, the dominant pore types and the temporal relationship between hydrocarbon emplacement and porosity evolution have not been well investigated previously due to the absence of studies on quantification and timing of porosity evolution. In order to well understand the mechanism of hydrocarbon accumulation and predict hydrocarbon-enriched zones, quantifying the spatial and temporal distribution of diagenetic alternations and of porosity is of significant importance.

The main objectives of this study are to analyze diagenetic modifications quantitatively, to estimate the volume of secondary pores and recognize the importance of dissolution pores, to determine the timing of important diagenetic processes that altered the petrophysical properties of the sandstones using a wide range of techniques, including petrography, fluid inclusion and isotopic studies, and thermal history modeling, and to establish porosity evolution through time and relate the porosity to each critical moment, especially to the timing of hydrocarbon emplacement.

2 Geological setting

The western Sichuan depression is located in the western part of the Sichuan Basin and structurally bounded by the Longmen Mountain thrust belt to the west and by the Qinling orogenic belt to the north (Meng et al. 2005) (Fig. 1). It underwent multi-stage tectonic movements from the Mesozoic, including Indosinian, Yanshanian, and Himalayan orogeny (Yang et al. 2012).

The western Sichuan depression contains numerous commercial gas accumulations in mainly Mesozoic reservoirs. More than ten tight gas sand fields have been discovered to date in this area. The Upper Triassic to Quaternary sediments deposited in western Sichuan are dominated by siliciclastic rocks of offshore marine to terrestrial origin with the thickness up to 8 km (Gu and Liu 1997; Xu et al. 1997; Yang et al. 2012).

The Middle Jurassic Shaximiao Formation is 600–750-m thick. It overlies the Middle Jurassic Qianfoya Formation and underlies the Upper Jurassic Suining Formation (Fig. 2). The Shaximiao Formation comprises interbedded siltstone and sandstone, interpreted as fluvial and lacustrine–deltaic deposits (Tan et al. 2008; Wang and Wang 2010; Ye and Lü 2010; Yang et al. 2012).

Most of the hydrocarbons in the Shaximiao Formation are thought to have been generated from the 5th member of the Xujiahe Formation (T_3x^5), and the time of peak hydrocarbon generation and migration was the late Early Cretaceous (100 Ma) (Cai and Liao 2000; Qin et al. 2007; Shen et al. 2008).

3 Samples and method

A total of 1225 core samples were selected from the Shaximiao Formation from 32 exploration wells. Thin sections were prepared for all samples using blue/red-dyed epoxy impregnation. Thin-section petrography was used to determine whole rock mineralogy, diagenetic relationships, porosity characteristics, and clay growth occurrence and habits in the pore space. The modal composition and porosity of 132 representative samples were obtained by counting 300 points in each thin section. Representative samples were viewed under a scanning electron microscope (SEM) equipped with an energy dispersive X-ray spectrometer (EDX) and cathodoluminescence imaging.

Pores were separated into residual primary intergranular pores, intragranular dissolution pores, and enlarged intergranular dissolution pores during the point counting of each thin section used in this study. Residual primary intergranular pores tend to be of straight–angular cross section and can thus be identified easily. However, it is difficult to estimate the proportion of secondary pores since enlarged intergranular dissolution pores include both primary and secondary pores. After obtaining the average volume of residual primary intergranular pores and intergranular dissolution pores based on point counts, the secondary contribution to the intergranular dissolution

Fig. 1 Location map and tectonic elements of the Sichuan Basin (simplified from Meng et al. 2005)

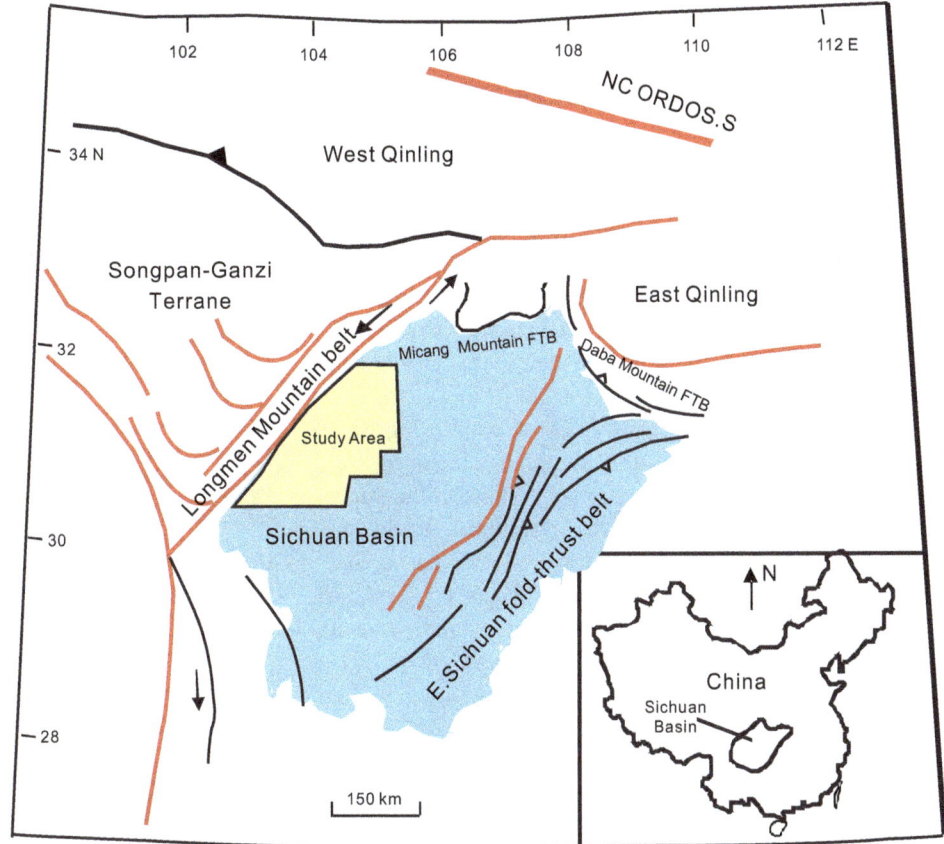

Fig. 2 Triassic to Cretaceous stratigraphic framework of the western Sichuan (adapted from Yang et al. 2012)

System	Series	Formation			Age, Ma	Hydrocarbon
Cretaceous						
Jurassic	Upper	Penglaizhen		J_3p	144	•
		Suining		J_3sn	154	•
	Middle	Shaximiao	Upper	J_2s	156	•
			Lower	J_2x	172	•
		Qianfoya		J_2q	175	
	Lower	Baitianba		J_1b	189	
Triassic	Upper	Xujiahe		T_3x	208	•

porosity was estimated as the difference between average enlarged intergranular dissolution porosity and average residual primary intergranular porosity.

In situ carbon and oxygen isotope analysis was performed using an Nd:YAG laser microprobe. Laser probe micro-sampling of C and O from carbonate cements for isotopic analysis was achieved by focusing a laser beam with a wavelength of 1064 nm and a diameter of 20 μm onto a sample situated in a vacuum chamber to ablate a small area on the sample and liberate CO_2 gas. After

purification, the CO_2 gas was led directly into a Finnigan MAT 252 mass spectrometer for isotopic analysis.

The temperature for calcite precipitation was calculated using the fractionation equation of Friedman and O'Neil (1977):

$$1000 \times \ln\alpha_{calcite-water} = 2.78 \times 10^6 \times T^{-2} - 2.89\,(SMOW) \quad (1)$$

$$\alpha_{calcite-water} = (1000 + \delta^{18}O_{calcite})/(1000 + \delta^{18}O_{water}) \quad (2)$$

Fourteen thick double-side polished thin sections were prepared for microthermometric measurements. Homogenization temperatures were measured using a Linkam THMS-600 heating/cooling stage. Only primary fluid inclusions with both aqueous and hydrocarbon phases were selected from authigenic quartz to determine their minimum precipitation temperatures (Liu et al. 2005; Lu and Guo 2000; Marfil et al. 2000).

X-ray diffraction (XRD) analyses were carried out on twenty-two bulk samples and <2 μm size fractions, which permitted quantification of quartz, feldspar, calcite, and clay contents.

The burial–thermal history was constructed using Schlumberger PetroMod 11 software. Input data included age, thickness, and lithology of stratigraphic units in the study area. The heat-flow model was developed based on knowledge of the tectonic history of the basin and calibrated against available vitrinite reflectance (R_o) data. The estimate of erosion was verified using two independent burial history techniques: sonic velocity and vitrinite reflectance (Shen et al. 2011).

4 Results

4.1 Sandstone petrology

The Shaximiao sandstones are generally well sorted, with a fine to medium grain size. Sandstones are mostly lithic arkoses (sample proportion: 59 %) and feldspathic litharenites (sample proportion: 19 %) (Fig. 3). Quartz is the predominant framework component in the sandstones, usually constituting 45 %–55 % of the whole rock volume with an average framework composition of $Q_{51}F_{28}R_{21}$. Plagioclase dominates over K-feldspar (Table 1). Rock fragments are sedimentary, metamorphic, and volcanic in origin. The detrital composition of sandstones varies with different provenance (Ye and Lü 2010). The sediments in the western region of depression were mainly sourced from the middle segment of Longmen Mountain and are mostly litharenites. The deposits in the middle and eastern parts were supplied from Micang Mountain and Daba Mountain and are generally sandstones with a high feldspar content.

Authigenic minerals comprise calcite, authigenic quartz, illite, chlorite, and kaolinite (Table 1). According to the point count data, cement volume ranges from 4.4 % to 28.7 %, with an average of 11.2 % (Table 2).

4.2 Diagenesis

The Shaximiao sandstones in western Sichuan are strongly influenced by mechanical compaction, cementation, and

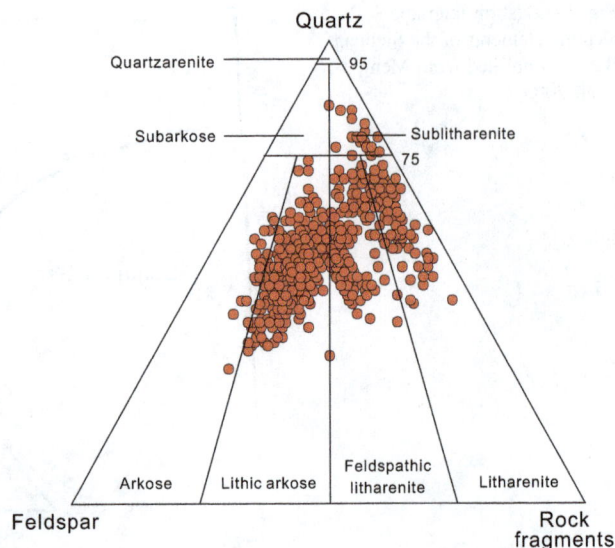

Fig. 3 Detrital compositions of the Shaximiao sandstones (nomenclature according to Folk's classification scheme, 1974)

Table 1 Statistical summary of XRD results (absolute values %)

	Minimum	Maximum	Mean	SD
Quartz	19	66	42.0	13.0
Plagioclase	5	64	38.0	15.9
K-feldspar	0	11	2.1	3.0
Clay	8	33	14.1	5.5
Calcite	0	18	3.8	4.3
Dolomite	0	1	0.1	0.3
Illite	0.4	14.5	2.4	3.0
Kaolinite	0	6.3	2.5	2.1
Chlorite	1.5	10.6	4.6	2.5

dissolution. Cementation was mainly by calcite, authigenic quartz, and clay minerals.

The sandstones had suffered a moderate degree of compaction, which is evidenced by rearrangement of detrital grains, squeezing of ductile grains, and the presence of straight contacts.

Diagenetic clay minerals include chlorite, kaolinite, and illite (Tables 1, 2). Chlorite is common in sandstones and occurs mainly in two forms: grain-coating (pore-lining) and pore-filling. SEM imaging indicates that grain-coating chlorites are made of thin and finely crystalline aggregates (thickness: 0.008–0.02 mm) perpendicularly to grain surfaces (Fig. 4a, c, d, f). Pore-filling chlorite typically consists of pseudohexagonal platelets and occurred after grain-coating chlorite (Fig. 4a, b, f). Kaolinite constitutes a trace to 57 % of the clay fraction in the sandstones and appears mostly as loose, micro-porous booklets (Fig. 4b). The

Table 2 Average values for the point-count authigenic minerals (%)	Authigenic quartz		Calcite		Chlorite		Kaolinite	Total
	Early	Late	Early	Late	Grain-coating	Pore-filling		
	1.7	0.8	1.7	1.0	4.5	1.0	0.5	11.2

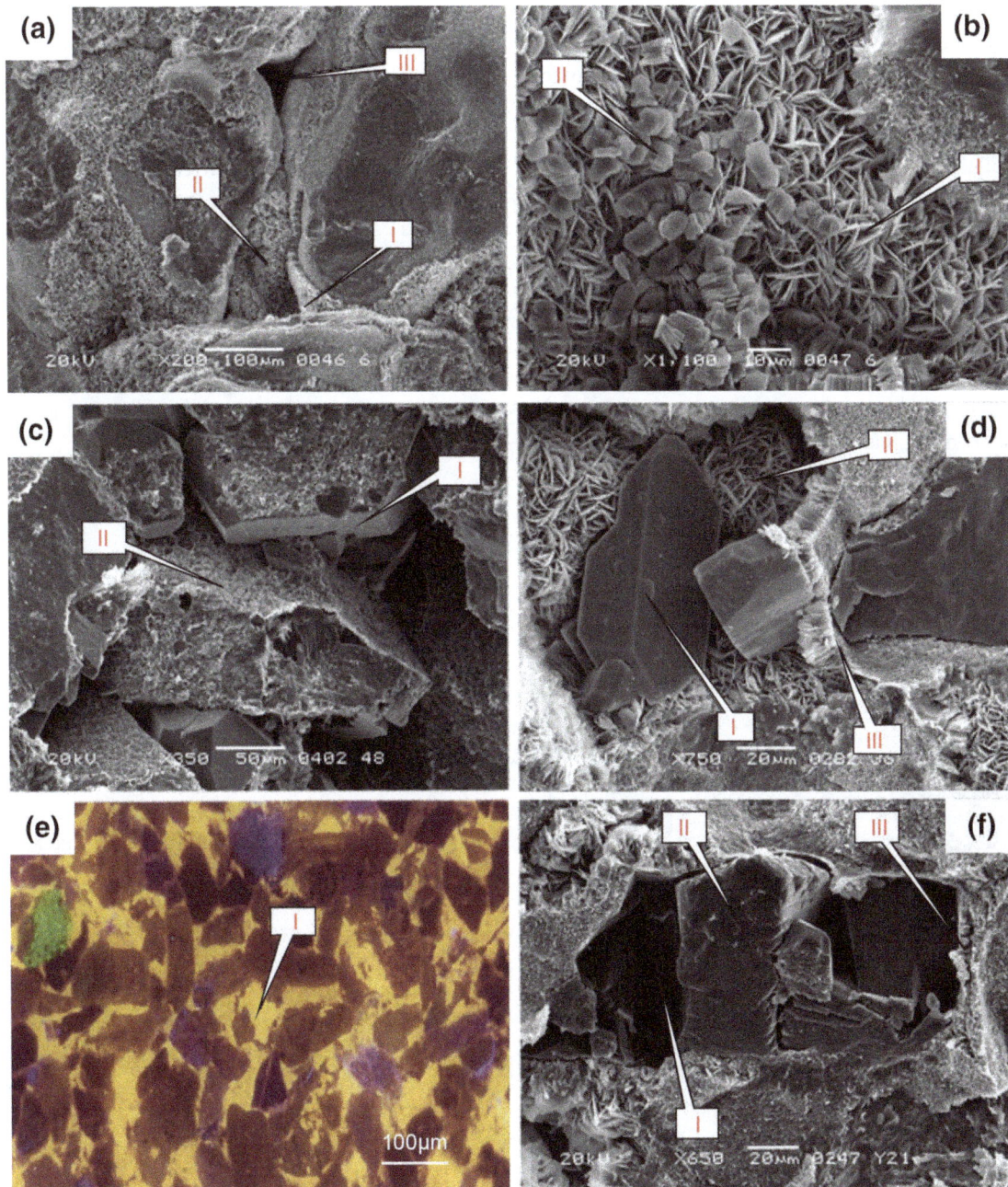

Fig. 4 Optical and SEM images of **a** grain-coating chlorite (I) occurred before pore-filling chlorite (II) and both partly filled primary intergranular pores (III) (Well CJ619, 2461.2 m). **b** Kaolinite (II) grew on pore-filling chlorite (I) (Well CJ619, 2461.2 m). **c** Quartz overgrowing (I) was developed from a quartz surface where grain-coating chlorite (II) was absent (Well CX605, 2824.34 m). **d** Authigenic quartz (I) grew on pore-filling chlorite (II); residual grain-coating chlorite is shown as III (Well GM31, 2587.34 m). **e** Cathodoluminescence image of early framework-stabilizing calcite (I) (Well CX168, 2332.70 m). **f** Calcite (II) partially filled secondary pores (I); remaining chlorite coating is shown as III (Well GJ6, 2819.41 m)

kaolinite cement typically has a fresh appearance and post-dates the authigenic quartz and chlorite (Fig. 4b). Fibrous illite is observed by SEM in many samples (trace to 14.5 %, av. 2.4 %).

Quartz cement (trace to 4 %, av. 2.5 %) occurs as syntaxial overgrowths on detrital quartz and as euhedral quartz (Fig. 4c, d). The average volume of early quartz overgrowths is relatively more than that of later euhedral overgrowths and outgrowths (Table 2).

Carbonate cements mainly include calcite and minor amounts of dolomite (Table 1). Calcite cement varies in abundance from trace to 20 % with an average of 2.7 % (Table 2). There are two forms of calcite: early diagenetic framework-stabilizing calcite (Type I) and late sparry calcite (Type II). Type I calcite fills intergranular pores and displays evidence of moderate to pervasive recrystallization into coarse, blocky calcite (Fig. 4e). The grained enclosed Type I calcite is loosely packed and shows no authigenic chlorite, indicating that it is the earliest formed authigenic mineral. Type II calcite fills intergranular secondary pores (Fig. 4f) and replaces feldspar and rock fragments. Type II calcite engulfs and thus post-dates quartz overgrowths and chlorite.

The dissolution of feldspar (mainly plagioclase) and early calcite cement is common in sandstones (Fig. 5). Feldspar dissolution has resulted in oversized and moldic pores which contain corroded remnants of feldspar. Partial dissolution of rock fragments has resulted in the formation of intragranular pores. Late authigenic quartz, calcite, and

kaolinite cements are interpreted to partially fill secondary porosity.

4.3 Quantity of dissolution pores

Quantification of the percentage of primary and secondary porosity is of great significance since it determines which factors or processes have the greatest impact on porosity. The Shaximiao sandstones contain primary intergranular porosity and secondary porosity (Fig. 5). The overall core–plug and thin-section porosity values range from 4.4 %–17.1 % (av. 11.1 %) to 4.4 %–20.2 % (av. 9.2 %), respectively (Table 3). The difference of about 1.8 % between core–plug and thin-section porosity values is attributed to the presence of micro-porosity within clay minerals.

Based on point count data, the percentage and contribution of primary/secondary pores to total visible porosity were calculated (Table 3), indicating that secondary pores are more abundant than primary pores (av. 5.5 % and 3.7 %, respectively), and most of samples are dominated by secondary pores.

4.4 Temperature of quartz cementation

Forty-seven homogenization temperatures (T_h) were measured in inclusions located at quartz grain-overgrowth boundaries. These inclusions have been termed Q1. They occur along the "dust-rims", delineating the surfaces of the detrital grains. The overall range of T_h in Q1 inclusions is 80–95 °C with an average of 85 °C, suggesting that significant quartz cementation started at around 80 °C (Fig. 6).

Seventy homogenization temperatures correspond to inclusions located in quartz outgrowths (termed as Q2) and range between 125 and 140 °C with an average of 130 °C (Fig. 6).

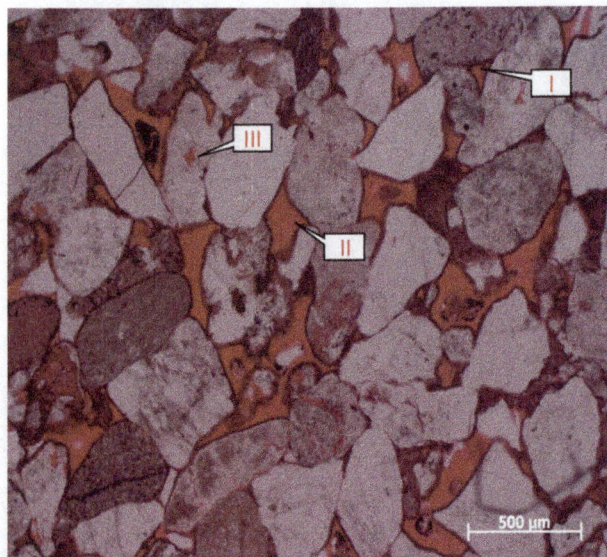

Fig. 5 Plane-polarized light thin-section photomicrograph showing primary intergranular pores (I), enlarged intergranular dissolution pores (II), and intragranular dissolution pores (III) (Well GM31, 2719.95 m)

Table 3 Statistical summary of the point-count primary and secondary porosity (%) and total thin-section and core–plug porosity (%)

	Minimum	Maximum	Mean	SD
Primary porosity				
Percent	0.5	10.0	3.7	1.6
Contribution	9.8	70.0	40.4	14.1
Secondary porosity				
Percent	1.8	12.5	5.5	2.2
Contribution	30	90.2	59.5	14.1
Total thin-section porosity	4.4	20.2	9.2	2.9
Total core–plug porosity	4.4	17.1	11.1	2.6

From 132 samples

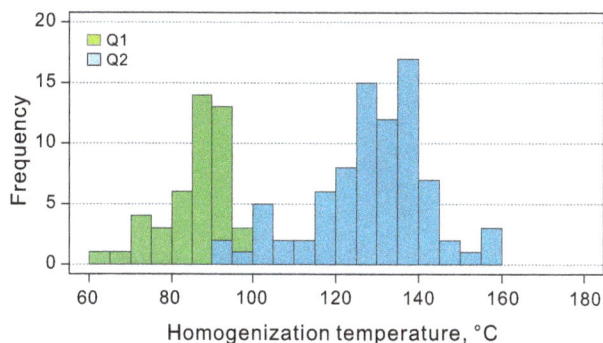

Fig. 6 Homogenization temperatures for 117 fluid inclusions in authigenic quartz cements

4.5 Temperature of calcite cementation

The oxygen isotope compositions of carbonates are widely used in paleothermometry (Friedman and O'Neil 1977; Gabitov et al. 2012; Zheng 2011).

The $\delta^{13}C$ and $\delta^{18}O$ values of calcite cements are listed in Table 4. The $\delta^{13}C$ values vary between -5.49 and -12.62 ‰ with an average of -8.7 ‰. The $\delta^{18}O$ values range from -15.39 to -16.84 ‰, and the $\delta^{18}O$ values of Type I calcite are very similar to those of Type II calcite (Table 4).

The Shaximiao sandstones were deposited under semi-arid and warm conditions (Wang and Wang 2010), and the $\delta^{18}O_{SMOW}$ value of -8 ‰ was assumed for early diagenetic formation water based on the results of Zhang et al. (2008). The current $\delta^{18}O_{SMOW}$ formation water composition (-4 ‰) (Shen et al. 2010; Li et al. 2008) was assumed for late diagenetic formation water. Using the fractionation equation of Friedman and O'Neil (1977), the temperatures for Type I and Type II calcite precipitation would be around 55–59 and 86–97 °C, respectively.

4.6 Burial and thermal history

In order to convert temperature and depth to time and thus date when various authigenic minerals precipitated, we have modeled the burial and thermal history in the study area. The modeling results shown in Fig. 7 indicate that the burial history of the Jurassic and Triassic sequences was characterized by an initial period of moderate to rapid subsidence from the Late Triassic to Late Jurassic (210–144 Ma). This was followed by a period of slow subsidence from Late Jurassic to Eocene (144–65 Ma). The subsidence phase was followed by uplift and erosion of 1000–2000 m of the sedimentary strata from Eocene to Neogene (65–5 Ma). Erosion estimates are based on sonic velocity and vitrinite reflectance (Shen et al. 2011).

5 Discussion

5.1 Paragenetic sequence

The Shaximiao sandstones have undergone complex diagenetic modifications. Petrographic examination allowed the relative timing of the main diagenetic events to be reconstructed (Fig. 8) and their influence on reservoir porosity to be also evaluated.

Grain-coating chlorite is poorly developed when grains are covered with early calcite cement. Quartz overgrowths were preferably developed where pore-lining chlorites are discontinuous or absent (Fig. 4c), indicating that chlorite coating formed before quartz overgrowth. Chlorite coating pre-dates quartz overgrowth, Type II calcite, and pore-filling chlorite (Fig. 4a, d) and is absent along the inter-granular straight contacts (Fig. 5), showing that sandstones have undergone mechanical compaction to a lesser extent. Fe and Mg ions required for the formation of authigenic chlorite were supplied by the alteration and dissolution of volcanic rock fragments, which are fairly common in the Shaximiao sandstones.

Authigenic quartz occurs as syntaxial overgrowths (Fig. 4c) and outgrowths (Fig. 4d). Quartz overgrowths are posterior to pore-lining chlorite authigenesis and pre-date pore-filling chlorite, whereas quartz outgrowths partially fill primary/secondary intergranular spaces and precipitated after pore-filling chlorite and before Type II calcite

Table 4 Isotopic compositions of calcite cements

Sample number	Description	$\delta^{13}C_{PDB}$, ‰	$\delta^{18}O_{PDB}$, ‰	Temperature, °C
370-11	Type I	−9.96	−15.45	55.6
168-58	Type I	−8.08	−15.39	55.2
168-68	Type I	−9.17	−16.02	59.4
168-11	Type II	−7.55	−16.84	97.0
168-39	Type II	−7.63	−16.34	92.4
168-97	Type II	−12.62	−16.19	91.1
168-118	Type II	−9.1	−15.6	86.0
167-3	Type II	−5.49	−16.06	90.0

Sweeney and Burnham (1990)_EASY%R_o=1.0, 2.0, 3.0 [%R_o]
Temperature=50.0, 100.0, 150.0, 200.0 [°C]
T
J
K
E
N
T_3
J_1
J_2
J_3
K_1
K_2
E_1
E_2
E_3
N_1
Depth, m
0
2000
4000
6000
t=100°C
R_o=1.0%
200
150
100
50
0
Time, Ma
K
J_3p
J_3sn
J_2s
J_2q
T_3x^5
T_3x^4
T_3x^3
T_3x^2

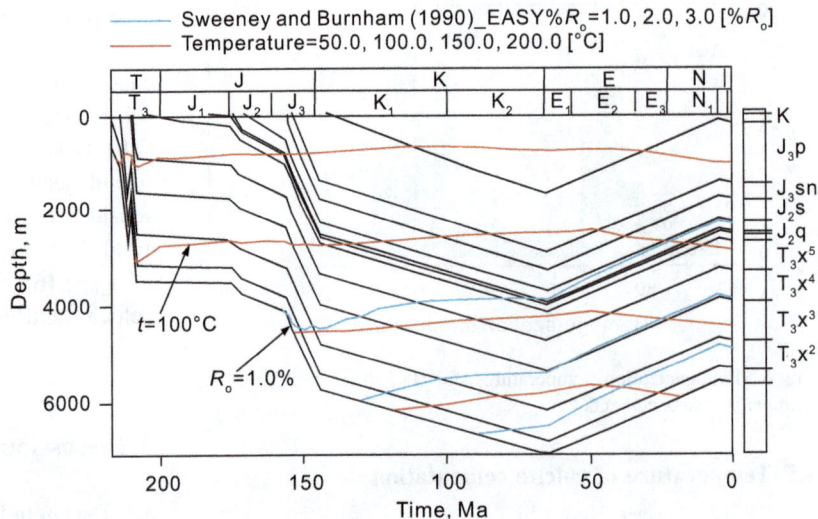

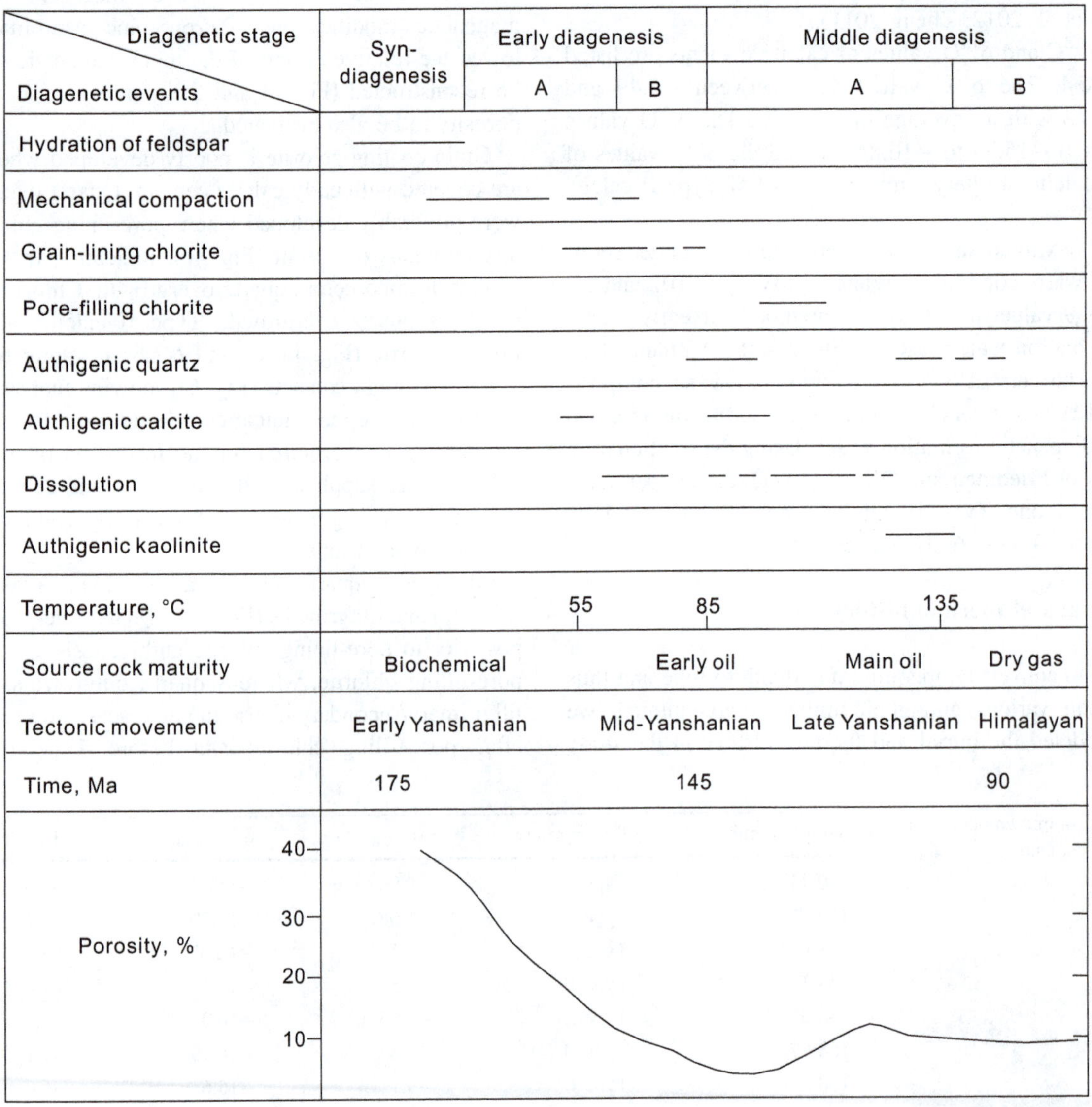

Fig. 7 Burial and thermal history of the study area. The heat-flow model was developed based on knowledge of the tectonic history of the basin and calibration to maturity data by comparing the R_o calibration data to the calculated values obtained using the Easy% R_o algorithm of Sweeney and Burnham (1990)

Diagenetic stage
Diagenetic events
Syn-diagenesis
Early diagenesis
Middle diagenesis
A
B
A
B
Hydration of feldspar
Mechanical compaction
Grain-lining chlorite
Pore-filling chlorite
Authigenic quartz
Authigenic calcite
Dissolution
Authigenic kaolinite
Temperature, °C
55
85
135
Source rock maturity
Biochemical
Early oil
Main oil
Dry gas
Tectonic movement
Early Yanshanian
Mid-Yanshanian
Late Yanshanian
Himalayan
Time, Ma
175
145
90
Porosity, %
40
30
20
10

Fig. 8 Paragenetic sequence and porosity evolution history of the Shaximiao sandstones

(Fig. 4d). Sources of silica include reaction of smectite to illite and dissolution of detrital feldspar minerals.

Type I calcite precipitated as concretionary masses during earliest diagenesis (55 °C, 150 Ma, Fig. 4e), and some samples display limited mechanical compaction due to early calcite cementation. Type II calcite fills dissolution pores (Fig. 4f), indicating that its precipitation was synchronous with or after late dissolution.

Detrital feldspars have undergone partial dissolution during eogenetic and mesogenetic stages. Remaining chlorite coating in early dissolution pores is generally deformed and exhibits vermiform structure, indicating that eogenetic dissolution pores were destroyed by subsequent compaction. However, remaining chlorite coating in late dissolution pores shows no deformation, demonstrating that no mechanical compaction occurred after late dissolution (Fig. 4f).

5.2 Porosity evolution

Alteration of sandstones mainly occurred during two paragenetic stages. Stage I is characterized by mechanical compaction and precipitation of early quartz, calcite, and chlorite. Stage II is characterized by dissolution and infilling of secondary pore spaces (Fig. 8).

5.2.1 Original porosity

The depositional porosity in sandstones was generally assumed to be 40 % (Houseknecht 1987). However, it has been demonstrated that original porosity is strongly influenced by depositional environment (Beard and Weyl 1973; Bloch and McGowen 1994). In this study, the original porosity was estimated using the empirical formula proposed by Beard and Weyl (1973).

Original porosity (OP)
$$= 20.91 + 22.9/\text{Trask's sorting coefficient (So)} \quad (3)$$

Trask's sorting coefficients for four sets of the Shaximiao reservoirs are 1.125, 1.13, 1.13, and 1.218, respectively. Thus, the original porosities were calculated to be 41.3 %, 41.2 %, 41.2 %, and 39.7 %, respectively, with an average of 40.9 %.

5.2.2 Stage I (<85 °C, 175–145 Ma)

Diagenetic alternations during Stage I include mechanical compaction and cementation by early quartz, calcite, and chlorite. Secondary pores formed by early dissolution were generally destroyed by continued compaction, and thus this process is not taken into account in this study.

Houseknecht (1987) and Ehrenberg (1989, 1995) developed a quantitative approach based on point count data to compute the original intergranular porosity loss due to compaction (COPL) and cementation (CEPL). However, the equations of Houseknecht (1987) and Ehrenberg (1989) assume that the secondary porosity and post-dissolution cement volumes are 0 %, which could cause COPL to be under-estimated if this assumption is far from reality. As discussed above, mechanical compaction has no or little impact on the Shaximiao sandstones after late dissolution. Thus, residual primary porosity (RPP) and pre-dissolution cement volumes (VEC) should be used for the calculation of COPL:

$$COPL = OP - RPP - VEC \quad (4)$$

where COPL is the original intergranular porosity loss due to compaction (%), OP is original porosity (%), RPP is residual primary porosity (%), and VEC is the total volume of early diagenetic cements (%).

During this stage, primary porosity was reduced from 40.9 % to 3.7 % by mechanical compaction and early diagenetic cementation through precipitation of quartz (1.7 %), calcite (1.7 %), and grain-coating chlorite (4.5 %) (Table 2; Fig. 8). Based on Eq. (4), COPL is then calculated to be 29.3 %. This suggests that mechanical compaction is a major factor of the loss of primary pores (Fig. 9), which is in good agreement with the rapid subsidence during this period (Fig. 7).

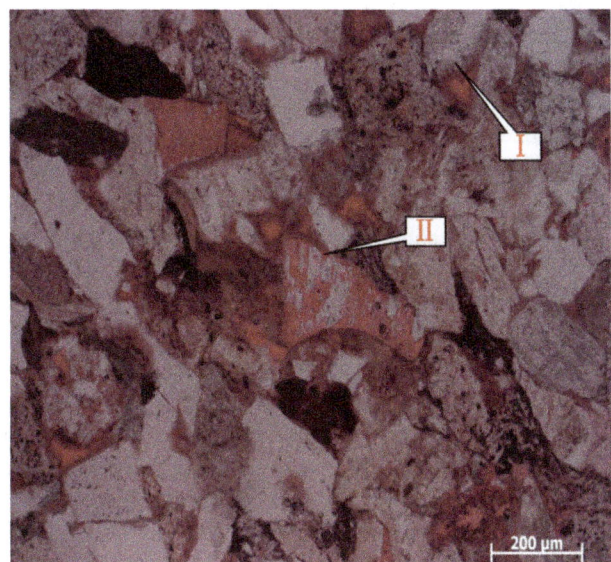

Fig. 9 Photomicrograph of remaining primary pore (I) and intragranular dissolution pore (II). Primary pores have been lost mainly by mechanical compaction. Well JS7, 1954.9 m

5.2.3 Stage II (>85 °C, 145–0 Ma)

This stage corresponds to a middle diagenetic stage and can be further divided into two sub-stages (Stage II_a and Stage II_b). Stage II_a (85–120 °C, 145–125 Ma) is dominated by late dissolution. Secondary pores formed in this stage by feldspar dissolution, and the source of acidic fluids is through the maturation of T_3x^5 source rocks. Organic acid-enriched fluids migrated upwards along faults and resulted in the formation of secondary pores in the Shaximiao sandstones. During stage II_b (>120 °C, 125–0 Ma), the dissolution pores were partially filled by late chlorite, quartz, calcite, and kaolinite cement (Table 2), leaving a residual secondary porosity of 5.5 % (Table 3). The total secondary porosity (TSP) formed during this period can be approximated as

$$TSP = RSP + VLC \qquad (5)$$

where TSP is the total secondary porosity (%), RSP is residual secondary porosity observed (%), and VLC is the total volume of late cements (%).

According to the point-count data, the volumes of late chlorite, quartz, calcite, and kaolinite are 1 %, 0.8 %, 1 %, and 0.5 %, respectively. Based on Eq. (5), the total secondary porosity is thus calculated to be 8.8 %.

It has widely been assumed that oil emplacement will retard the precipitation of quartz and calcite cements (Luo et al. 2009). Hence, the porosity of hydrocarbon-saturated sandstones would be 12.5 % (9.2 % (current thin-section porosity) + 3.3 % (late cement volumes) = 12.5 %), thus providing a good quality reservoir for hydrocarbon migration and accumulation (Fig. 8).

6 Conclusions

(1) The Shaximiao sandstones in western Sichuan have undergone complex diagenetic modifications, including a moderate degree of compaction, cementation by authigenic quartz (early and late), calcite (early and late), and chlorite (grain-coating and pore-filling), and dissolution.

(2) The point count data show that secondary pores (av. 5.5 %) are more abundant than remaining primary pores (av. 3.7 %), and most of samples are dominated by secondary pores.

(3) Analysis of porosity evolution indicates that alteration of sandstones mainly occurred during two paragenetic stages (Stage I and Stage II). Mechanical compaction and cementation by early chlorite, calcite, and quartz typically decrease the depositional porosity (40.9 %) by an average of 37.2 % leaving porosity of 3.7 % after stage I (<85 °C, 175–145 Ma). The original

intergranular porosity loss due to compaction is calculated to be 29.3 %, suggesting that mechanical compaction is the most significant diagenetic process in primary porosity destruction. Stage II can be further divided into two sub-stages (Stage II_a and Stage II_b). During stage II_a (85–120 °C, 145–125 Ma), acidic fluids generated by the thermal maturation of organic matter in T_3x^5 source rocks flowed into the Shaximiao sandstones and resulted in dissolution of feldspar and rock fragments, which enhanced porosity by 8.8 %, and the porosity after this stage increased from 3.7 % to 12.5 %. During stage II_b (>120 °C, 125–0 Ma), the precipitation of late chlorite, calcite, quartz, and kaolinite destroyed 3.3 % porosity, leaving porosity of 9.2 % in the rock today.

Acknowledgments This work was financially supported by the National Science Foundation of China (No. 41172119) and the Important National Science & Technology Specific Project (2011ZX05002-004-001). The authors acknowledge China Petroleum & Chemical Corporation (Sinopec) for permission to publish this paper.

References

Ajdukiewicz JM, Lander RH. Sandstone reservoir quality prediction: the state of the art. AAPG Bull. 2010;94(8):1083–91.

Beard DC, Weyl PK. Influence of texture on porosity and permeability of unconsolidated sand. AAPG Bull. 1973;57(2):349–69.

Bjørlykke K, Jahren J. Open or closed geochemical systems during diagenesis in sedimentary basins: constraints on mass transfer during diagenesis and the prediction of porosity in sandstone and carbonate reservoirs. AAPG Bull. 2012;96(12):2193–214.

Bloch S, McGowen JH. Influence of depositional environment on reservoir quality prediction. In: Wilson MD, editor. reservoir quality assessment and prediction in clastic rocks, vol. 30. Tulsa: SEPM Short Course; 1994. p. 41–57.

Cai KP, Liao SM. A research on the gas source of Jurassic gas reservoirs in western Sichuan Basin. Nat Gas Ind. 2000;20(1):36–41 (in Chinese).

Dai JY, Zhang YW, Xiong QH, et al. Effects of diagenesis on reservoir property and quality, a case study of the Cainan Oilfield in the east of Zhun Ga'er Basin. Pet Explor Dev. 2003;30(4):54–5 (in Chinese).

Ehrenberg SN. Measuring sandstone compaction from modal analysis of thin sections: how to do it and what the results mean. J Sediment Res. 1995;65A(2):369–79.

Ehrenberg SN. Assessing the relative importance of compaction process and cementation to reduction of porosity in sandstones; discussion; compaction and porosity evolution of Pliocene sandstones, Ventura Basin, California: discussion. AAPG Bull. 1989;73(10):1274–6.

Friedman I, O'Neil JR. Compilation of stable isotopic fractionation factors of geochemical interest. Reston: US Geological Survey Professional Paper; 1977.

Gabitov RI, Watson EB, Sadekov A. Oxygen isotope fractionation between calcite and fluid as a function of growth rate and temperature: an in situ study. Chem Geol. 2012;306–307:92–102.

Gu XD, Liu XH. Stratigraphy in the Sichuan Basin. Wuhan: China University of Geosciences Press; 1997 (in Chinese).

Houseknecht DW. Assessing the relative importance of compaction processes and cementation to reduction of porosity in sandstones. AAPG Bull. 1987;71(6):633–42.

Li W, Zhao KB, Liu CX. Hydrogeology study of petroliferous basins. Beijing: Geological Publishing House; 2008. p. 179–83 (in Chinese).

Liu DL, Tao SZ, Zhang BM. Application and questions about ascertaining oil-gas pools age with inclusion. Nat Gas Geosci. 2005;16(1):16–9 (in Chinese).

Liu MJ, Liu ZS, Xiao M, et al. Paleoporosity and critical porosity in the accumulation period and their impacts on hydrocarbon accumulation—a case study of the middle Es_3 member of the Paleogene Formation in the Niuzhuang Sag, Dongying Depression, Southeastern Bohai Bay Basin, East China. Pet Sci. 2014;11(4):495–507.

Lu HZ, Guo DJ. Progress and trends of researches on fluid inclusions. Geol Rev. 2000;46(4):385–92 (in Chinese).

Luo JL, Morad S, Salem A, et al. Impact of diagenesis on reservoir-quality evolution in fluvial and lacustrine-deltaic sandstones: evidence from Jurassic and Triassic sandstones from the Ordos basin, China. J Pet Geol. 2009;32(1):79–102.

Lü ZX, Ye SJ, Qing C. Characteristics and evaluation of the upper reservoir of Shaximiao Formation in Xiaoquan gas field in west Sichuan. Nat Gas Ind. 2000;20(5):15–8 (in Chinese).

Marfil R, Rossi C, Lozano RP. Quartz cementation in Cretaceous and Jurassic reservoir sandstones from the Salam oil field, Western Desert, Egypt: constraints on temperature and timing of formation from fluid inclusions. In: Worden RH, Morad S, et al., editors. Quartz cementation in sandstones., Special Publication Number 29 of the International Association of SedimentologistsOxford: Blackwell Science; 2000. p. 163–82.

Meng QR, Wan E, Hu JM. Mesozoic sedimentary evolution of the northwest Sichuan Basin: implication for continued clockwise rotation of the South China block. Geol Soc Am Bull. 2005;117(3–4):396–410.

Prodanović M, Bryant SL, Davis JS. Numerical simulation of diagenetic alteration and its effect on residual gas in tight gas sandstones. Transp Porous Media. 2013;96(1):39–62.

Qin SF, Dai JX, Wang LS. Different origins of natural gas in secondary gas pool in western Sichuan foreland basin. Geochimica. 2007;36(4):368–74 (in Chinese).

Shen ZM, Gong YJ, Liu SB, et al. A discussion on genesis of the Upper Triassic Xujiahe Formation water in Xinchang area, western Sichuan depression. Geol Rev. 2010;56(1):82–8 (in Chinese).

Shen ZM, Liu T, Lü ZX, et al. A comparison study on the gas source of Jurassic natural gas in the western Sichuan depression. Geol J China Univ. 2008;14(4):577–82 (in Chinese).

Shen ZM, Liu Y, Liu SB. Denudation recovery of Himalayan periods in middle section of western Sichuan depression. Comput Tech Geophys Geochem Explor. 2011;33(2):189 (in Chinese).

Stroker TM, Harris NB, Elliott WC, et al. Diagenesis of a tight gas sand reservoir: upper cretaceous mesa verde group, Piceance Basin, Colorado. Mar Pet Geol. 2013;40:48–68.

Sweeney JJ, Burnham AK. Evaluation of a simple model of vitrinite reflectance based on chemical kinetics. AAPG Bull. 1990;74:1559–70.

Tan WC, Hou MC, Dong GY, et al. Research on depositional system of Middle Jurassic Shaximiao Formation in western Sichuan foreland basin. J East China Inst Technol. 2008;31(4):336–43 (in Chinese).

Taylor KG, Gawthorpe RL, Pannon-Howell S. Basin-scale diagenetic alteration of shoreface sandstone in the Upper Cretaceous Spring Canyon and Aberdeen Members, Blackhawk Formation, Book Cliffs, Utah. Sediment Geol. 2004;172(1–2):99–115.

Taylor TR, Giles MR, Hathon LA, et al. Sandstone diagenesis and reservoir quality prediction: models, myths, and reality. AAPG Bulletin. 2010;94(8):1093–132.

Wang DY, Wang J. Sedimentary facies and its distribution of the Middle Jurassic Shaximiao Formation in the West Sichuan Foreland Basin. Acta Geol Sichuan. 2010;30(3):255–9 (in Chinese).

Wood JR, Byres AP. Alteration and emerging methodologies in geochemical and empirical modeling. In: Wilson MD, editor. Reservoir quality assessment and prediction in clastic rocks, vol. 30. Tulsa: SEPM Short Course; 1994. p. 395–400.

Xu XS, Liu BJ, Zhao YG, et al. Sequence stratigraphy and basin-mountain transformation in the western margin of upper Yangtze landmass during the Permian to Triassic. Beijing: Geological Publishing House; 1997 (in Chinese).

Yang KM, Zhu HQ, Ye J, et al. The geological characteristics of tight sandstone gas reservoirs in West Sichuan Basin. Beijing: Science Press; 2012 (in Chinese).

Ye SJ, Lü ZX. Reservoir characterization and factors influencing reservoir characteristics of the lower Shaximiao Formation in Xinchang gas field, western Sichuan. J Mineral Petrol. 2010;30(3):96–104 (in Chinese).

Zhang JL, Zhang PH, Xie J, et al. Diagenesis of clastic reservoirs: advances and prospects. Adv Earth Sci. 2013;28(9):957–67 (in Chinese).

Zhang L, Chen ZY, Nie ZL, et al. Correlation between $\delta^{18}O$ in precipitation and surface air temperature on different time-scale in China. Nucl Tech. 2008;31(9):715–20 (in Chinese).

Zheng YF. On the theoretical calculations of oxygen isotope fractionation factors for carbonate-water systems. Geochem J. 2011;45(4):341–54.

5

Numerical simulation of hydraulic fracture propagation in tight oil reservoirs by volumetric fracturing

Shi-Cheng Zhang[1] · Xin Lei[1] · Yu-Shi Zhou[1] · Guo-Qing Xu[1]

Abstract Volumetric fracturing is a primary stimulation technology for economical and effective exploitation of tight oil reservoirs. The main mechanism is to connect natural fractures to generate a fracture network system which can enhance the stimulated reservoir volume. By using the combined finite and discrete element method, a model was built to describe hydraulic fracture propagation in tight oil reservoirs. Considering the effect of horizontal stress difference, number and spacing of perforation clusters, injection rate, and the density of natural fractures on fracture propagation, we used this model to simulate the fracture propagation in a tight formation of a certain oilfield. Simulation results show that when the horizontal stress difference is lower than 5 MPa, it is beneficial to form a complex fracture network system. If the horizontal stress difference is higher than 6 MPa, it is easy to form a planar fracture system; with high horizontal stress difference, increasing the number of perforation clusters is beneficial to open and connect more natural fractures, and to improve the complexity of fracture network and the stimulated reservoir volume (SRV). As the injection rate increases, the effect of volumetric fracturing may be improved; the density of natural fractures may only have a great influence on the effect of volume stimulation in a low horizontal stress difference.

Keywords Tight oil reservoir · Volumetric fracturing · Fracture propagation · Horizontal stress difference · Stimulated reservoir volume

1 Introduction

Due to ultralow matrix permeability, multistage fracturing of horizontal wells is recognized as the main stimulation technology for an economical and effective approach to recover oil and gas from tight reservoirs (Zhao et al. 2012; Li et al. 2013). The main mechanism (King 2010) is to open natural fractures and expand them until shear sliding occurs in the process of hydraulic fracturing. The multistage, multi-perforation clusters per fracturing treatment in a horizontal wellbore and the natural fractures may create a fracture network system that may enhance the stimulated reservoir volume (SRV) and improve both the initial production and the ultimate recovery factor (Mayerhofer et al. 2008; Cipolla et al. 2009). There are many factors which influence the fracture propagation, like the horizontal stress difference, density of natural fractures, injection rate, number of perforation clusters, and the spacing of clusters (Yost et al. 1988; Palmer et al. 2007; Cipolla et al. 2011; Olson and Wu 2012).

Many researchers have studied the complex propagation of hydraulic fractures induced by volumetric fracturing with different numerical simulation approaches. Dahi-Taleghani and Olson (2009), Dahi-Taleghani (2010) and Keshavarzi et al. (2012) used a two-dimensional finite element method to simulate the complex fracture propagation. In this model, a uniform and constant net pressure is loaded on the surface of the hydraulic fractures. Olson (2008) and Olson and Dahi-Taleghani (2009) presented a pseudo-three-dimensional complex fracture network model

✉ Shi-Cheng Zhang
zhangsc@cup.edu.cn

[1] College of Petroleum Engineering, University of Petroleum, Beijing 102249, China

Edited by Yan-Hua Sun

based on the displacement discontinuity method, which considered the injection of non-Newtonian fluids, Carter filtration, random non-planar propagation and fracture height extension in three layers. Zhao and Young (2009) used a two-dimensional particle discrete element method, where the model consists of cohesive particles and pore space between these particles. The pore pressure will increase with the injection of the fracturing fluid, and it will remove the cohesion between particles. The simulated natural fractures are non-cohesive or weakly cohesive. Nagel et al. (2011) and Zangeneh et al. (2012) used a discrete element method to simulate the complex fracture network system. In this model, the rock mass is divided by multiple joints. The fractures only propagate along the joint network. No new hydraulic fractures occur and grow except the initial natural fractures. In recent years, the simulation technology of a complex fracture network has been developed, which combines with a micro-seismic imaging system to represent the complexity of hydraulic fractures. There are two primary models. One is the wire mesh model (Xu and Ghassemi 2009; Xu et al. 2010; Meyer and Bazan 2011), which can effectively simulate the complexity of fractures and the spacing between perpendicular fractures; another one is the unconventional fracture model (Weng et al. 2011), which describes complex geological conditions and evaluates the propagation of complex fractures more strictly. However, the extended finite element and boundary element method may not apply various hydraulic pressures on the fracture surface, and also does not consider the impact of seepage and leak off of fracturing fluids; the discrete element method restricts the path of hydraulic fractures.

In this study, we use a mixed finite element and discrete element method to build a model for predicting propagation of fractures induced by volumetric fracturing in a tight oil reservoir. By using this model, we mainly examine the impact of horizontal stress difference, number and spacing of perforation clusters, injection rate, and density of natural fractures on fracture patterns. This research may have a significant effect on future hydraulic fracturing design of tight oil reservoirs.

2 Numerical simulation and analysis

2.1 Fracture propagation model

The rock deformation is based on a linear elastic fracture mechanism. The governing equation consists of a stress equilibrium equation and a fracturing fluid flow equation (continuity equation). In this model, the coupling of fluid flow in fractures and rock deformation of the matrix block is based on the continuum discrete element method. The

domain is discretized into many matrix blocks, which are linked by virtual springs. The breakage of a spring represents the failure of the rock. There is a fracture element between two blocks to calculate the flow of fracturing fluids and the distribution of the hydraulic pressure. As an external load, the hydraulic pressure will be applied on the surface of fractures. We use the finite element method to solve the deformation of continuous blocks, and use the discrete element method to solve the breakage of springs. The spring breakage is based on the maximum tensile stress criterion and the Mohr–Coulomb criterion. Figure 1 indicates the calculation model. Due to the ultralow matrix permeability of tight oil reservoirs, the seepage and leak off of fracturing fluids can be ignored. This model mainly considers some key parameters, including rock mechanical properties, in situ stress, reservoir pressure, natural fracture characteristics, injection rate, and the fracturing fluid viscosity.

The key point of the successful volumetric fracturing is to connect the natural fractures in the reservoir. Many studies show that the horizontal stress difference is the major geological factor contributing to form a complex fracture network system. Laboratory fracturing experiments (Blanton 1982; Warpinski and Teufel 1987; Gu and Weng 2010) indicate that a horizontal stress difference of lower than 4 MPa is beneficial to open the natural fractures and form complex fractures; and a horizontal stress difference of higher than 8 MPa is unsuitable for opening the natural fractures, it is easier to form a planar fracture. 4–8 MPa is the transition zone from complex fractures to planar fractures.

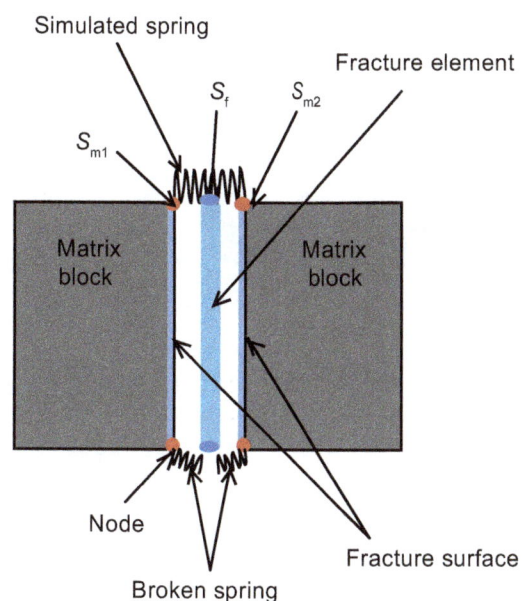

Fig. 1 Sketch of calculation model

By using the real reservoir data and fracturing parameters, we simulated the volumetric fracturing in a tight oil formation of a certain oilfield, and examined the impact of the horizontal stress difference, linear density of natural fractures, number and spacing of perforation clusters on fracture patterns. Figure 2 shows the model of a horizontal well with single-stage fracturing and 4 perforation clusters. The horizontal well trajectory is perpendicular to the maximum horizontal stress. The creation of the natural fractures is from random sampling. The angle of natural factures is real reservoir data, which varies from 0 to 30° oriented to the maximum horizontal stress.

2.2 Calculation of the stimulated reservoir volume (SRV)

The conception of the SRV is presented by Mayerhofer et al. (2008). According to the production data from the Barnett shale gas, the bigger the volume of the fracture network, the better the effect on production after fracturing. The primary calculation method for the SRV is to divide the micro-seismic monitor data into several blocks, which are presented as bands, and to add up the volume of all blocks, shown as Fig. 3. For ease of calculation, the calculation of the SRV can be simplified as the calculation of the stimulated reservoir area (SRA), which is a permeability enhanced area. In that case, the solution can be the calculation of any closed region. The main calculation method includes the ellipse method, boundary analytical method, and the probability method. In this study, we use the ellipse method to calculate the stimulated reservoir area, shown as Fig. 4. We only accumulate the effective closed region area of the fracture network, which is connected the fracture network, as the part of the stimulated reservoir area. The single plane fractures cannot be

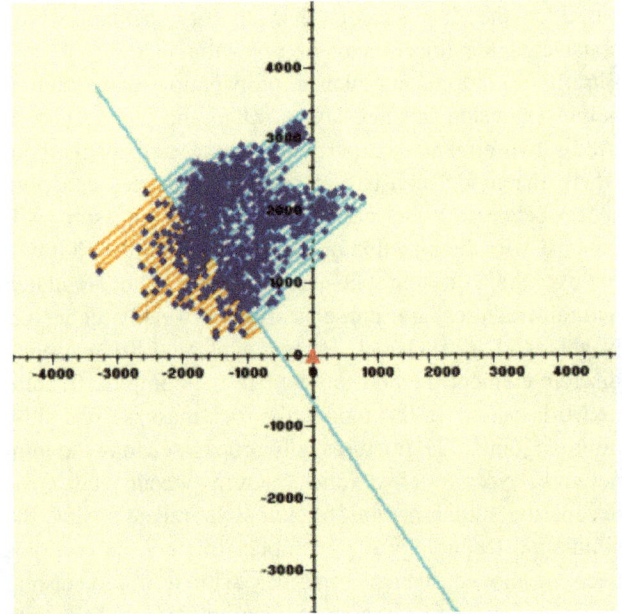

Fig. 3 Estimating SRA from micro-seismic mapping data

included, as shown in Fig. 5. First we calculate the stimulated reservoir area, and assume the fracture height equaling the reservoir thickness. Then we calculate the SRV with Eq. (1).

$$SRV = SRA \times H_\mathrm{f} = \sum_{i=1}^{n} A_i H_\mathrm{p}, \tag{1}$$

where H_f is the fracture height; SRA is the stimulated reservoir area; H_p is the reservoir thickness.

2.3 Impact of the horizontal stress difference

In this simulation, the linear density of the natural fractures is 0.12 m/m^2, the number of the perforation clusters is 4, the perforation cluster spacing is 20 m, and the injection rate is 15 m^3/min. Figure 6 shows the fracture network geometry at horizontal stress differences of 3, 6, and 9 MPa, respectively. Figure 6a shows under a horizontal stress difference of 3 MPa, once the hydraulic fracture meets a natural fracture, hydraulic fractures could easily deflect and connect with more natural fractures. The stimulated reservoir area (SRA) is about 12,450 m^2, and the average fracture length is 411 m. Under a horizontal stress difference of 6 MPa, the fracture network is narrower. The SRA is about 6470 m^2, and the average fracture length is 458 m, shown as Fig. 6b. Under a horizontal stress difference of 9 MPa, when the hydraulic fracture encounters a natural fracture, instead of deflecting, the hydraulic fracture will more easily pass through the natural fracture, which will cause the number of connected natural

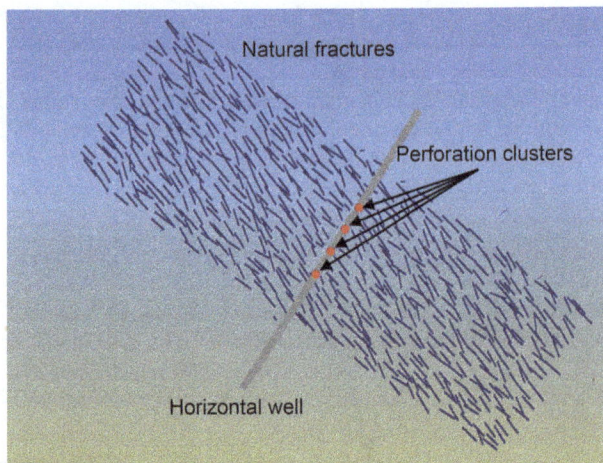

Fig. 2 Single-stage fractured horizontal well with 4 perforation clusters

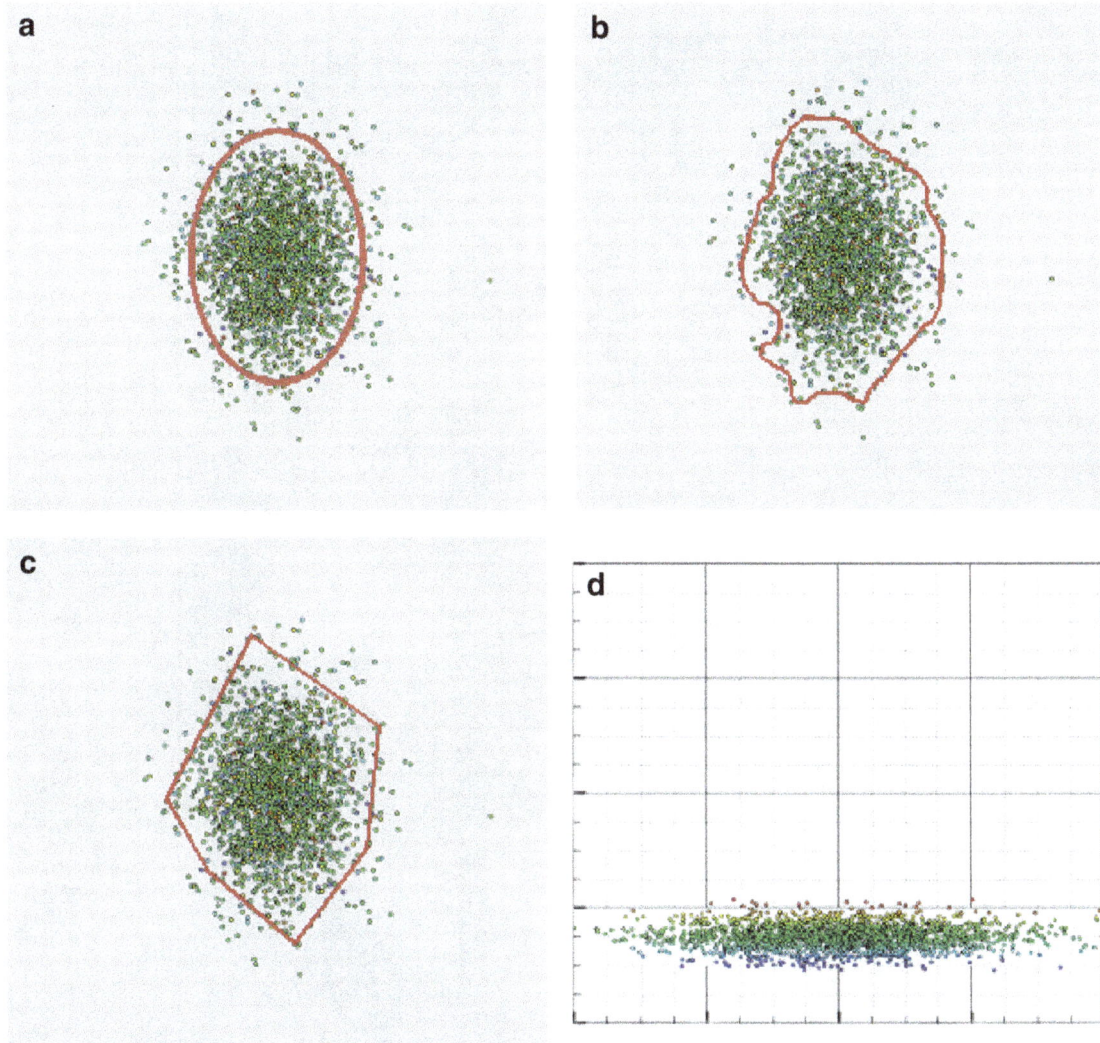

Fig. 4 Methods for calculating the SRA. **a** Ellipse method; **b** Probability method; **c** Boundary analytical method; **d** Profile map

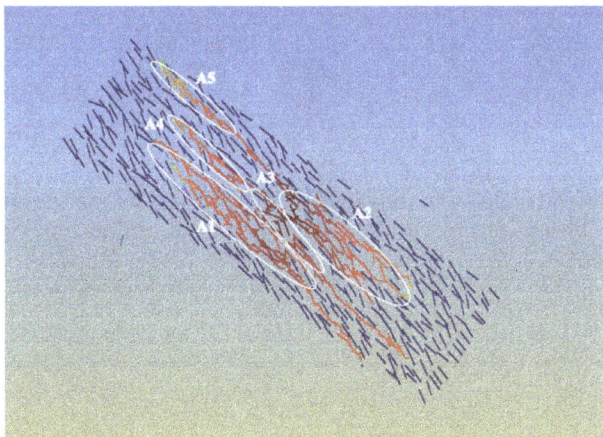

Fig. 5 Calculation of the stimulated reservoir area by the ellipse method

fractures to reduce. The SRA is about 3100 m², and the average fracture length is 493 m, shown as in Fig. 6c.

With a lower horizontal stress difference, the natural fractures could easily be opened by hydraulic fractures, and then form branch fractures. Finally, the hydraulic fractures connect with the natural fractures to form a complex fracture network system. With an increase in the horizontal stress difference, the SRV is reduced and the average fracture length is increased. When the horizontal stress difference is higher than 6 MPa, the fracture geometry changes from a complex fracture network to planar fractures, which causes a reduction in the SRV and an increase in the fracture length. The horizontal stress difference of the target formation is 3–5 MPa. It is beneficial to form a complex fracture network by volumetric fracturing.

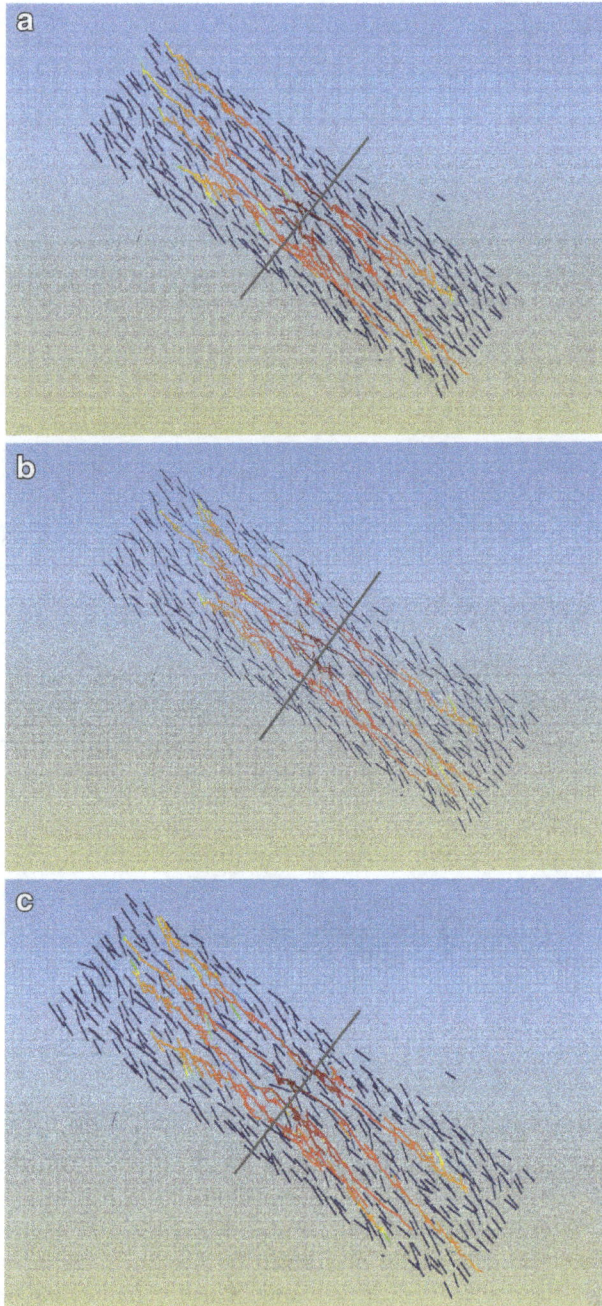

Fig. 6 Fracture geometry under various horizontal stress differences. **a** 3 MPa; **b** 6 MPa; **c** 9 MPa

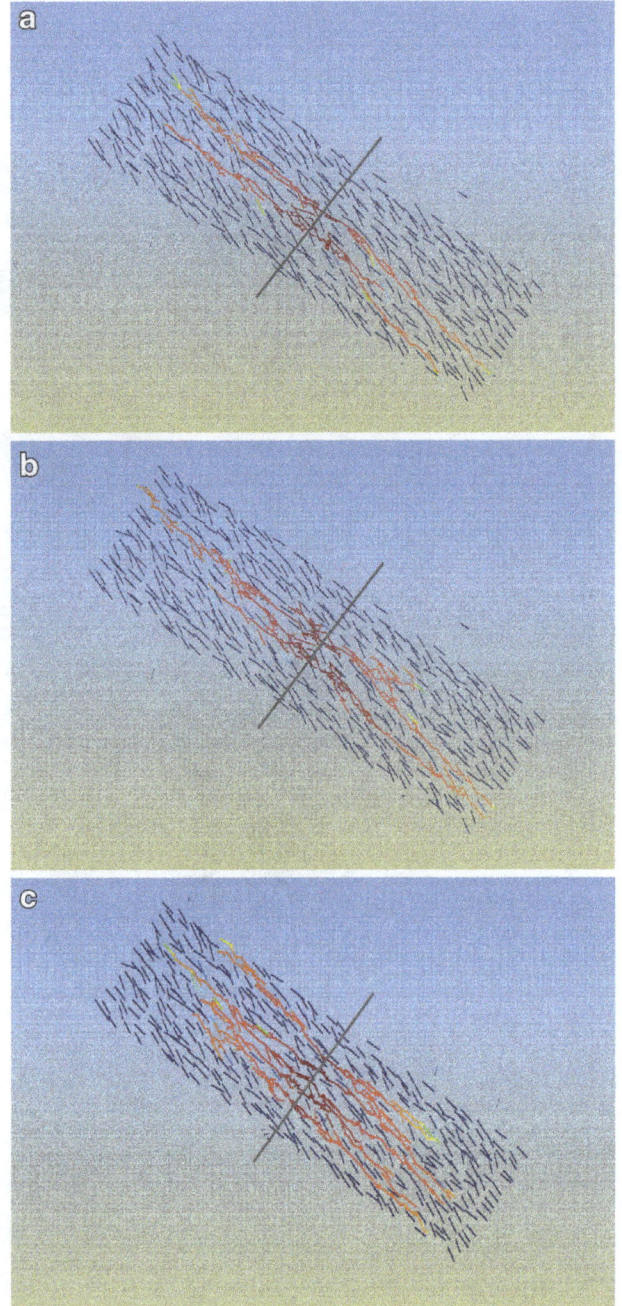

Fig. 7 Fracture geometry under various perforations. **a** 2 clusters, 6 m^3/min injection rate; **b** 2 clusters, 8 m^3/min injection rate; **c** 5 clusters, 15 m cluster spacing, 15 m^3/min injection rate

2.4 Impact of the number and spacing of perforation clusters

The number and spacing of perforation clusters has a direct influence on the fracture propagation geometry of horizontal well fracturing (Cheng 2009). For a tight oil reservoir, 2 perforation clusters, 20 m cluster spacing and 6–8 m^3/min injection rate cannot meet the requirement of stimulated reservoir volume (SRV). The SRA is only

1243 m^2 (Fig. 7a). The directions of the hydraulic fractures are the same as those of the natural fractures. The number of branch fractures is too low to form a complex fracture network system. Once the number of the perforation clusters is increased to 5 and the cluster spacing is reduced to 15 m (Fig. 7c), more natural fractures are induced to open and connect with each other to form a fracture network, shown as Fig. 7c. The SRA increases from 12,450 m^2 in 4

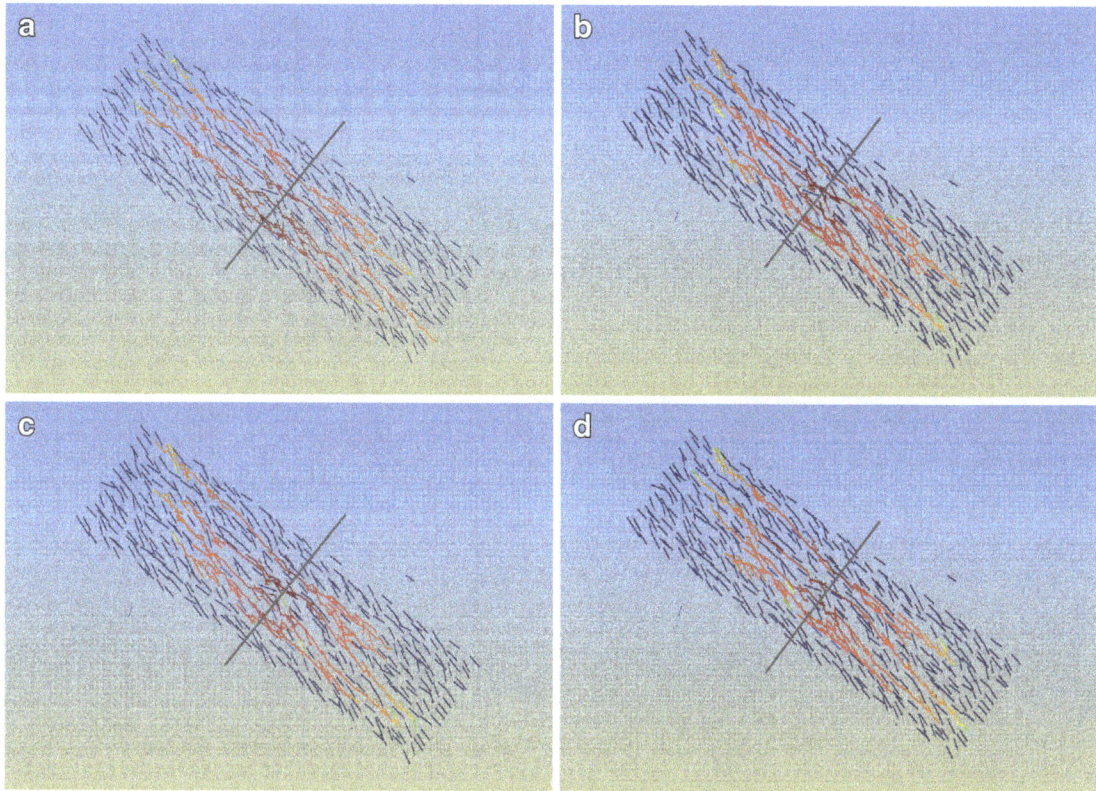

Fig. 8 Fracture geometry at various injection rates. **a** 10 m^3/min; **b** 12 m^3/min; **c** 14 m^3/min; **d** 15 m^3/min

clusters (20 m cluster spacing) to 15,040 m^2, and the average fracture length reduces from 411 to 379 m.

Increasing the number of perforation clusters and reducing the cluster spacing are beneficial to improve the complexity of the fracture network under the condition of low horizontal stress difference (Manchanda et al. 2012). However, propagation of multiple fractures at a low horizontal stress difference will lead to interaction among fractures (Bunger et al. 2011), which causes compressed middle fractures. Due to the small fracture width, as the hydraulic fracture propagates, the volume of liquid that is injected into the fractures will be reduced. Therefore, it is hard to inject the proppant into the fractures thus leading to sand plugging (Olsen et al. 2009). On the other hand, if the horizontal stress difference is high, the fracture propagation becomes difficult if the horizontal well has a close fracture spacing. As a result, the number of fractures will be reduced. Therefore, a reduction in the effective stimulated area which is caused by short perforation cluster spacing should be avoided.

2.5 Impact of the injection rate

Injection rate is one of the most important engineering factors in tight oil reservoir stimulation (King et al. 2008). The simulation results show that a high injection rate is beneficial to improve the complexity of the fracture network. Figure 8 indicates the geometry of the fracture network after volumetric fracturing at different injection rates. The horizontal stress difference is 3 MPa, the number of the perforation clusters is 4 and the cluster spacing is 20 m. The simulation results show that the higher the injection rate, the bigger the SRA. When the injection rate increases from 10 to 15 m^3/min, the SRA increases from 8080 to 14,910 m^2 correspondingly.

2.6 Impact of the linear density of natural fractures

If the linear density of natural fractures increases from 0.12 m/m^2 (Fig. 6) to 0.14 m/m^2, more stimulated natural fractures are developed. With 4 perforation clusters, 20 m cluster spacing and 15 m^3/min injection rate, the SRA will increase by a large margin. When the horizontal stress difference is 3 MPa, the SRA increases from 12,450 to 15,870 m^2, an increases of about 27.5 % (Fig. 9a). When the horizontal stress difference is 6 MPa, the SRA increases from 6470 to 6900 m^2, only 6.6 % larger (Fig. 9b). This result shows the linear density of the natural fractures only has a large impact on the volumetric fracturing in a low horizontal stress difference (Wu and Pollard 2002). Under a higher horizontal stress difference, the hydraulic fractures may hardly open and connect with the natural

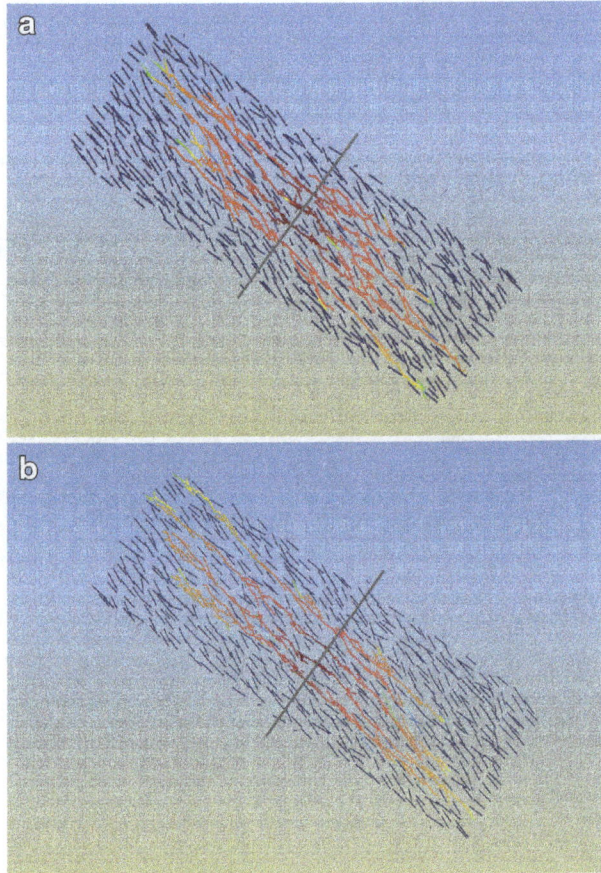

Fig. 9 Fracture geometry under higher linear density of natural fractures. **a** Horizontal stress difference of 3 MPa; **b** Horizontal stress difference of 6 MPa

fractures even if the linear density is high and volumetric stimulation will be difficult.

3 Field application

By using the physical properties and fracturing parameters (Table 1) of a tight oil reservoir in a certain oilfield, we simulated the fracture geometry after volumetric fracturing. Then the SRA was calculated and the simulated result was compared with the real micro-seismic monitoring result to verify the accuracy of our model.

Figure 10 shows a micro-seismic monitoring diagram, which is the fracture geometry of a horizontal well after 8-stage fracturing. Table 2 shows the results of micro-

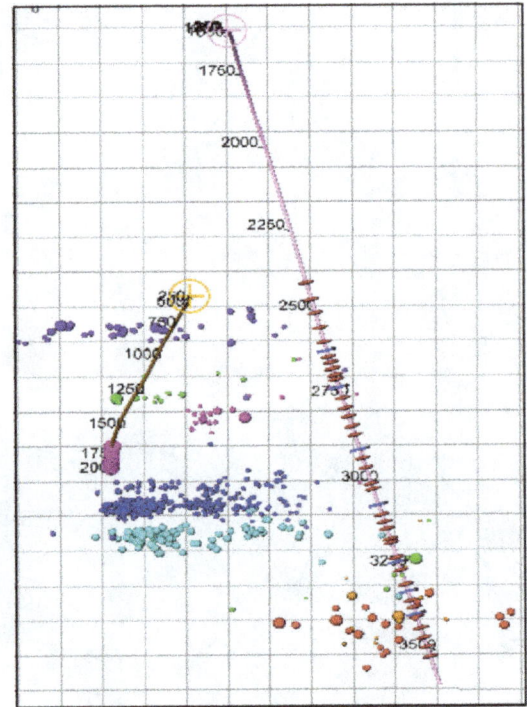

Fig. 10 Micro-seismic monitoring diagram

seismic monitoring, including the band length, band width, and the fracture height. According to these results, the SRV is 12×10^6 m^3. Figure 11 indicates the simulated fracture geometry, the SRV is 10.9×10^6 m^3. The error between the real fracturing and simulation results is only 8.2 %. This result verifies the accuracy of the model proposed in this paper.

4 Conclusions

(1) By using a mixed finite element and discrete element method, a model has been built to predict the propagation of fractures induced by fracturing in a tight oil reservoir. The influence of horizontal stress difference, number and spacing of perforation clusters, injection rate, and the linear density of natural fractures on fracture propagation has been studied with this model.

(2) When the horizontal stress difference is lower than 5 MPa, it is beneficial to form a complex fracture network system; when the horizontal stress

Table 1 Reservoir physical properties and fracturing parameters

Target formation	Permeability, 10^{-3} μm^2	Porosity, %	Horizontal stress difference, MPa	Perforation clusters	Cluster spacing, m	Injection rate, m^3/min
Tight oil	0.27	7.5	3	5	30	15

Table 2 Comparison of simulation and micro-seismic monitoring results

Formation	Calculation method	Stage number	Band length, m	Fracture height, m	Band width, m	SRV, m³	Error, %
Tight oil	Micro-seismic monitoring result	8	330	82.5	165	12×10^6	8.2
			198	25	171		
			283	22.9	72.3		
			340	26.3	129		
			373	34.7	177		
			171	41	96.5		
			304	30.3	127		
			217	98	42.4		
	Simulation result	8	300	30	152	10.9×10^6	

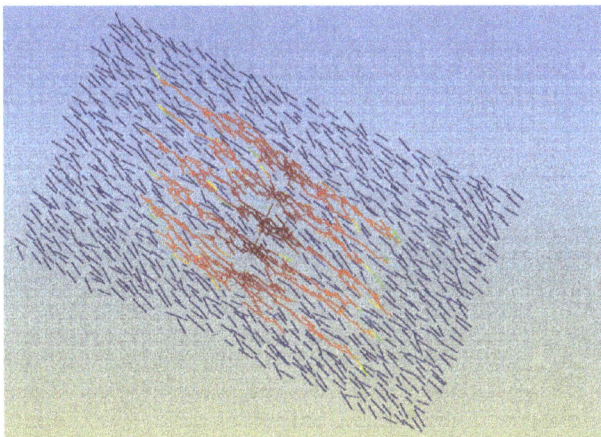

Fig. 11 Simulated fracture geometry

difference is higher than 6 MPa, it is easy to form a planar fracture system. The density of natural fractures only has a great influence on the effect of volumetric stimulation when there is a low horizontal stress difference.

(3) When there are low horizontal stress differences, increasing perforation clusters or reducing cluster spacing has a little impact on increasing the stimulated reservoir volume (SRV). The interaction among fractures is serious. Fractures at some regions will deflect and coalesce. With high horizontal stress differences, increasing the number of perforation clusters is beneficial to open and connect more natural fractures, and to improve the complexity of the fracture network and the SRV. As the injection rate increases, the effect of volumetric fracturing may be improved.

References

Blanton TL. An experimental study of interaction between hydraulically induced and pre-existing fracturing. In: SPE unconventional gas recovery symposium, Pittsburgh, 1982. doi: 10.2118/10847-MS.

Bunger AP, Jeffrey RG, Kear J, Zhang X. Experimental investigation of the interaction among closely spaced hydraulic fractures. In: 45th U.S. rock mechanics/geomechanics symposium, San. 2011. ARMA 11-318.

Cipolla CL, Lolon EP, Dzubin B. Evaluating stimulation effectiveness in unconventional gas reservoirs. In: SPE annual technical meeting, New Orleans, Louisiana. 2009. doi:10.2118/124843-MS.

Cipolla CL Weng X, Mack M, et al. Integrating microseismic mapping and complex fracture model to characterize fracture complexity. In: SPE hydraulic fracturing technology conference, The Woodlands, Texas, USA. 2011. doi:10.2118/140185-MS.

Cheng Y. Boundary element analysis of the stress distribution around multiple fractures: implications for the spacing of perforation clusters of hydraulically fractured horizontal wells. In: SPE Eastern Regional meeting, Charleston, West Virginia, USA, 2009. doi:10.2118/125769-MS.

Dahi-Taleghani A, Olson J E. Numerical model of multi-stranded hydraulic fracture propagation: accounting for the interaction between induced and natural fractures. In: SPE annual technical conference and exhibition, New Orleans, Louisiana, USA, 2009. doi:10.2118/124884-PA.

Dahi-Taleghani A. Fracture re-initiation as a possible branching mechanism during hydraulic fracturing. In: 44th U.S. rock mechanics symposium and 5th U.S.–Canada rock mechanics symposium, Salt Lake City, Utah, 2010. ARMA 10-278, 2010.

Gu H, Weng X. Criterion for fractures crossing frictional interfaces at non-orthogonal angles. In: The 44th US rock mechanics symposium and 5th U.S.–Canada rock mechanics symposium, Salt Lake City, 2010. ARMA 10-198.

King GE, Haile L, Shuss JA. Increasing fracture path complexity and controlling downward fracture growth in the Barnett shale. In: SPE shale gas production conference, Fort Worth, Texas, USA, 2008. doi:10.2118/119896-MS.

King GE. Thirty years of gas shale fracturing: what have we learned? J Petrol Technol. 2010;62(11):88–90.

Keshavarzi R, Mohammadi S, Bayesteh H. Hydraulic fracture propagation in unconventional reservoirs: the role of natural fractures. In: 46th U.S. rock mechanics/geomechanics symposium, Chicago, Illinois, 2012. ARMA 12-129.

Li XW, Zhang KS, Fan FL, et al. Study and experiments of volumetric fracturing in low pressure tight formation of the Ordos Basin. J Oil Gas Technol. 2013;35(3):142–6, 152 (in Chinese).

Mayerhofer MJ, Lolon EP, Warpinski NR, et al. What is stimulated rock volume? In: SPE shale gas production conference, Fort Worth, Texas, USA, 2008. doi:10.2118/119896-MS.

Meyer BR, Bazan LW. A discrete fracture network model for hydraulically-induced fractures: theory, parametric and case studies. In: SPE hydraulic fracturing conference and exhibition, Woodlands, Texas, 2011. doi:10.2118/140514-MS.

Manchanda R, Roussel NP, Sharma, MM. Factors influencing fracture trajectories and fracturing pressure data in a horizontal completion. In: 46th U.S. Rock mechanics/geomechanics symposium, Chicago, Illinois, 2012. ARMA 12-633, 2012.

Nagel N, Gil I, Sanchez-Nagel M, Damjanac B. Simulating hydraulic fracturing in real fractured rock—overcoming the limits of pseudo3D models. In: SPE hydraulic fracturing technology conference and exhibition, The Woodlands, Texas, 2011. doi:10.2118/140480-MS.

Olsen T, Bratton T, Thiercelin M. Quantifying proppant transport for complex fractures in unconventional formations. In: SPE hydraulic fracturing technology conference, The Woodlands, Texas, USA, 2009. doi:10.2118/119300-MS.

Olson JE. Multi-fracture propagation model: applications to hydraulic fracturing in shales and tight gas sands. In: 42nd U.S. symposium on rock mechanics, San Francisco, California, 2008. ARMA 08-327.

Olson JE, Dahi-Taleghani A. Model simultaneous growth of multiple hydraulic fractures and their interaction with natural fractures. In: SPE hydraulic fracturing technology conference, The Woodlands, Texas, USA, 2009. doi:10.2118/119739-MS.

Olson JE, Wu K. Sequential versus simultaneous multi-zone fracturing in horizontal wells: insights from a non-planar, multi-frac numerical model. In: SPE hydraulic fracturing technology conference, The Woodlands, Texas, USA, 2012. doi:10.2118/152602-MS.

Palmer ID, Moschovidis ZA, Cameron JR. Model shear failure and stimulation of the Barnett shale after hydraulic fracturing. In: SPE hydraulic fracturing technology conference, College Station, Texas, USA, 2007. doi:10.2118/106113-MS.

Xu W, Ghassemi A. Poroelastic analysis of hydraulic fracture propagation. In: 43rd U.S. rock mechanics symposium & 4th U.S.–Canada rock mechanics symposium, Asheville, North Carolina, 2009. ARMA 09-129.

Xu W, Thiercelin M, Ganguly U. Wiremesh: a novel shale fracture simulator. In: International oil and gas conference and exhibition in China, Beijing, China, 2010. doi:10.2118/132218-MS.

Warpinski N, Teufel LW. Influence of geologic discontinuities on hydraulic fracture propagation. J Petrol Technol. 1987;39(2):20–9. doi:10.2118/13224-PA.

Wu H, Pollard D. Imaging 3-D fracture networks around boreholes. AAPG Bull. 2002;86(4):593–604.

Weng X, Kresse O, Cohen C. Model of hydraulic fracture network propagation in a naturally fractured formation. SPE Prod Oper. 2011;26(4):368–80. doi:10.2118/140253-PA.

Yost AB, Overby WK, Wilkins DA. Hydraulic fracturing of a horizontal well in a naturally fractured reservoir: gas study for multiple fracture design. In: SPE gas technology symposium, Dallas, Texas, 1988. doi:10.2118/17759-MS.

Zangeneh N, Eberhardt E, Bustin RM. Application of the distinct-element method to investigate the influence of natural fractures and in situ stresses on hydrofrac propagation. In: 46th US rock mechanics/geomechanics symposium, Chicago, IL, USA, 2012. ARMA 12-223.

Zhao XP, Young RP. Numerical simulation of seismicity induced by hydraulic fracturing in naturally fractured reservoirs. In: SPE annual technical conference and exhibition, New Orleans, Louisiana, 2009. doi:10.2118/124690-MS.

Zhao ZZ, Du JH, Zou CN, et al. Tight oil and Gas. Beijing: Petroleum Industry Press; 2012 (in Chinese).

6

Genetic types and geochemical characteristics of natural gases in the Jiyang Depression, China

Wen-Tao Li · Yang Gao · Chun-Yan Geng

Abstract Natural gases were widely distributed in the Jiyang Depression with complicated component composition, and it is difficult to identify their genesis. Based on investigation of gas composition, carbon isotope ratios, light hydrocarbon properties, as well as geological analysis, natural gases in the Jiyang Depression are classified into two types, one is organic gas and the other is abiogenic gas. Abiogenic gas is mainly magmatogenic or mantle-derived CO_2. Organic gases are further divided into coal-type gas, oil-type gas, and biogas according to their kerogen types and formation mechanisms. The oil-type gases are divided into mature oil-type gas (oil-associated gas) and highly mature oil-type gas. The highly mature oil-type gases can be subdivided into oil-cracking gas and kerogen thermal degradation gas. Identification factors for each kind of hydrocarbon gas were summarized. Based on genesis analysis results, the genetic types of gases buried in different depths were discussed. Results showed that shallow gases (<1,500 m) are mainly mature oil-type gases, biogas, or secondary gases. Secondary gases are rich in methane because of chromatographic separation during migration and secondary biodegradation. Secondary biodegradation leads to richness of heavy carbon isotope ratios in methane and propane. Genesis of middle depth gases (1,500–3,500 m) is dominated by mature oil-type gases.
Deep gases (3,500–5,500 m) are mainly kerogen thermal degradation gas, oil-cracking gas, and coal-type gas.

Keywords Genetic types · Natural gases · Jiyang Depression · Light hydrocarbon properties · Carbon isotope ratios · Identification factors

1 Introduction

The Jiyang Depression is located in the north part of Shandong Province in China, on the fluvial plain and the delta where the Huanghe River runs into the Bohai Sea. Tectonically, the Jiyang Depression is located in the southeast part of the Bohai Bay Basin. It is a big terrestrial depression and ranks as one of the most prolific petroliferous area (Li et al. 2003). Since the discovery of the Shengli Oilfield in 1960, 50×10^8 t of OOIP and $2,500 \times 10^8$ m^3 OGIP have been proved, at the same time, 10.7×10^8 tons of oil and 460×10^8 m^3 gases have been produced.

Five sets of source rocks were developed in the Jiyang Depression, and they are distributed in the Carboniferous-Permian, the second member of the Kongdian Formation (Ek$_2$), and the fourth, third, and first members of the Eogene Shahejie Formation (Es$_4$, Es$_3$, and Es$_1$). The kerogen in those source rocks is mainly sapropelic type, and some of them are humic type. After a series of tectonic movements, these source rocks vary greatly in depth and evolution histories which influence gas generation and accumulation in many aspects, such as gas components, genesis, etc. (Zhang 1991). Thirteen commercial gas bearing layers have been discovered in the Neogene Minghuazhen and Guantao Formations, the Eogene Shahejie Formation, and the Paleozoic Carboniferous-Permian and Ordovician in the Jiyang Depression (Fig. 1). Gas reservoirs occurred widely at a depth from 192

W.-T. Li
School of Energy Resources, China University of Geosciences, Beijing 100083, China

W.-T. Li · Y. Gao (✉) · C.-Y. Geng
Geoscience Research Institute, Shengli Oilfield Company SINOPEC, Dongying 257015, Shandong, China
e-mail: swap124@163.com

Edited by Jie Hao

to 4,750 m. In these reservoirs, gas compositions vary greatly from hydrocarbon gas to abiogenic gas. As for hydrocarbon gas, the paraffin hydrocarbon composition and carbon isotope ratios varied dramatically. There are several different genesis models such as oil-type gas, coal-type gas, biogas, and inorganic mantle source gas, etc. (Gao et al. 2011; Zhou 2004; Luo et al. 2008). Natural gases usually occur as normal gas reservoirs, tight sandstone gas reservoirs, shale gas reservoirs, and coal-bed methane.

To make a thorough investigation of the gas genesis in the Jiyang Depression, the authors collected abundant data from exploration wells with commercial gas flow including 472 sets of natural gas composition data, 293 sets of carbon isotope ratio data (both hydrocarbon gas and carbon dioxide), and 69 sets of light hydrocarbon properties data (Fig. 2). According to gas component contents, carbon isotope ratios and light

hydrocarbon properties, combined with geological analysis, natural gases in the Jiyang Depression are divided into two categories namely organic gas and abiogenic gas. Organic gas was further divided into coal-type gas, oil-type gas, and biogas according to kerogen type and formation mechanism. The oil-type gases were finally divided into mature oil-type gas (oil-associated gas) and highly mature oil-type gas (including oil-cracking gas and kerogen thermal degradation gas) (Schoell 1980). The geochemical properties of each kind of natural gas were discussed, respectively.

2 Abiogenic gas

Abiogenic gas in the Jiyang Depression is mainly CO_2, and its distribution is controlled by great deep faults (Tang

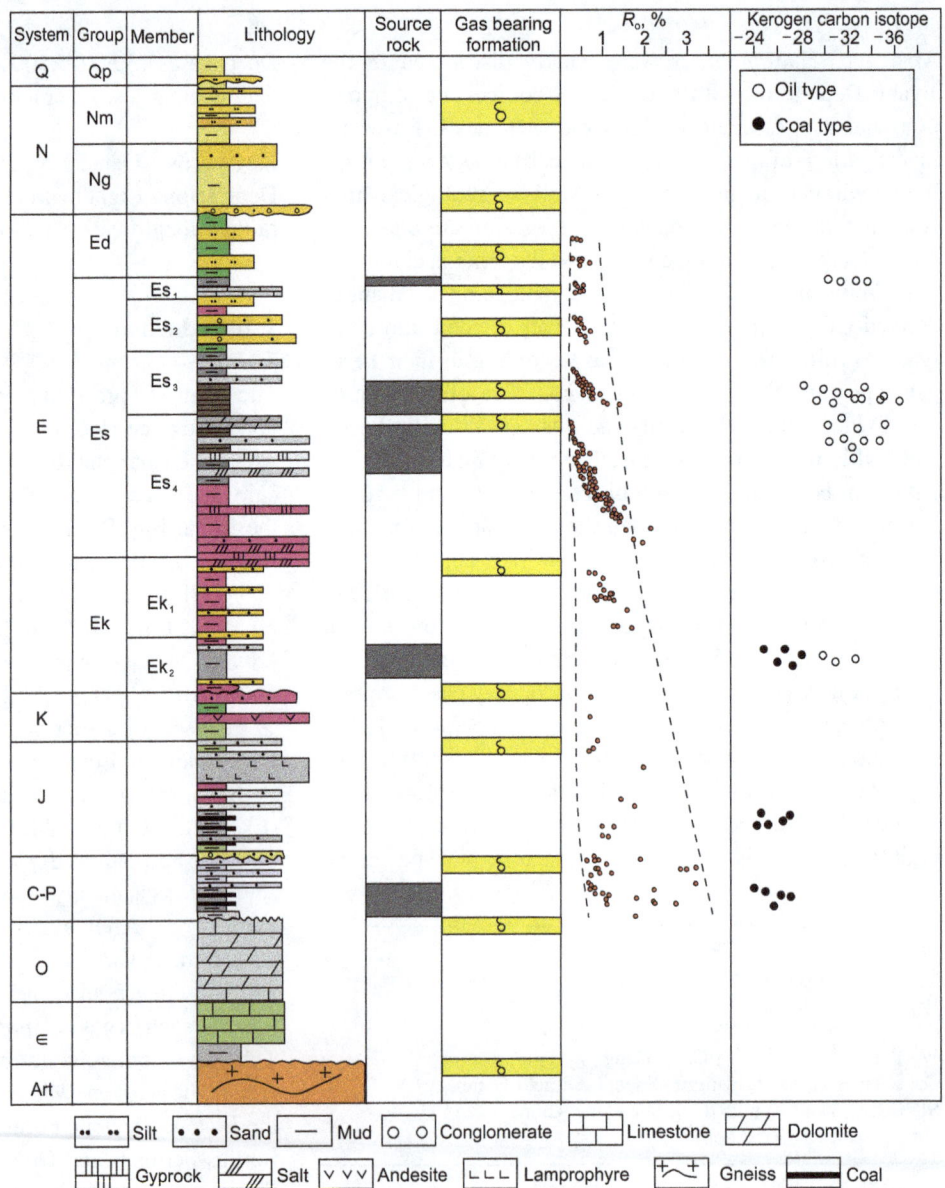

Fig. 1 Strata histogram and gas bearing layers in the Jiyang Depression

Fig. 2 Structural framework and typical gas producing wells in the Jiyang Depression

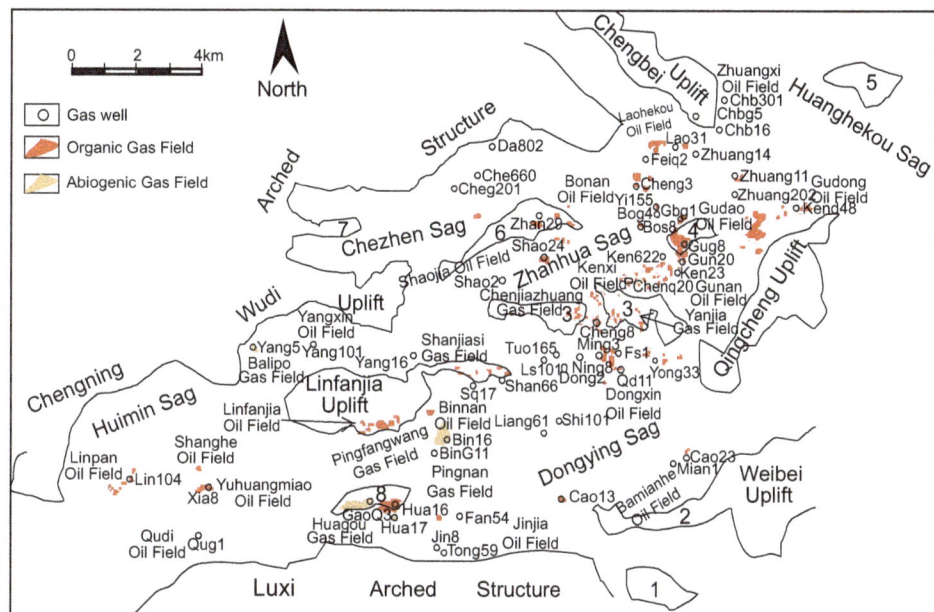

1-Shouguang Uplift; 2-Guangrao Uplift; 3-Chenjiazhuang Uplift; 4-Gudao Uplift; 5-Bonan Uplift; 6-Yihezhuang Uplift; 7-Qingyun Uplift; 8-Gaoqing Uplift;

et al. 2002). This type of gas is mainly found in Pingfangwang, Pingnan, and Huagou gas fields in the western part of the Dongying Sag and the Balipo gas field in the northern part of the Huimin Sag. Vertically, CO_2 is mainly distributed in the Shahejie Formation of Eogene (Es for short), Neogene, and Ordovician. The CO_2 content of such reservoirs ranges from 55.5 % sto 100 % and averages 82.4 %. Hydrocarbon gases were mixed into CO_2 reservoirs in varying degrees. The methane content in those reservoirs ranges from 0 % to 37.2 % with an average of 13.3 %, while the heavy hydrocarbon (C_{2+}) content was low with an average value of 1.4 % (Table 1).

There are usually three genetic types of natural CO_2, namely magma degassing, decomposition of carbon rich crustal rock, and decomposition of organic matter. Studies have confirmed that $\delta^{13}C_{CO2}$ can be used to identify its genesis. It is generally believed that $\delta^{13}C_{CO2} > -8$ ‰ indicates inorganic genesis, and $\delta^{13}C_{CO2} < -10$ ‰ indicates organic genesis (Zhang 1991; Dai 1993). In the Jiyang Depression, gas reservoirs with high CO_2 content (>60 %) usually have heavy carbon isotope ratios ranging from -9.8 ‰ to -3.4 ‰ (PDB). The $\delta^{13}C_{CO2}$ of most samples was higher than -7 ‰, and this can be classified as inorganic genesis. The CO_2 content in hydrocarbon gas reservoirs was usually lower than 10 % with $\delta^{13}C_{CO2}$ less than -8 ‰, and it can be deduced that the CO_2 came from decarboxylation of organic matter (Fig. 3).

Previous studies have shown that helium of different genesis has different isotopic compositions, and the $^3He/^4He$ values of atmosphere, earth mantle, and crust are,

respectively, 1.4×10^{-6}, 1.1×10^{-5}, and 2×10^{-8} (Sun et al. 1996). As shown in Table 2, the $^3He/^4He$ value of CO_2 reservoirs in the Pingfangwang, Huagou, and Yangxin gas fields in the Jiyang Depression was high (3.55–4.49×10^{-6}), and R/Ra was 2.5–3.2, indicating a mixed He origin of mantle genesis and crust genesis (Wang et al. 2013). The isotopic analysis of rare gases and CO_2 indicated that the highly concentrated CO_2 gas reservoir in the Jiyang Depression originated from magma–mantle degassing (Hunt et al. 2012).

3 Organic gases

Organic hydrocarbon gases are produced from sedimentary organic matter due to a series of biological-geochemical reactions. Organic matter of different types and in different thermal evolution stages will produce hydrocarbon gases with different component compositions and isotopic compositions. C_1/C_{1-5} and $\delta^{13}C_1$ of hydrocarbon gases in the Jiyang Depression changed regularly with depth, and could be divided into three categories according to the reservoir depth (Fig. 4):

(1) Buried less than 1,500 m: the C_1/C_{1-5} value is usually high and ranges from 0.8 % to 1.0, 90 % of the natural gases are dry gas ($C_1/C_{1-5} > 0.95$) with a methane content more than 95 %, while the values of $\delta^{13}C_1$ can be separated into two groups: the values of group one are between -40 ‰ and -50 ‰, and the values of group two are less than -55 ‰.

Table 1 Geochemical characteristics of CO_2 in the Jiyang Depression

Area	Well	Formation	Depth, m	$\delta^{13}C_{CO2}$, ‰ (PDB)	$\delta^{13}C_1$, ‰ (PDB)	CH_4, %	C_{2+}, %	N_2, %	CO_2, %
Pingfangwang Gas Field	Bin1	E	1890.0–1898.0	−6.1	−45.8	3.53	2.09	0	94.13
	Ping13-2	E	1453.6–1483.2	−4.7	−52.7	26.43	2.79	1.07	68.85
	Ping13-4	E	1450.8–1486.4	−4.4	−51.7	19.04	4.85	1.21	74.92
	Ping14-3	E	1467.0–4684.6	−4.3	−51.8	18.17	3.30	0.61	77.93
	Ping4	E	1459.4–1461.4	−5.4	−50.9	20.89	3.33	0.46	75.33
	Ping9-3	E	1462.6–1489.2	−4.5	−51.6	22.46	3.40	0.25	73.87
	Pingq12	E	1470.5–1472.5	−4.4	−51.9	21.63	3.39	0.63	74.2
	Pingq12-61	E	1452.4–1487.6	−4.5	−51.8	17.13	3.19	0.38	79.17
	Pingq4	E	1459.4–1474.5	−5.4	−51.7	20.89	3.32	0.46	75.33
Pingnan Gas Field	Bin11	E	1980.2–2250.0	−5.9	−47.6	1.31	1.06	0	97.31
	Bing11	O	2301.0–2307.0	−6.3	−47.6	1.52	0.74	0.25	97.06
Binnan Oilfield	Bin4	E	1510.0–1568.0	−9.8	−49.4	32.62	4.91	0	60.72
	Bin4-13-1	E	1453.0–1455.0	−5.1	−52.4	22.71	3.76	0.85	72.68
	Bin4-6-6	E	1469.7–1474.7	−4.6	−51.7	23.52	3.53	0.33	72.5
Huagou Gas Field	Gao10	N	824.3–838.9	−5.2		0	0	0	99.99
	Gao3	N	833.4–834.8	−4.4	−35	0.07	0	0	97.87
	Gao53	N	811.4–818.0	−6.8		0.04	0	0	99.96
	Hua17	E	1965.1–1980.0	−3.4	−54.39	7.47	0.51	2.05	89.70
	Hua17	E	2000.0–2009.6	−3.4	−54	10.99	0.35	9.03	79.56
Balipo Gas Field	Yang25	E	2793.9–2805.0	−4.4	−42.51	0.52	0	3.83	95.64
	Yang5	O	2380.4–2386.0			0.76	0	0	99.24

E Eogene, N Neogene, O Ordovician

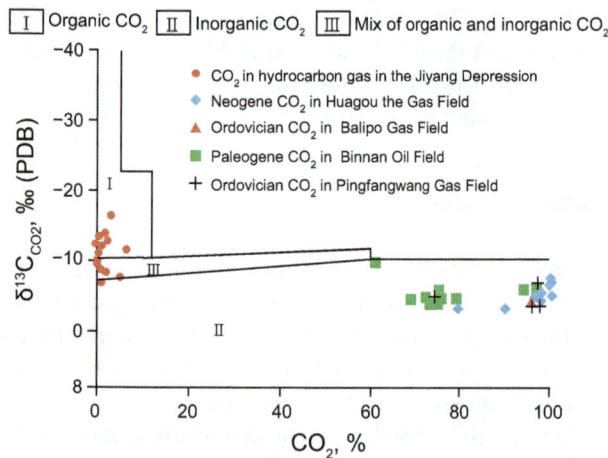

Fig. 3 Identifying inorganic and organic CO_2 with $\delta^{13}C_{CO2}$—CO_2 relationships (Dai 1993)

(2) Buried between 1,500 and 3,500 m: the heavy hydrocarbon content is usually high, and C_1/C_{1-5} ranges from 0.4 to 0.9. Most gases are associated with oil, and their $\delta^{13}C_1$ ranges from −45 ‰ to −52 ‰.

(3) Buried between 3,500 and 5,500 m: the value of C_1/C_{1-5} ranges from 0.6 to 1.0. Compared with the natural gas buried between 1,500 and 3,500 m, the value of C_1/C_{1-5} is higher, and the $\delta^{13}C_1$ is also heavier with a value of −30 ‰ to −50 ‰.

The carbon isotope composition can be used to determine the natural gas genesis as concluded below. Under conditions of similar maturity, hydrocarbon gases generated from sapropelic-type kerogen usually had heavier $\delta^{13}C_1$ than gases generated from humic kerogen; under the condition of similar kerogen type, the natural gases of high thermal evolution degree tend to have heavy $\delta^{13}C_1$. Due to multiple reasons such as various kerogen types, different thermal evolution degrees, and secondary changes, the distribution characteristics of $\delta^{13}C_1$ show that $\delta^{13}C_1$ values of natural gases at middle depth are much lower, but those at shallow and deep depths are higher (Fig. 4).

To identify the genesis of natural gases, the three categories of natural gases (shallow gas <1,500 m, middle gas 1,500–3,500 m, and deep gas >3,500 m) were put into the genesis identification template built by Dai (1993). As shown in Fig. 5, the genesis of shallow gas is complicated with biogas and oil-associated gas dominating, while there is still a small portion of shallow gas having a $\delta^{13}C_1$ value of −40 ‰ to −55 ‰ with high C_1/C_{1-5} and C_1/C_{2+3} values higher than 500, which indicate secondary changes.

Table 2 Characteristics of rare gas isotope in CO_2 gas reservoirs in the Jiyang Depression

Area	Well	Depth, m	$^3He/^4He$, 10^{-8}	R/Ra	$^{40}Ar/^{36}Ar$	$^4He/^{20}Ne$
Binnan Oilfield	Bin4-6-6	1469.7–1474.7	387	2.76	1,791	934
Huagou Gas Field	Hua17	1965.1–1980	445	3.18	770	
	Hua17	2000–2009.6	449	3.18	1,054	
Pingfangwang Gas Field	Ping13-2	1453.6–1483.2	359	2.56	1,220	493
	Ping13-4	1450.8–1486.4	355	2.54	1,722	
	Ping14-3	1467–1484.6	447	3.19	1,378	110
	Ping9-3	1462.6–1489.2	387	2.76	317	467
	Pingq12	1470.5–1472.5	385	2.75	1,051	
	Pingq12-61	1452.4–1487.6	361	2.58	1,478	495
	Pingq4	1459.4–1474.5	385	2.75	1,758	221
Yangxin Oilfield	Yang25	2793.9–2805	412	2.94		

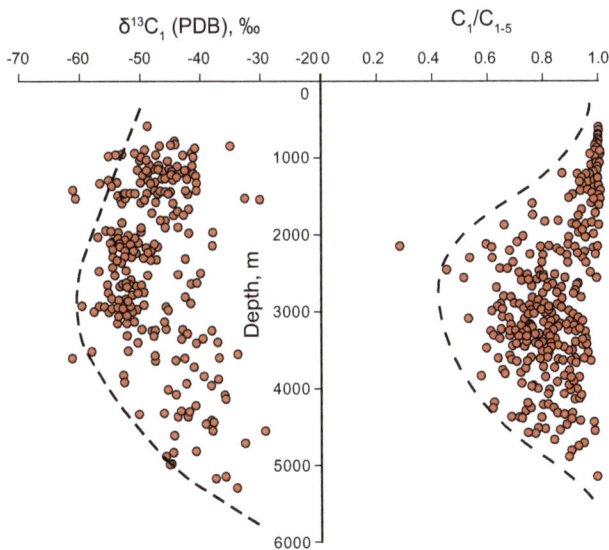

Fig. 4 Variation of $\delta^{13}C_1$ and C_1/C_{1-5} of hydrocarbon gases with depth in the Jiyang Depression

The genesis of middle gas is comparatively simple and dominated by oil-associated gas. As for deep gas, there are several genesis options such as highly mature oil-cracking gas, associated gas with condensate oil, coal-type gas, and the mixture of two or more of them (Fig. 5).

Carbon isotope ratios of methane, ethane, and propane can also be used to identify the genesis of natural gases (Lin et al. 2011). As shown in Fig. 6, shallow buried hydrocarbon gases in the Jiyang Depression were located in the area of oil-associated gas, but the ethane and propane carbon isotope ratios of some samples were abnormally heavy, which might be caused by secondary changes such as biodegradation. The middle buried hydrocarbon gases are mainly oil-associated gas mixed with a small amount of coal-type gas. The deep natural gases were coal-type gas, oil-type gas, as well as a mixture of the both.

In general, hydrocarbon gases in the Jiyang Depression include biogas, oil-associated gas, highly mature oil-type gas, coal-type gas, and their geochemical characteristics are separately discussed below.

3.1 Biogas

Biogas is defined as hydrocarbon gas or nonhydrocarbon gas that is produced due to biochemical reactions of fermentative bacteria and methanogens in the process of degradation of organic matter in source rocks or crude oil (Gao et al. 2010). Naturally existed biogas was formed due to two kinds of processes: one is methyl-type fermentation ($CH_3COOH \rightarrow CH_4 + CO_2$) and the other is carbonate reduction ($CO_2 + 4H_2 \rightarrow CH_4 + 2H_2O$). Limited by the survival temperature of methanogens (0–75 °C) (Li et al. 2008), biogas was mainly developed in shallow-middle buried horizons. The present geothermal gradient in the Jiyang Depression is about 3 °C/100 m, and the surface temperature was about 17 °C, so biogas in the Jiyang Depression tends to occur above 2,000 m. Biogas reservoirs have already been discovered and they are scattered in the Huagou and Yangxin gas fields.

The composition of biogas is fairly simple and is mainly methane. Heavy hydrocarbon (C_{2+}) contents are extremely low (usually less than 0.5 %), the value of C_1/C_{1-5} is higher than 0.995, and there are also low levels of nonhydrocarbon components (mainly N_2, CO_2). $\delta^{13}C_1$ ranged from −55 ‰ to −60.9 ‰ (Table 3) (Hu et al. 2010). Since very little ethane and propane exist in biogas, it is difficult to measure their corresponding carbon isotope ratios.

Biogas is mainly developed in the first member of the Eogene Shahejie Formation (Es₁ for short) in the Yangxin and Huagou gas fields. Source rocks in Es₁ were buried in less than 2,000 m, and were at an immature stage with an R_o value of 0.3 %–0.6 %, the formation temperature was about 55–75 °C, which provided favorable conditions for

Fig. 5 Genesis identification template for hydrocarbon gas in the Jiyang Depression (Dai 1993)

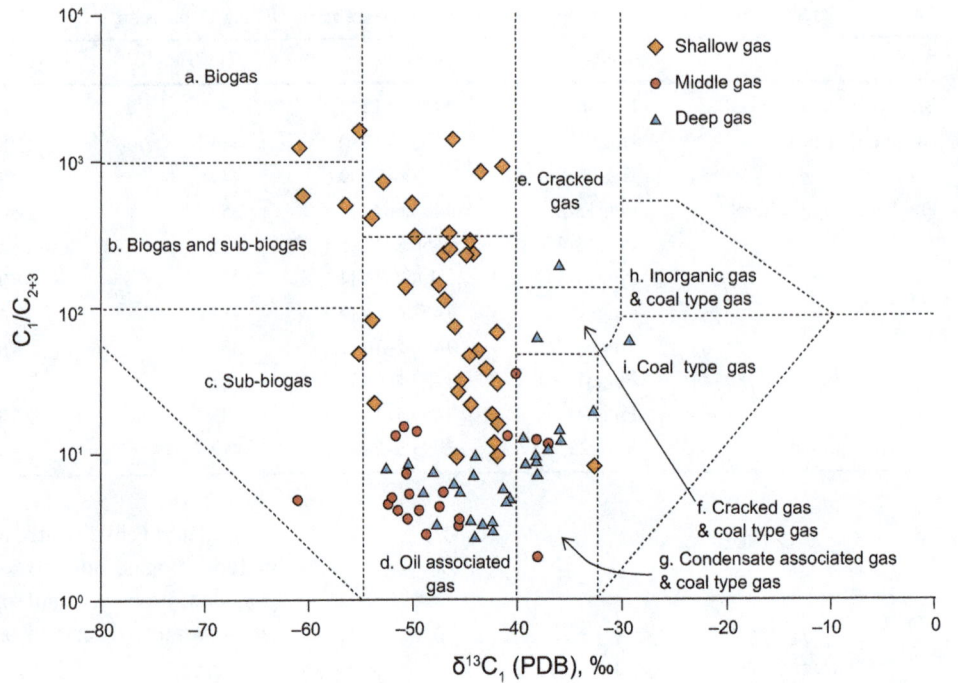

Fig. 6 Characteristics of $\delta^{13}C_1$, $\delta^{13}C_2$, and $\delta^{13}C_3$ in the Jiyang Depression (Dai 1993)

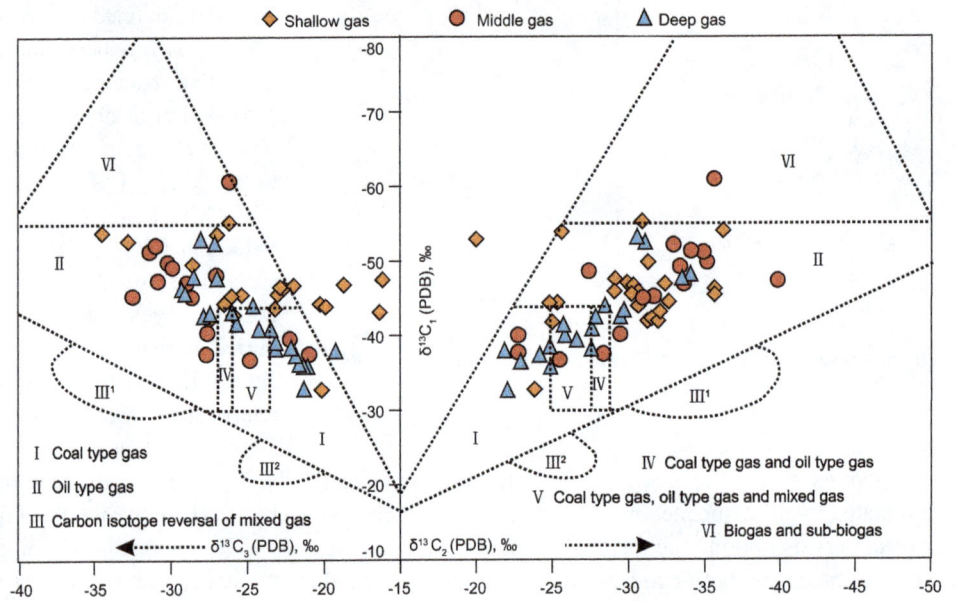

Table 3 Geochemical properties of biogas in the Jiyang Depression

Area	Formation	Well	Depth, m	CO_2, %	N_2, %	CH_4, %	C_{2+}, %	$\delta^{13}C_1$, ‰ (PDB)
Yangxin Gas Field	Es_1	Yang101	1504.2–1529.6	0.39	1.04	98.4	0.17	−60.6
	Es_1	Yang16	1319.0–1325.0	2.98	2.17	94.66	0.19	−56.5
	Es_1	Yang21	1412.0–1415.6	0.1	1.69	97.16	0.08	−60.9
Huagou Gas Field	Es_1	Hua4	1276.0–1307.0	0.17	7.38	89.25	0.17	−55.4
	Es_1	Hua171	1453.0	13.9	6.68	79.10	0.14	−55.0

survival of methanogens. Anaerobes and methanogens have already been detected in the formation water in this area, and this confirmed that natural gas occurring in this interval is biogas.

3.2 Mature oil-type gas (oil-associated gas)

Mature oil-type gas is generated by sapropelic-type source rocks in mature stage ($R_o = 0.6\ \%-1.3\ \%$). Since sapropelic-type source rocks tended to generate more oil than gas during its mature stage, this kind of gas usually occurred as dissolved gas in oil reservoirs. Sometimes gas would exsolve from oil due to changes in temperature and pressure, and a gas cap would be formed.

Mature oil-type gas is the most important kind of natural gas in the Jiyang Depression. Most shallow gases and middle gases as well as part of deep gases are of this kind, and the reserves of this kind of gas resource account for 3/4 of all the proved gas reserves in place. Mature oil-type gas usually occurred in the third member and the fourth member of the Eogene Shahejie Formation (Es$_3$, Es$_4$), and sometimes in buried hills in the Paleozoic Carboniferous or Ordovician (e.g., Zhuangxi Oilfield). This kind of gas always occurred associated with oil reservoirs and gas was produced together with oil.

The methane content of mature oil-type gas varied greatly and ranged from 25.6 % to 99.6 %. The methane content of most oil-associated gas (81 %) was about 60 %–90 %, the heavy hydrocarbon content ranged from 0 % to 70.9 %, and C$_1$/C$_{1-5}$ ranged from 0.6 to 0.99. δ^{13}C$_1$ of mature oil-type gas in the Jiyang Depression ranged from -38 ‰ to -55 ‰, δ^{13}C$_2$ ranged from -26.3 ‰ to -34.9 ‰, δ^{13}C$_3$ ranged from -25.6 ‰ to -32.1 ‰, δ^{13}C$_4$ ranged from -25.6 ‰ to -32.1 ‰, and they were arranged in the order of δ^{13}C$_1 < \delta^{13}$C$_2 < \delta^{13}$C$_3 < \delta^{13}$C$_4$ (Table 4).

δ^{13}C$_1$ and δ^{13}C$_2$ of mature oil-type gas in the Jiyang Depression correlated well with depth, so it is possible to calculate the maturity using gas carbon isotope ratios. Comparison between gas samples and source rock samples was carried out, and the relationship between δ^{13}C$_1$ and R_o was established:

$$\delta^{13}C_1 = 6.942 \ln R_o - 45.254\ (R_o = 0.4 - 1.3), \quad (1)$$

where δ^{13}C$_1$ is the methane carbon isotope ratios of mature oil-type gas, ‰; R_o is vitrinite reflectance, %.

3.3 Highly mature oil-type gas

Highly mature oil-type gas was generated by sapropel-type source rock in highly mature stage ($R_o > 1.3\ \%$) (Zhao et al. 2013). There are two options for the genesis of highly mature oil-type gas, one is kerogen thermal degradation gas

which means that sapropel-type kerogen degrades into natural gas at high temperature; and the other is oil-cracking gas which means that oil cracks into natural gas at high temperature (Lu et al. 2006).

Compared with mature oil-type gas, kerogen thermal degradation gas usually has a higher value of C$_1$/C$_{1-5}$, and heavier δ^{13}C$_1$ and δ^{13}C$_2$. Methane comprises 70.78 %–88.6 % of kerogen thermal degradation gas and heavy hydrocarbons about 5 %–29 %, usually in the range of 10 %–15 %. The value of C$_1$/C$_{1-5}$ ranged from 0.7 to 0.9 and was a little higher than that of mature oil-type gas. δ^{13}C$_1$ ranged from -43.9 ‰ to -33.9 ‰, δ^{13}C$_2$ ranged from -27.6 ‰ to -28.7 ‰, δ^{13}C$_3$ ranged from -23.3 ‰ to -25.9 ‰, and δ^{13}C$_4$ ranged from -25.0 ‰ to -26.6 ‰. There was an apparent reversal of δ^{13}C$_3$ and δ^{13}C$_4$ (Table 5). There are multiple reasons for the isotope reversal such as mixture of gases from different kerogen types, mixture of gases from the same kerogen type but of different maturities, inorganic originated hydrocarbon gas, and biodegradation gas (Burruss and Laughrey 2010). Analysis of the reservoir forming processes indicated that the discovered highly mature oil-type gases originated from the same source rocks, i.e., the fourth member of the Eogene Shahejie Formation (Es$_4$), which was deeply buried with little chance of undergoing biodegradation. Therefore, it can be inferred that the reversal in δ^{13}C$_3$ and δ^{13}C$_4$ was caused by the mixing of gases of different maturities.

Only oil-cracking gas was discovered in the Minfeng area (Chen et al. 2014), where the fourth member of the Shahejie Formation (Es$_4$) was deeply buried, and the temperature might exceed 210 °C in its maximum depth. According to the experiment carried out by Luo et al. (2008), crude oil would crack into gases when the temperature exceeded 160 °C. Compared with kerogen thermal degradation gas with the similar maturity, δ^{13}C$_1$ and δ^{13}C$_2$ of oil-cracking gas were fairly light and, respectively, ranged from -48.4 ‰ to -50.4 ‰ and from -33 ‰ to -34 ‰ (Song et al. 2009; Tian et al. 2009).

Based on comparison of light hydrocarbon compounds in oil-cracking gas and kerogen thermal degradation gas, Hu et al. (2005) put forward that MCC$_6$/nC$_7$ and (2-MC$_6$ + 3-MC$_6$)/nC$_6$ of oil-cracking gas were higher than those of kerogen thermal degradation gases (MCC$_6$ means methylcyclohexane, 2-MC$_6$ means 2-methylhexane, 3-MC$_6$ means 3-methylhexane, nC$_7$ means n-heptane, nC$_6$ means n-hexane). Based on simulation experiments, Wang (2005) discovered that there were differences in MCC$_6$/CC$_6$, MCC$_6$/nC$_7$, and (2-MC$_6$ + 3-MC$_6$)/nC$_6$ between these two kinds of highly mature oil-type gases (CC$_6$ means cyclohexane). The content of thermally stable compounds in kerogen thermal degradation gas was higher than that in oil-cracking gas. In the Jiyang Depression,

Table 4 Geochemical characteristics of typical mature oil-type gases in the Jiyang Depression

Area	Formation	Well	Depth, m	$\delta^{13}C_1$, ‰(PDB)	$\delta^{13}C_2$, ‰ (PDB)	$\delta^{13}C_3$, ‰ (PDB)	$\delta^{13}C_4$, ‰ (PDB)	CH_4, %	C_{2+}, %	N_2, %	CO_2, %
Chengdao Oilfield	N	Chengbs19	1308.2	−53.9	−36.2	−34.9	−32.1	98.03	1.2	0.37	0.2
	E	Chengb12	2144.5	−38.0	−28.3	−27.8	−27.1	60.63	38.79	0.62	0.63
	Pz	Chengb242	2936.6	−45.7	−31.2	−28.9	−28.1	70.97	26.37	0.29	2.26
Linpan Oilfield	N	Lin2-6	1582.8	−44.5	−32.5	−26.7	−27.7	96.6	2.728	0.594	0.177
Yanjia Gas Field	E	Yan22	1573.0	−47.4	−33.5	−29.2	−27.7	70.77	25.39	0.39	3.45
Dongfenggang Oilfield	E	Che57	4067.0	−44.2				81.8	11.86	0.06	6.28
Shengtuo Oilfield	E	Ning3	1805.6	−45.8	−35.6	−29.4	−28.4	83.79	13.98	0	0.4
	E	Tuo113	1948.6	−53.8				91.86	6.18	0	0.34
	E	Tuo165	3391.1	−50.3	−35.1	−30.5	−29.0	65.95	19.02	2.79	12.24
Gubei Oilfield	E	Gub1	2138.5	−47.1				75.36	17.03	0.77	6.84
Bonan Oilfield	E	Xiny12	2454.9	−52.3	−33	−31.2	−29.8	74.78	20.5	0	4.3
	E	Yi37	3220.3	−45.6	−31.7	−32.9	−28.3	66.2	22.45	0	9.55
	E	Yi170	3817.6	−52.6	−31.2	−27.3	−27.2	84.44	12.4	0.14	2.72
	E	Bos4	3911.5	−52.7	−30.8	−28.2	−28.1	83.38	14.59	1.53	1.4
Gudong Oilfield	E	Gud9	2506.4	−48.7	−27.4	−27.3	−25.6	60.05	20.99	24	2.52
Guangli Oilfield	E	Lai10	2665.1	−50.6				81.35	15.39	0	0.46
Liangjialou Oilfield	E	Liang60	2844.8	−52.3				73.12	20.23	0	3.61
	E	Ling35	3119.9	−50.8				89.84	5.93	0	1.28
Lijin Oilfield	E	Li54	2904.3	−49.5				70.11	22.23	0	5.93
Xianhe Oilfield	E	Niu23	3289.8	−52.1				73.2	20.13	0	4.05
	E	Wang53	3389.0	−50.4				67.63	23.94	0	4.99
Dawangzhuang Oilfield	O	Dag23	1738.3	−44.4				90.18	7.23	0	2.04
Chengdong Oilfield	P	Chengk1	2588.0	−51.5	−34.1	−31.7	−29.8	73.92	25.53	0.18	0.71
Yong'anzhen Oilfield	E	Yong12-21		−47.6	−39.9	−31.3	−28.5	98.55	1.28	1.109	0.663
Zhuangxi Oilfield	E	Zhuang202	2644.5	−51.6	−34.9	−31.7	−29.4	86.22	7.49	0	1.13
	E	Zhuang50	3228.2	−49.7	−33.5	−30.1	−28.2	89.44	7.47	0	1.73
	E	Zhuang74	3634.5	−47.6	−33.3	−28.9	−27.5	68.44	23.49	0	4.81
	O	Zhuangg10	3627.2	−44.1				63.74	28.75	0	3.56
	O	Zhuangg21	3929.1	−42.4	−27.9	−27.7	−26.4	67.15	25.73	0	3.13
	O	Zhuangg4	4013.5	−45.4				77.5	16.97	0	4.55
	∈-Anz	Zhuangg25	4277.6	−43.2	−29.7	−26.3	−27.7	71.34	25.48	0	1.36
	O	Zhuangg14	4318.5	−46.1	−32.0	−29.6		76.77	13.21	0	9.24
	O	Zhuangg13	4367.5	−42.2	−29.5	−28.1	−27.2	69.54	27.75	0	1.87
	O	Zhuangg18	4582.4	−44.4				71.18	24.65	0.59	3.08
	∈	Zhuangg17	4886.2	−45.8	−31.9	−29.2	−28.5	85.99	10.25	0.58	3.09

MCC_6/nC_7 of oil-cracking gas was higher than 1.0, (2-MC_6 + 3-MC_6)/nC_6 of oil-cracking gas was higher than 0.4, which were higher than those of kerogen thermal degradation gas and mature oil-type gas, while MCC_6/nC_7 of oil-cracking gas was less than 0.8 which was lower than that of kerogen thermal degradation gas and mature oil-type gas (Fig. 7).

3.4 Coal-type gas

Coal-type gas is defined as natural gas generated by coal or humic kerogen due to biochemical and chemical action. Coal-type gas discovered in the Jiyang Depression was mainly developed in the Paleozoic Ordovician and Carboniferous—Permian in the Gubei buried hill belt, the

Table 5 Geochemical properties and genesis of typical highly mature oil-type gas in the Jiyang Depression

Area	Formation	Well	Depth, m	$\delta^{13}C_1$, ‰ (PDB)	$\delta^{13}C_2$, ‰ (PDB)	$\delta^{13}C_3$, ‰ (PDB)	$\delta^{13}C_4$, ‰ (PDB)	C_1/C_{1-5}	CH_4, %	C_{2+}, %	N_2, %	CO_2, %	Genetic type
Bonan Oilfield	O	Bo601	5007.0–5009.0	−43.8	−28.7	−25.8	−26.1	0.716	70.78	29.22			Kerogen thermal degradation gas
	E	Bos5	4491.9–4587.3	−38.0				0.859	79.55	13.07	0.43	7.03	
	O	Bos6	4165.5–4246.0	−40.8	−27.6	−24.5		0.799	74.98	19.27	0.51	4.77	
Laohekou Oilfield	Pz	Chengb39	4173.0–4320.0	−41.3	−27.6	−25.9	−25.7	0.835	75.02	15.47	2.04	7.48	
Shengtuo Oilfield	E	Tuo765	4354.1–4386.0	−43.9	−28.6	−24.9	−26.6	0.880	87.82	12.03	0	0.15	
Lijin Oilfield	E	Xinlis1	4271.2–4371.0	−41.8				0.905	87.82	9.27	0.04	2.87	
Zhuangxi Oilfield	O	Zhuangg23	3897.0–3988.5	−38.3	−27.6	−23.3	−26.2	0.890	86.38	10.65	1.14	1.53	
Dongxin Oilfield	E	Feng8	4314.0–4316.0	−49.0				0.798	62.92	16.61	1.25	19.16	Oil-cracking gas
	E	Fengs1	4316.0–4343.0	−48.4	−33.0	−26.8	−25.0	0.862	46.73	8.12	0.39	44.76	
	E	Fengs1	4400.0–4402.0	−50.4	−34.0	−27.3	−25.6	0.876	81.01	11.59	1.6	5.74	
	E	Fengs1		−48.0				0.860	71.39	11.75	0.53	16.28	

fourth member of the Shahejie Formation in the Bonan deep sag, and the Shahejie Formation in the Qudi Oilfield in the Huimin Sag. Coal-type gas in the Gubei buried hill and Qudi Oilfield was generated by coal and humic kerogen in the Shanxi Formation and Taiyuan Formation in Carboniferous—Permian, while that in the Bonan Sag (Well Yi115 and Yi121) was generated by humic kerogen in the upper part of Es_4.

The methane content of coal-type gas ranged from 75 % to 92 %, and the heavy hydrocarbon content varied greatly from 0.51 % to 19.5 %. C_1/C_{1-5} ranged from 0.8 to 0.99 and the value of most samples exceeded 0.9. C_1/C_{1-5} of coal-type gas is usually higher than that of oil-type gas with a similar maturity. $\delta^{13}C_1$ of coal-type gas in the Jiyang Depression ranged from −32.6 ‰ to −41.0 ‰, $\delta^{13}C_2$ ranged from −22.0 ‰ to −27.6 ‰ (Table 6). There was a slight reversal in $\delta^{13}C_3$ and $\delta^{13}C_4$ and this might be caused by mixing with oil-type gas. It is pointed out that $\delta^{13}C_2$ of coal-type gas in China is usually higher than −28 ‰ (Song et al. 2012; Dai et al. 2012; Wang et al. 2010), and in the Jiyang Depression, the carbon isotope ratios of coal-type gas are located in the "I" area of the "V" shaped $\delta^{13}C_1$-$\delta^{13}C_2$-$\delta^{13}C_3$ template (Fig. 6).

C_7 light hydrocarbon information can also be used to distinguish coal-type gas from oil-type gas. The C_7 system is composed of three kinds of compounds: normal heptane (nC_7), methylcyclohexane (MCC_6), and multi-structured dimethylcyclopentane ($\sum DMCC_5$). MCC_6 mainly came from higher plants and was a major component of C_7 system in coal-type gas, and $\sum DMCC_5$ mainly came from aquatic organisms and was a major component of C_7 system in oil-type gas (Song and Zhang 2004).

As shown in Fig. 8, coal-type gas differed significantly from oil-type gas. $MCC_6/\sum C_7$ of the coal-type gas exceeded 50 %, while $\sum DMCC_5/\sum C_7$ was less than 40 %; as for oil-type gas, $nC_7/\sum C_7$ exceeded 30 %, $MCC_6/\sum C_7$ ranged from 20 to 40 %.

4 Secondary changes of natural gas

Most shallow gas in the Jiyang Depression was originally dissolved gas that escaped from oil when temperature and pressure changed due to migration of oil along faults or sand bodies. This kind of natural gas was located in the "d" area (oil-associated gas) of "$\delta^{13}C_{CH4}$−$C_1/(C_2 + C_3)$ template" in Fig. 5, and in "II" area of "V" shaped "$\delta^{13}C_1$−$\delta^{13}C_2$−$\delta^{13}C_3$ template" in Fig. 6, and was typical mature oil-type gas.

There is a kind of shallow gas whose hydrocarbon carbon isotope ratios are similar to mature oil-type gas, but its heavy hydrocarbon content is extremely low, the methane content is very high (>95 %), C_1/C_{1-5} is higher than 0.95,

Table 6 Geochemical properties of typical coal-type gas in the Jiyang Depression

Area	Formation	Well	Depth, m	$\delta^{13}C_1$, ‰ (PDB)	$\delta^{13}C_2$, ‰ (PDB)	$\delta^{13}C_3$, ‰ (PDB)	$\delta^{13}C_4$, ‰ (PDB)	CH_4, %	C_{2+}, %	N_2, %	CO_2, %
Gudao Oilfield	C–P	Bo93	3230.0–3249.4	−38.1	−22.7	−21.25	−21.8	88.99	7.81		2.29
	O	BoG4	4375.0–4460.0	−38.2	−24.9	−22.5	−23.6	85.32	10.02	0	4.66
	O	BoG403	3850.5–3889.3	−37.1	−24.2	−22.0	−23.5	85.16	8.68	0.83	5.31
	Es	BoS3	4450.1–4472.4	−39.1	−26.7	−23.4	−23.9	83.15	10.8	0.73	5.05
	P	GBG1	4020.6–4139.5	−35.9	−23.1	−21.2	−21.2	88.44	6.48	0.55	4.54
	C–P	GBG1	4120.6–4139.0	−35.8	−22.9	−21.5	−20.8	82.52	10.09	0.74	6.66
	C–P	GBG2	3689.0–3731.0	−41.0	−25.8	−23.6	−23.6	75.87	19.52	0.96	3.65
	Mz	Yi132	3374.0–3387.0	−37.0	−25.3	−25.0	−25.5	87.01	7.9	1.98	2.18
	P	Yi155	4696.3–4706.7	−32.7	−22.0	−21.5	−21.0	87.64	4.85		6.64
Bonan Oilfield	Es	Yi115	5110.4–5164.4	−35.9	−24.9	−21.8		80.18	0.51	0.05	19.27
	Es	Yi121	4426.1–4438.4	−38.0	−22.0	−19.3	−20.6	91.36	1.46	0	7.09
	O	BoS6	4165.5–4246.0	−40.8	−27.6	−24.5		74.98	19.27	0.51	4.77
Qudi Oilfield	Es	QuG1	1514.0–1520.0	−32.6	−23.9	−20.3	−20.2	77.25	9.53	11.99	0.93

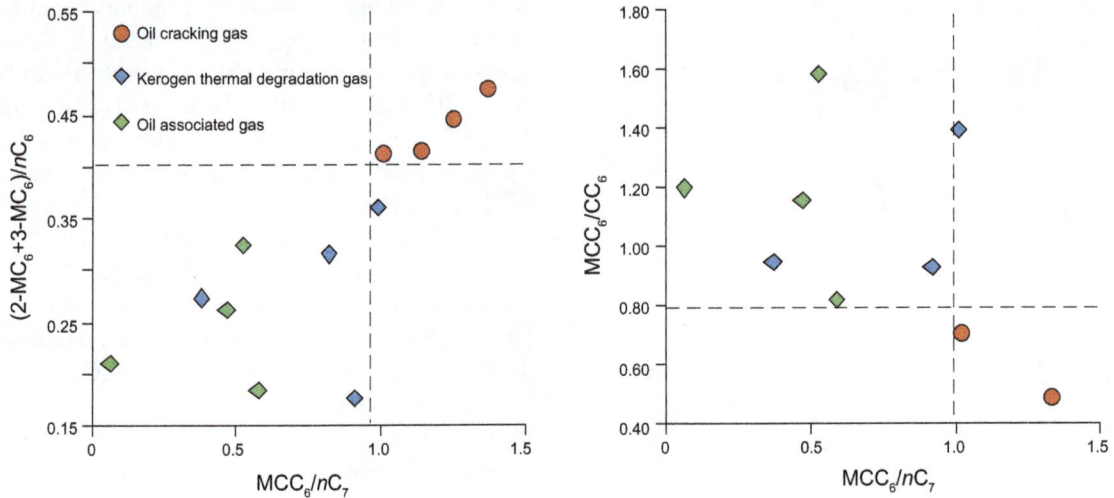

Fig. 7 Light hydrocarbon property differences between oil cracking gas and kerogen thermal degradation gas

and is located above the "d" area of the "$\delta^{13}C_{CH4}-C_1/(C_2+C_3)$ template" (Table 7; Fig. 5). Such characteristics are caused by composition changes during long distance migration of natural gas. Gas migration experiments in porous sandstone core samples indicated that, with increasing migration distance, the methane content tended to increase while the heavy hydrocarbon content (C_{2+}) decreased correspondingly. Furthermore, the carbon isotopes of hydrocarbon differentiated slightly and this means that carbon isotopes became lighter with an increase of migration distance (variation range usually less than −2 ‰). Therefore, it is believed that this kind of natural gas with a high content of methane was dissolved gas that escaped from oil after long-distance migration.

There is another kind of shallow gas whose methane carbon isotope ratios are heavier than those of mature oil-type gas, and $\delta^{13}C_2$ and $\delta^{13}C_3$ are extremely heavy. $\delta^{13}C_3$ of

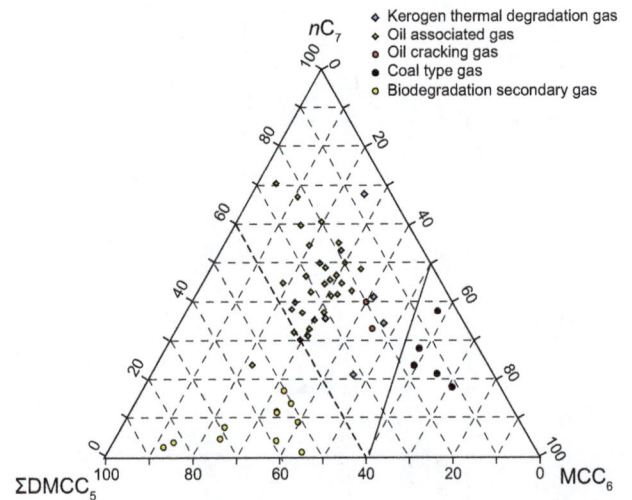

Fig. 8 Triangular template of C_7 light hydrocarbon in different kinds of natural gases in the Jiyang Depression

Table 7 Geochemical properties and genesis of shallow gas (part) in the Jiyang Depression

Area	Well	Formation	Depth, m	δ13C1, ‰ (PDB)	δ13C2, ‰ (PDB)	δ13C3, ‰ (PDB)	δ13C4, ‰ (PDB)	CH4, %	N2, %	Genesis		
Caoqiao Oilfield	Cao104	Es	1258–1265.6	−46.4	−35.7	−23	−22.12	97.87	0.4	1.23	0.42	High methane secondary gases
Huagou Gas Field	Hua16	N	828.1–831.1	−46.6	−30.4	−22.64	−26.1	98.99	0.31	0.71	0	
	Hua6-4	N	790–830	−44.7	−25.2			96.5	0.43	\	1.93	
Gudong Oilfield	Gud22-3	N	1214.8–1219.6	−43.7	−30.6	−23.3	−22.4	97.17	2.41	0.26	0.13	
	Gud29-416	N	1246–1270.4	−45.9	−29.1	−23	−21.1	94.99	1.6	1.52	1.83	
	Gud3-015	N	1203.7–1207.4	−45.4	−30.7	−26.1	−23.3	95.19	3.92	0.33	0.54	
	Gud31-15	N	1196.2–1238.2	−45.6	−30.3	−25.6	−23.6	93.51	5.01	0.83	0.51	
	Gud3-517	N	1303.4–1315.4	−49.9	−31.4	−28.8	−29.2	96.79	0.37	2.26	0.49	
	Gud13-N11	N	1252.8–1263.6	−41.9	−32.1	−27.7	−24.8	93.57	5.2	0.45	0.71	Biodegradation secondary gases
	Gud13-P513	N	1380.5–1562	−42.2	−31.9	−27.5	−25	88.25	10.59	0.5	0.56	
	Gud2-2	N	1191–1204	−41.9	−31.1	−27.8	−25.5	90.42	8.1	0	1.27	
	Gud22-N3	N	1261.6–1294	−42.3	−31.3	−26	−23.9	83.12	5.42	0.28	1.15	
	Gud2-5	N	1175.2–1204.2	−41.8	−31.4	−27.7	−25.2	73.28	12.42	12.41	1.63	
Chenjiazhuang Oilfield	Chenq11	N	935–942	−52.9	−33.11	−20.08	−18.49	95.02	0.152	4.706	0.122	
	Chenq8	N	945–948	−53.87	−25.7	−27.14	−29.03	88.436	0.315	10.49	0.752	
Shanjiasi Oilfield	Shan66	Anz	1076–1100	−46.79	−29.89	−18.92	−24.11	95.85	0.41	0	1.61	
Huagou Gas Field	Gao41-5	Mz	1965.1–1984	−41.8	−25.14	−8.52	−22.13	97.384	3.592		0.82	
	Gao42	Es	945–952	−42.95	−32.03	−16.49	−23.09	89.138	5.328	3.096	5.1	
	Hua6	N	818–819.6	−44.28	−25.24	−20.33	−27.3	99.26	0.44	0.28	0.02	
	Hua6-2	N	743.8–783.4	−44.35	−24.91	−20.11	−24.85	96.7	0.43		1.92	
Linpan Oilfield	Lin2-4	N	1398.8–1426	−47.57	−29.27	−16.18	−20.81	96.624	0.719	2.496	0.16	
Yuhuangmiao Oilfield	Xia8	Ed	1457.7–1464	−46.95	−32.48	−22.03	−22.76	98	0.9	0	0	

normal mature oil-type gas ranged from -26 ‰ to -34 ‰, while $\delta^{13}C_3$ of this kind of natural gas might as heavy as -8.5 ‰. $\delta^{13}C_3$ of most natural gas ranged from -8 ‰ to -22 ‰, and carbon isotope ratios of light hydrocarbons arranged in the order of $\delta^{13}C_1 < \delta^{13}C_2 < \delta^{13}C_3 > \delta^{13}C_4$. $\delta^{13}C_1$ of some samples was about 2 ‰–7 ‰ heavier than that of mature oil-type gas. Take the Gudong Oilfield as an example, $\delta^{13}C_1$ of mature oil-type gas (Well Gud3-517) was about -49.9 ‰, $\delta^{13}C_2$ was about -31.4 ‰, but $\delta^{13}C_1$ and $\delta^{13}C_2$ of the sample from the same horizon and similar depth (Gud2-2) were, respectively, -41.9 ‰ and -31.1 ‰, that is to say $\delta^{13}C_1$ was about 8 ‰ heavier than that in Well Gud3-517.

Analysis indicated that the main reason that caused abnormal carbon isotope ratios was biodegradation. James and Burns (1984) analyzed the carbon isotope ratios of light hydrocarbons of biodegradation natural gases in Australia and Canada, and discovered that $\delta^{13}C_3$ was abnormally heavy. They deduced that since propane is soluble in water, it is readily biodegradable. Secondary biodegradation shallow gas in the Jiyang Depression exhibited the similar characteristics.

Stahl (1980) carried out bacterial degradation experiments, and pointed out that long-chained paraffin hydrocarbon is more easily degradable than short-chained ones, and normal paraffin hydrocarbon is more easily degradable than isomeric ones. As shown in Fig. 8, $nC_7/\sum C_7$ of biodegradation shallow gas was less than 20 %, and was obviously less than that of mature oil-type gas.

Leythaeuser et al. (1979) studied biodegradation of oil using light hydrocarbon data, and summarized typical characteristics: the content of normal paraffin hydrocarbon was low, while the contents of isomeric ones (such as 3,3-DMC_5; 2,3,3-TMC_4; 2,2-DMC_5; 2,4-DMC_5 and 2,2-DMC_4) were high (DMC_5 means dimethylpentane, TMC_4 means triptane, DMC_4 means dimethylbutane). Based on analysis of light hydrocarbons in shallow gas in the Jiyang Depression, Zhang (1991) pointed out that the relationship between 2,4-DMC_5/nC_6 and the heptane index can be used to distinguish biodegradation oil-type gas from other oil-type gases. As shown in Fig. 9, the value of 2,4-DMC_5/nC_6 for biodegradation gas was usually higher than 0.5, and the heptane index was usually less than 5; in contrast, the heptane index usually ranged from 20 to 50, and 2,4-DMC_5/nC_6 for other oil-type gases was less than 0.1.

5 Identification factors

Based on discussion above, the identification factors for different kinds of natural gases are summarized in Table 8,

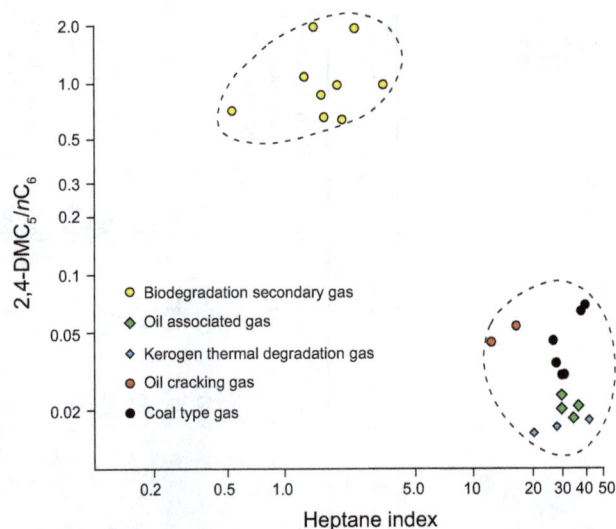

Fig. 9 2,4-DMC_5/nC_6–heptane index template of natural gases in the Jiyang Depression (Zhang 1991)

$$\text{Heptane index} = 100 \times nC_7/(CC_6 + 2\text{-}MC_6 + 2,3\text{-}DMC_5 + 1,1\text{-}DMC_5 + 3\text{-}MC_6 + 1,C,3\text{-}DMCC_5 + nC_7 + MCC_6)$$

and with the help of gas compositions, carbon isotope ratios of paraffin hydrocarbon and CO_2, and light hydrocarbon index, it is feasible to identify the genesis of natural gases in the Jiyang Depression.

Take mature oil-type gas as reference, biogas has a high methane content, and $\delta^{13}C_1$ was less than -55 ‰; highly mature oil-type gases are divided into kerogen thermal degradation gas and oil-cracking gas. They both have a high value of C_1/C_{1-5} and heavy methane carbon isotope ratios, and they can be distinguished by the (2-MC_6 + 3-MC_6)/nC_6–MCC_6/nC_7 template. Ethane carbon isotope ratios of coal-type gas in the Jiyang Depression are usually higher than -28 ‰ and the compositions of C_7 can be used to effectively distinguish coal-type gas from oil-type gas.

Heavy hydrocarbons usually reduce in the process of gas migration. The C_1/C_{2+3} value of methane rich secondary gas might exceed 280, and its carbon isotope compositions and light hydrocarbon compositions are similar to those of mature oil-type gas. Secondary biodegradation gas is featured by heavy carbon isotope ratios of $\delta^{13}C_1$ or $\delta^{13}C_3$, and light hydrocarbon isotope ratios arrange in the order of $\delta^{13}C_1 < \delta^{13}C_2 < \delta^{13}C_3 > \delta^{13}C_4$. Influenced by biodegradation, the normal paraffin hydrocarbon content is low. The triangular template of C_7 and 2,4-DMC_5/nC_6–heptane index template can be used to distinguish the secondary biodegradation gas from other natural gases.

Table 8 Identification factors of hydrocarbon gases with different genesis in the Jiyang Depression

Genetic types		Gas composition			Isotope composition					Light hydrocarbon characteristics
		CO_2, %	C_1/C_{1-5}	C_1/C_{2+3}	$\delta^{13}C_1$, ‰	$\delta^{13}C_2$, ‰	$\delta^{13}C_3$, ‰	$\delta^{13}C_4$, ‰	Isotope series	
CO_2 gas		>50 %	/	/	/	/	/	/	/	/
Hydrocarbon gases	Biogas	<10 %	>0.995	/	<−55	/	/	/	/	/
	Biodegradation secondary gas		0.95–1	17–207	−38 to −48	−20 to −34	−13 to −24	−14 to −25	$\delta^{13}C_1 < \delta^{13}C_2 < \delta^{13}C_3 > \delta^{13}C_4$	$nC_7/\sum C_7 < 0.2$; 2,4-$DMC_5/nC_6 < 0.5$; heptane index <5;
	High methane secondary gas		>0.95	>280	−42 to −55	−29 to −38	−26 to −34	−25 to −32	$\delta^{13}C_1 < \delta^{13}C_2 < \delta^{13}C_3 < \delta^{13}C_4$	$nC_7/\sum C_7 = 30\text{-}60$; $MCC_6/\sum C_7 = 20\text{-}40$; 2,4-$DMC_5/nC_6 < 0.1$; heptane index = 30–40;
	Oil-associated gas		0.6–0.99	2–39	−42 to −55	−29 to −38	−26 to −34	−25 to −32	$\delta^{13}C_1 < \delta^{13}C_2 < \delta^{13}C_3 < \delta^{13}C_4$	$nC_7/\sum C_7 = 30\text{-}60$; $MCC_6/\sum C_7 = 20\text{-}40$; 2,4-$DMC_5/nC_6 < 0.1$; heptane index = 30–40;
	Kerogen thermal degradation gas		0.7–0.95	5–46	−33 to −44	−27 to −33	−23 to −26	−26 to −28	$\delta^{13}C_1 < \delta^{13}C_2 < \delta^{13}C_3 < \delta^{13}C_4$	$MCC_6/nC_7 < 1.0$; (2-MC_6 + 3-MC_6)/$nC_6 < 0.4$; $MCC_6/CC_6 > 0.8$; $DMCC_5/\sum C_7 = 40\text{-}60$;
	Oil-cracking gas		0.7–0.9	2–55	−44 to −52	−28 to −34	−24 to −28	−24 to −26	$\delta^{13}C_1 < \delta^{13}C_2 < \delta^{13}C_3 < \delta^{13}C_4$	$MCC_6/nC_7 > 1.0$; (2-MC_6 + 3-MC_6)/$nC_6 > 0.4$; $MCC_6/CC_6 < 0.8$; $DMCC_5/\sum C_7 = 30\text{-}40$;
	Coal-type gas		0.8–0.99	10–190	−29 to −41	−17 to −25	−19 to −24	−20 to −25	$\delta^{13}C_1 < \delta^{13}C_2 < \delta^{13}C_3 < \delta^{13}C_4$	$MCC_6/\sum C_7 > 50$ %; $\sum DMCC_5/\sum C_7 < 40$ %; 2,4-$DMC_5/nC_6 = 0.03\text{—}0.1$; heptane index = 30–50

6 Conclusions

1) Based on analysis of gases compositions, carbon isotope ratios, light hydrocarbon properties, combined with geological analysis, natural gases in the Jiyang Depression were classified into two categories namely hydrocarbon gas and abiogenic gas. The abiogenic gas was mainly magmatogenic or mantle derived CO_2. Hydrocarbon gases were further divided into coal-type gas, oil-type gas, and biogas according to the kerogen types and formation mechanisms. The oil-type gases were divided into mature oil-type gas (oil-associated gas), highly mature oil-type gas. Highly mature oil-type gases were subdivided into oil-cracking gas and kerogen thermal degradation gas.

2) Analysis results showed that shallow gases (buried less than 1,500 m) are mainly mature oil-type gases, secondary gas is rich in methane after chromatographic separation during migration and secondary mature oil-type gas after biodegradation is featured by rich in ^{13}C in methane and ethane. Meanwhile, biogas is another kind of shallow gas. The genesis of middle gases buried in the depth of 1,500–3,500 m was simple and was dominated by mature oil-type gases. Deep gases buried in the depth of 3,500–5,500 m were usually kerogen thermal degradation gas, oil-cracking gas, and coal-type gas.

3) Due to chromatographic effects, the methane content increases and heavy hydrocarbons decrease during the progress of migration. Secondary biodegradation gas was featured by heavy carbon isotope ratios of $\delta^{13}C_1$ or $\delta^{13}C_3$, and light hydrocarbon isotope ratios arranged in the order of $\delta^{13}C_1 < \delta^{13}C_2 < \delta^{13}C_3 > \delta^{13}C_4$. Influenced by biodegradation, the normal paraffin hydrocarbon content was low. Triangular template of C_7 and $2,4\text{-}DMC_5/nC_6$—heptane index template can be used to distinguish secondary biodegradation gas from other natural gases.

References

Burruss RC, Laughrey CD. Carbon and hydrogen isotopic reversals in deep basin gas: evidence for limits to the stability of hydrocarbons. Org Geochem. 2010;41(12):1285–96.

Chen ZH, Zhang SC, Zha M. Geochemical evolution during the cracking of crude oil into gas under different pressure systems. Sci China: Earth Sci. 2014;57(3):480–90.

Dai JX. The carbon and hydrogen isotope characteristics and identification of different kinds of natural gases. Nat Gas Geosci. 1993;2:1–40 (in Chinese).

Dai JX, Ni YY, Zou CN. Stable carbon and hydrogen isotopes of natural gases sourced from the Xujiahe Formation in the Sichuan Basin, China. Org Geochem. 2012;43:103–11.

Gao Y, Jin Q, Zhu GY. Genetic types and distribution of shallow-buried natural gases. Pet Sci. 2010;7(3):347–54.

Gao Y, Jin Q, Shuai YH, et al. Genetic types and accumulation conditions of biogas in Bohaiwan Basin. Nat Gas Geosci. 2011;22(3):407–14 (in Chinese).

Hu GY, Luo X, Li ZS, et al. Geochemical characteristics and origin of light hydrocarbons in biogenic gas. Sci China: Earth Sci. 2010;53(6):832–43.

Hu GY, Xiao ZY, Luo X, et al. Light hydrocarbon composition difference between two kinds of cracked gases and its application. Nat Gas Ind. 2005;25(9):23–5 (in Chinese).

Hunt AG, Darrah TH, Poreda RJ. Determining the source and genetic fingerprint of natural gases using noble gas geochemistry: A northern Appalachian Basin case study. AAPG Bull. 2012;96(10):1785–811.

James AT, Burns BJ. Microbial alteration of subsurface natural gas accumulations. AAPG Bull. 1984;68(8):957–60.

Leythaeuser D, Schaefer RG, Cornford C, et al. Generation and migration of light hydrocarbons (C_2-C_7) in sedimentary basins. Org Geochem. 1979;1(4):191–204.

Li J, Hu GY, Zhang Y, et al. Study and application of carbon isotope fractionation during the reduction process from CO_2 to CH_4. Earth Sci Front. 2008;15(5):357–63 (in Chinese).

Li PL, Jin ZJ, Zhang SW, et al. The present research status and progress of petroleum exploration in the Jiyang Depression. Pet Explor Dev. 2003;30(3):1–4 (in Chinese).

Lin HX, Cheng FQ, Jin Q. Fractionation mechanism of natural gas components and isotopic compositions and sample analysis. Nat Gas Geosci. 2011;22(2):195–200 (in Chinese).

Lu SF, Li JJ, Xue HT, et al. Chemical kinetics of carbon isotope fractionation of oil-cracking methane and its initial application. J Jilin Univ (Earth Science Edition). 2006;36(5):825–9 (in Chinese).

Luo X, Wang YB, Li J, et al. Origin of gas in deep Jiyang Depression. Nat Gas Ind. 2008;28(9):13–6 (in Chinese).

Schoell M. The hydrogen and carbon isotopic composition of methane from natural gases of various origins. Geochim Cosmochim Acta. 1980;44(5):649–61.

Song GQ, Jin Q, Wang L, et al. Study on kinetics for generating natural gas of Shahejie Formation in deep-buried sags of Dongying Depression. Acta Petrolei Sinica. 2009;30(5):672–7 (in Chinese).

Song MS, Zhang XC. Discussion on deep gas geochemical characteristics and genesis of Bonan Sag, Jiyang Depression. Nat Gas Geosci. 2004;15(6):646–9 (in Chinese).

Song Y, Liu SB, Zhang Q, et al. Coalbed methane genesis, occurrence and accumulation in China. Pet Sci. 2012;9(3):269–80.

Stahl WJ. Compositional changes and $^{13}C/^{12}C$ fractionations during the degradation of hydrocarbons by bacteria. Geochim Cosmochim Acta. 1980;44(11):1903–7.

Sun ML, Chen JF, Liao YS. Helium isotopic characteristics, genesis of CO_2 in natural gases and distribution of Tertiary magamatite in the Jiyang Depression. Geochimica. 1996;25(5):475–80 (in Chinese).

Tang DZ, Liu HX, Li XM, et al. Probe into deep-seated structural factors of abiogenic gas accumulation and storage in Jiyang Depression. Earth Sci: J China Univ Geosci. 2002;27(1):30–4 (in Chinese).

Tian H, Xiao XM, Yang LG, et al. Pyrolysis of oil at high temperatures: gas potentials, chemical and carbon isotopic signatures. Chin Sci Bull. 2009;54(7):1217–24.

Wang G L. Accumulation conditions of natural gas in cratonic area, Tarim Basin. Ph.D. Thesis. Chengdu: Southwest Petroleum University. 2005. 43–47.

Wang P, Shen ZM, Liu SB, et al. Geochemical characteristics of noble gases in natural gas and their application in tracing natural gas migration in the middle part of the western Sichuan Depression, China. Pet Sci. 2013;10(3):327–35.

Wang YP, Dai JX, Zhao CY, et al. Genetic origin of Mesozoic natural gases in the Ordos Basin (China): comparison of carbon and hydrogen isotopes and pyrolytic results. Org Geochem. 2010;41(9):1045–8.

Zhang LY. Identifying criteria of natural gases in the Jiyang Depression. Pet Geol Exper. 1991;13(4):355–69 (in Chinese).

Zhao XZ, Jin Q, Jin FM, et al. Origin and accumulation of high-maturity oil and gas in deep parts of the Baxian Depression, Bohai Bay Basin, China. Pet Sci. 2013;10(3):303–13.

Zhou JL. Gas accumulation analysis of upper Paleozoic coal in the Jiyang Depression. J Earth Sci Environ. 2004;26(2):47–50 (in Chinese).

7

Influence of strong preformed particle gels on low permeable formations in mature reservoirs

Mahmoud O. Elsharafi[1] · Baojun Bai[2]

Abstract In mature reservoirs, the success of preformed particle gel (PPG) treatment rests primarily on the ability of the PPG to reduce and/or plug the high permeable formations, but not damage the low permeable formations. Static test models (filtration test model and pressure test model) were used to determine the effect of PPG on low permeable formations. This work used a strong preformed particle gel, Daqing (DQ) gel made by a Chinese company. The particle gel sizes were ranged from 30 to 120 mesh for this work. PPGs are sized in a millimeter or micrometer, which can absorb over a hundred times their weight in liquids. The gel strength was approximately 6500 Pa for a completely swollen PPG with 1 % (weight percentage) NaCl solution (brine). 0.05 %, 1 %, and 10 % NaCl solutions were used in experiments. Sandstone core permeability was measured before and after PPG treatments. The relationship between cumulative filtration volumes versus filtration times was determined. The results indicate that DQ gels of a particle size of 30–80 mesh did not damage the cores of a low permeability of 3–25 mD. The DQ gels of a smaller particle size ranging from 100 to 120 mesh damaged the core and a cake was formed on the core surface. The results also indicate that more damage occurred when a high load pressure (400 psi) was applied on the high permeability cores (290–310 mD). The penetration of the particle gels into the low permeable formations can be decreased by the best selection of gel types, particle sizes, and brine concentrations.

Keywords Formation damage · Mature reservoirs · Preformed particle gel · Low permeable formations

1 Introduction

Water production is the main problem in oil/gas well operations as reservoirs mature (Seright 2003; Bai et al. 2008). Veil et al. (2004) reported that nearing the end of oil/gas production lives, water production can be 98 % of the material brought to the surface. Water production makes oil/gas wells unproductive and economically wasteful, which can cause early shut-in wells and decreased oil/gas production. Also, more water production can increase the costs of removing both scale and corrosion, and separating water from hydrocarbon. These costs increase as the water production increases (Dalrymple 1997). Worldwide, approximately three barrels of water are produced daily with each barrel of oil (Wiedeman 1996; Bailey et al. 2000). The situation is even worse in the United States, where more than 10 barrels of water are produced for each barrel of oil (Nemec 2014). Bailey et al. (2000) reported that the cost of treating and removing the surplus water production is expected to be 40 billion U.S. dollars (USD) globally. Hence, conformance control treatments and water shut-off are important in the oil industry.

Fluid flow in porous media is affected by the reservoir heterogeneity. It affects the selection of production plans, reservoir management, and oil recovery methods. Reservoir heterogeneity is the most important cause of increased water production and decreased oil production. Water

✉ Mahmoud O. Elsharafi
mahmoud.elsharafi@mwsu.edu

[1] McCoy School of Engineering, Midwestern State University, Wichita Falls, TX 76308, USA

[2] Department of Geological Sciences & Engineering, Missouri University of Science and Technology, Rolla, MO 65409, USA

Edited by Yan-Hua Sun

flooding is widely used in the oil industry to maintain reservoir pressure. Numerous reservoirs have cracks and channels due to mineral dissolution or sand production in the duration of water flooding (Liu et al. 2010). Reservoirs with high permeability zones and fractures are relatively common in mature reservoirs (Bai et al. 2008; Liu et al. 2010). To solve the problem of reservoir heterogeneity, gel treatments are widely used in oilfields (Zitha and Darwish 1999; Thomas et al. 2000; Bai et al. 2007a, b; Wang et al. 2008; Al-Muntasheri and Zitha 2009; Wu et al. 2011).

A chemical method extensively uses in situ gel treatment for both water shut-off and conformance control. In the in situ gel treatment process, a mixture of polymer and cross-linker (gelant, the liquid formulation of the in situ gel composition is called a gelant) is injected into the formation, and the gel forms under reservoir conditions (Liu et al. 2010). In situ gel penetrates into the high permeable formation and creates gel to reduce or plug the high permeable formation. Therefore, the gel is affected by the rock and fluid properties. Preformed particle gels (PPGs) are used to overcome different drawbacks inevitable in in situ gelation systems. These drawbacks include the inability to control gelation time, the uncertainty of gelling due to shear degradation, gelant composition changes caused by chromatographic fractionation effect and dilution by formation water. PPG is created on the surface and injected as gel particles (Chauveteau et al. 2001; Pritchett et al. 2003; Frampton et al. 2004; Rousseau et al. 2005; Sydansk et al. 2005; Bai et al. 2007a, b; Zaitoun et al. 2007; Wu et al. 2011). PPGs are a better selection than in situ gels from the point of controlling the particle sizes for different reservoir characterization. Presently, preformed gels contain preformed bulk gels, partially preformed gels, and particle gels (Chauveteau et al. 2000, 2001; Bai et al. 2007a, b, 2008).

For conformance control treatments, all particle gels used in oilfields are superabsorbent polymers (SAP). Particle gels contain PPGs, microgels, swelling micrometer-sized polymers (Bright Water[R]), and a pH-sensitive cross-linked polymer (Coste et al. 2000; Chauveteau et al. 2000, 2001; Al-Anaza and Sharma 2002; Pritchett et al. 2003; Frampton et al. 2004; Huh et al. 2005; Rousseau et al. 2005; Bai et al. 2007a, b, 2008; Zaitoun et al. 2007; Roussennac and Tosschi 2010; Zhang et al. 2010; Jia et al. 2011; Juntail et al. 2011a, b). Table 1 displays different types of particle gels used in the oil industry, related studies and the researchers who developed these gels, particle size, and applications. PPGs, microgels, and BrightWater[R] have all been used as water shut-off materials in mature reservoirs. pH-sensitive polymers are used to solve potential problems caused by polymer flooding, such as high injection pressure with associated pumping costs, the creation of unwanted injection well fractures, and the mechanical degradation of polymers due to high shear near the wellbore (Al-Anaza and

Sharma 2002; Huh et al. 2005, 2009; Al-Muntasheri et al. 2009, 2010). During matrix acidizing treatments for in situ-gelled acids, numerous acid methods have been used to improve acid diversion in heterogeneous reservoirs (Gomaa and Naser-El-Din 2010; Gomaa et al. 2011; Rabie et al. 2011, 2012; Reddy 2014). Legemah et al. (2014) reported the impact of different crosslinkers on the fluid properties while using low polymer loading as fracturing fluids.

PPGs have a collection of compositions which could absorb more than a hundred times their weight in solutions (Bai et al. 2007a, b, 2008; Zhang and Bai 2011; James 2011). In addition, they do not easily release the absorbed fluids under pressure (Bai et al. 2007a, b, 2008; Zhang and Bai 2011; James 2011). Bai et al. (2012) reported that the PPGs could absorb an enormous quantity of water because water has a large quantity of hydrogen ions. The water absorption volume is affected by sodium chloride (NaCl) concentrations (Bai et al. 2012). Sun et al. (2014) reported that the main element of PPG is the potassium salt of a cross-linked polyacrylic acid/polyacrylamide copolymer.

Liu et al. (2010) reported that in China, PPGs are widely used to decrease the theft zones in production and/or injection wells. Lately, to control CO_2 breakthrough for CO_2-flooded zones, both Occidental Oil Company and Kinder-Morgan Company used similar materials with good results (Smith et al. 2006; Pyziak and Smith 2007; Larkin and Creel 2008).

A lot of researchers have studied the propagation and blocking effectiveness of PPGs in both high permeable reservoirs and fractured layers (Bai et al. 2007a, b; Zhang and Bai 2011). Elsharafi and Bai (2012) determined that the best PPG treatments occurred when the PPG could simply penetrate through the high permeable layers without damaging the low permeable formations. Elsharafi and Bai (2015) found that the permeability of the gel pack in the fluid channels depended upon the particle strength, particle size, brine concentration, and the load pressure. The relationship among the cumulative injection volumes against time is needed to determine the damage to low permeable reservoirs (Ershaghi et al. 1986; Vetter et al. 1987). The cumulative injection volume value was used because if the core were damaged the flow rate will not be constant. As a result, the curves (the relationship between the cumulative filtration volume and the injection time) will not be straight lines. Filtration tests have been used in the past to study the damage of cores fully saturated with brine, oil, or residual oil while injecting suspended particles, oily water, or a combination of both into these cores (Hsi et al. 1994; Coleman and McLelland 1994; Al-Abduwani et al. 2005; Ali et al. 2009). The main purpose of this paper is to study the effect of strong PPGs on low permeable zones/areas in mature reservoirs, including the effects of rock permeability, salinity, particle size of PPGs, PPG types, pore throat size in cores, gel strength, and the injection pump

Table 1 Various particle gels used in the oil industry

Particle gels	Related studies	Developer	Particle size	Applications
Preformed particle gel (PPG)	Coste et al. (2000), Bai et al. (2007a, b, 2008), Jia et al. (2011), Zhang and Bai (2011), Juntail et al. (2011a, b)	PetroChina and Missouri S&T	Millimeter (10 µm–mms)	4000 plus injection wells
Microgels	Chauveteau et al. (2000, 2001), Rousseau et al. (2005), Zaitoun et al. (2007)	IFP	Microgel (1–10 µm)	10 plus gas wells
Bright water[R]	Pritchett et al. (2003), Frampton et al. (2004), Roussennac and Toschi (2010)	Nalco, Chevron, and BP	Submicron (<1 µm)	60 plus injectors

pressure. This paper is also needed to find out how to minimize the formation damage when the PPGs were used for water shut-off or conformance control treatment in mature reservoirs. This work used both filtration tests and load pressure tests to find out if the PPGs will damage the low permeable zones/areas. The damage or penetration caused by PPGs on unswept low permeability oil-rich zones could be effectively controlled by controlling particle gel strength, gel type, particle size, and brine concentration. This paper's results can be used to properly select the gel particles that will not damage the formation for the best particle gel treatment.

2 Experimental

2.1 Materials

2.1.1 Performed particle gel

Daqing (DQ) gel, a type of preformed particle gel (PPG), was used in all filtration tests and load pressure experiments. Table 2 lists the typical characteristics of the DQ gel. The percentage in Table 2 for gel and NaCl are weight percent. We used gels of various particle sizes of 30, 50–60, 80, and 100–120 mesh to determine the influence of PPG sizes on the reduction in the formation permeability. Figure 1 illustrates different particle sizes of the DQ gel fully swollen with 1 % brine.

2.1.2 Sodium chloride solutions (brines)

Brines used in all filtration tests and load pressure experiments were prepared by dissolving sodium chloride (NaCl)

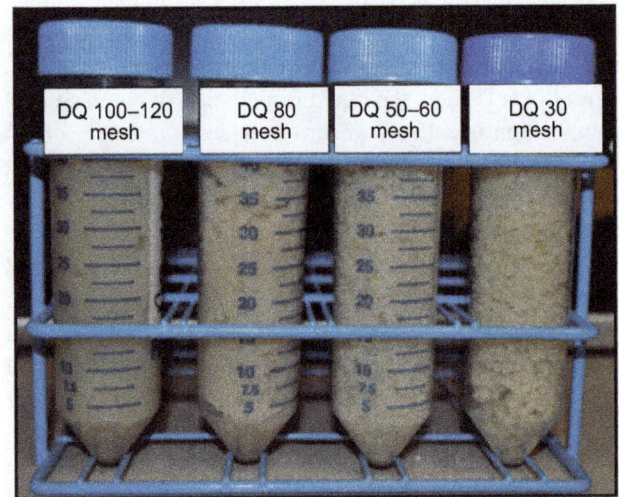

Fig. 1 Different particle sizes of the DQ gel fully swollen with 1 % brine

in deionized water. Different NaCl solutions (0.05, 1, and 10 wt% NaCl solutions at laboratory conditions) were carefully chosen to make preformed particle gels. The NaCl concentration expressively influences the swelling ratio and the strength of PPGs. Figure 2 illustrates the influence of the NaCl concentration on the ultimate swelling ratio of PPGs. Table 3 indicates the PPG strength variance at different NaCl concentrations before and after being compressed.

2.1.3 Sandstone core preparation

Twenty samples were collected from different sandstone sources (Missouri sandstone, Roubidoux sandstone, and Berea sandstone). The length (L) of all samples was 1.5 in. (3.7 cm) and the diameter (d) of all samples was 1.5 in.

Table 2 Typical characteristics of the DQ gel (after Zhang and Bai 2011)

Properties	Value
Absorption deionized water, g/g	>15
Apparent bulk density, g/L	850
Moisture content, %	0.96
pH value	6.5–7.0 (±0.5; 1 % gel in 0.9 % brine)

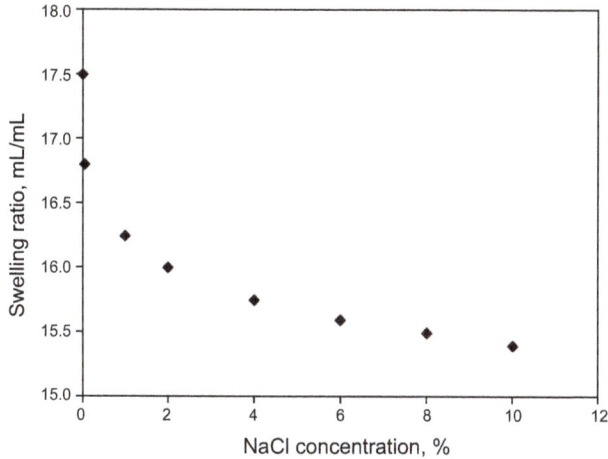

Fig. 2 Effect of the NaCl concentration on the swelling ratio of PPGs

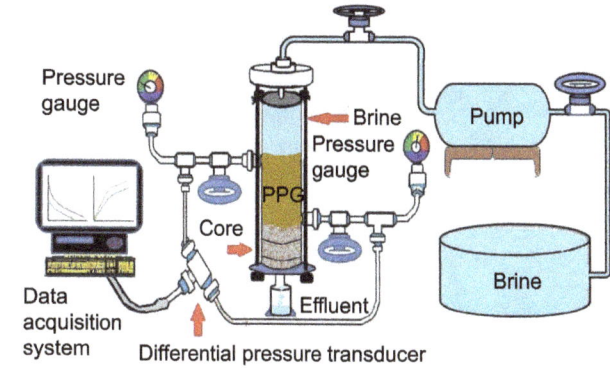

Fig. 3 Schematic of the filtration test model

(3.7 cm). To make sure that the dimensions of all samples were exact, a caliper was used to measure the core length and diameter. The samples were placed in an oven at 120 °C for 24 h. After that, the samples were evacuated and then saturated to 100 % with brines of chosen concentrations.

2.2 Experimental setup

2.2.1 Filtration test model

Figure 3 displays an experiment setup, which was primarily composed of a Teledyne ISCO model 500D syringe pump and one filtration test model. The syringe pump was used to inject brine into the filtration test model. The filtration test model was set up with a transparent round tube that contains a core sample fixed inside the round tube by two O-rings. In addition, a heavy duty glue was used to seal the space between the round tube and the core sample. Two lids were used (one above the round tube and one below the round tube). Bolts, nuts, and shims were used to fasten the two lids on the round tube. The first lid above the round tube had a hole which represented an inlet for the injected brine. The second lid, which was located below the round tube, had a hole which represented an outlet to discharge brine. To measure the pressures around both the sandstone

core and the PPG sample, pressure gauges were used. A differential pressure transducer was attached to the data acquisition unit to measure the differential pressure around the gel pack. The filtration test model did not use a piston.

2.2.2 Load pressure model

Figure 4 displays a load pressure model which was used to determine the influence of PPGs on the core damage after being compressed by a piston. The difference between this model and the filtration test model is that the load pressure model used a piston. The piston was placed on the top of PPGs inside the round tube. Brine was used to fill the space above the piston inside the round tube. To measure the pressure on the bottom of the piston (load pressure), two pressure gauges were used, one under the piston and another above the sandstone core. A differential pressure transducer was connected between the two pressure gauges to measure the differential pressure around the PPG sample.

2.3 Experimental procedures

2.3.1 Filtration test

The sandstone core sample was evacuated and saturated with the desired brine, and then the porosity of the sandstone core sample (ϕ) was determined. The desired sandstone core sample was fixed on the bottom portion of the transparent filtration test model. Brine was injected into the

Table 3 Effect of the NaCl concentration on PPG strength before and after being compressed

No.	Type of gel	Particle size, mesh	Gap, mm	NaCl concentration, %	PPG strength before being compressed G'_b, Pa	PPG strength after being compressed G'_a, Pa
1	DQ gel	30	1.5	0.05	4089	5994
2	DQ gel	30	1.5	0.25	4328	6358
3	DQ gel	30	1.5	1	4486	6583
4	DQ gel	30	1.5	10	4603	7368

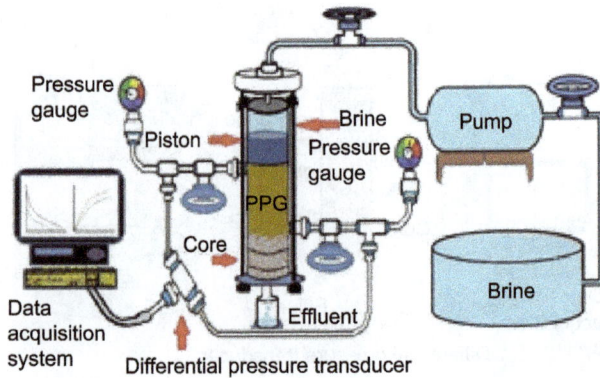

Fig. 4 Schematic of the load pressure model

filtration model to determine the core permeability before filtration tests. Completely swollen PPGs were poured into approximately half of the transparent round tube on the top of the core surface and the other space inside the round tube was filled with brine. Brine was injected at different constant pressures of 10, 50, 10, 100, 10, 200, 10, 400, 10 psi, and each constant pressure was run for 30 min as shown in Table 4, or until 500 mL of brine (pump volume) was injected into the core sample. The reason for repeating the 10 psi pressure test, was to find out the further damage to the sandstone core sample while using various injection pressures. A cumulative flow rate was measured at 1, 2, 3, 4, 5, 7.5, 10, 15, 20, 25, and 30 min for each pressure used. After 4 h and 30 min from the first core permeability measurements, PPGs were poured out from the round tube and brine was injected to determine the permeability of the core sample.

2.3.2 Load pressure test

A round piston was placed on the top of the PPGs inside the round tube after each filtration test. The PPGs were compacted by the round piston using brine as an injection liquid, with up to 300 psi pump pressure. Compacted PPGs

Table 4 Steps for various injection pressures versus time

Step	Injection pressure, psi	Time, min
1	10	30
2	50	30
3	10	30
4	100	30
5	10	30
6	200	30
7	10	30
8	400	30
9	10	30

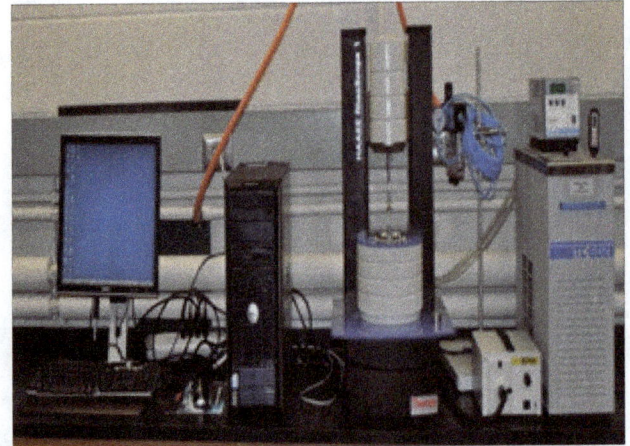

Fig. 5 Instrument for measuring gel strength

were poured out from the round tube, and brine was injected to measure the permeability of the core sample.

2.4 Measurement of PPG strength

Gel strength measurements were important. Particle gel strength measurements were taken to determine the PPG strength. These measurements indicated which particle gel was a weak gel and which was strong. These measurements also were used to determine which PPG could be selected without damaging the unswept oil-bearing zone. To determine the PPG strength (G'), a rheometer, KAAKE RheoScope1 (Thermo Scientific) was used as shown in Fig. 5. PPG strength measurements were taken, before and after the PPG was compacted by the piston at room temperature to find out the influence of compression on the PPG strength. The measurement model was fixed for oscillations with a frequency of 1.000 Hz, and stress of 1.0 Pa. The sensor which was used for gel strength measurement was PP35 Ti Po LO2 016, with a gap of 1.5 mm. PPG strength measurements were measured for each PPG sample during 60 s.

3 Calculation

3.1 Permeability of sandstone rocks

The linear Darcy equation was used to calculate the permeability (k) of different sandstone samples as shown in Eq. (1):

$$k = \frac{Q\mu L}{0.78 d^2 \Delta p},$$ (1)

where Q is the fluid flow rate, cm^3/s; μ is the brine viscosity, cP; L is the sandstone core length, cm; Δp is the differential pressure, atm; d is the diameter of the sandstone core, cm; and the physical meaning of the constant 0.78 is $\pi/4$.

3.2 Pore throat of sandstone rocks

Numerous attempts have been made to relate the pore diameter of a solid to intrinsic, more readily measurable properties, such as porosity and permeability (Elgmati et al. 2011). The Kozeny model describes the flow of fluids across straight cylindrical channels in a rock bed by combining Darcy's and Poiseuille's laws (Elgmati et al. 2011). This study used a calculation method to find out the average pore throat size (d_o) of the various sandstone core samples. The relationship between the porosity, sandstone rock permeability, and the average pore throat diameter could be described by Eq. (2) (Hong 1985; Lei et al. 2010):

$$d_o = \sqrt{\frac{32\tau k}{\phi}}, \tag{2}$$

where k is the sandstone rock permeability, μm^2; ϕ is the sandstone rock porosity, %; and τ is the tortuosity constant, dimensionless. This analysis assumes that the tortuosity coefficient is equal to 1. The pore diameter of the sandstone core sample was calculated with the simplified Kozeny formula. Table 5 shows the PPG sizes and the ratio of the particle diameter to the pore throat diameter of different

samples (d_p/d_o). The d_p/d_o values are 84–390, 26–127, and 17–84 for the core samples of permeability of 5–25, 110–115, and 290–310, respectively.

4 Results and discussion

4.1 Results of filtration tests

The results included the influence of particle size of PPGs, core permeability, and NaCl concentration on the damage to different sandstone core samples. The outcomes also contained the alteration of each core permeability after gel injection. Altered constant injection pressures were used to determine the relationship between the various cumulative volumes against time (filtration curves). The curve shape could be used to find out the sandstone rock damage.

4.1.1 Influence of PPG particle size

Several PPGs (30, 50–60, 80, and 100–120 mesh) were used to determine the influence of particle size on sandstone rock damage.

Figure 6 shows experimental results of the cumulative filtration volume versus the filtration time. The permeabilities of various cores were 10.65 mD for Fig. 6a, 20.45 mD for Fig. 6b, 12.35 mD for Fig. 6c, and 9.75 mD

Table 5 Properties of each sandstone sample used in experiments with various PPG sizes and NaCl solutions

No.	Type of sandstone	Porosity ϕ, %	Permeability k, mD	NaCl concentration, %	d_p, mesh	d_p, μm	d_o, μm	d_p/d_o
1	Missouri sandstone	14.0	10.65	1	30	595	1.55	383.84
2	Berea sandstone	15.0	20.45	1	50–60	250–297	2.07	120.77–143.47
3	Berea sandstone	15.0	12.35	1	80	177	1.61	109.75
4	Missouri sandstone	14.0	9.75	1	100–120	125–149	1.48	84.45–100.67
5	Roubidoux sandstone	16.0	114.20	1	30	595	4.74	125.30
6	Roubidoux sandstone	16.0	111.80	1	50–60	273	4.69	53.21
7	Roubidoux sandstone	16.0	110.00	1	100–120	137	4.66	26.82
8	Roubidoux sandstone	18.5	306.00	1	30	595	7.22	82.31
9	Roubidoux sandstone	18.5	300.28	1	50–60	250–297	7.16	34.91–41.48
10	Roubidoux sandstone	18.5	294.22	1	80	177	7.08	24.97
11	Roubidoux sandstone	18.5	293.20	1	100–120	125–149	7.07	17.68–21.07
12	Berea sandstone	14.0	10.80	0.05	30	595	1.56	381.16
13	Berea sandstone	14.0	10.65	1	30	595	1.55	383.84
14	Berea sandstone	14.0	10.35	10	30	595	1.52	389.37
15	Roubidoux sandstone	16.0	111.80	0.05	30	595	4.69	126.64
16	Roubidoux sandstone	16.0	113.23	1	30	595	4.72	125.84
17	Roubidoux sandstone	16.0	114.70	10	30	595	4.75	125.03
18	Roubidoux sandstone	18.5	300.50	0.05	30	595	7.16	83.06
19	Roubidoux sandstone	18.5	306.00	1	30	595	7.22	82.31
20	Roubidoux sandstone	18.5	305.00	10	30	595	7.21	82.44

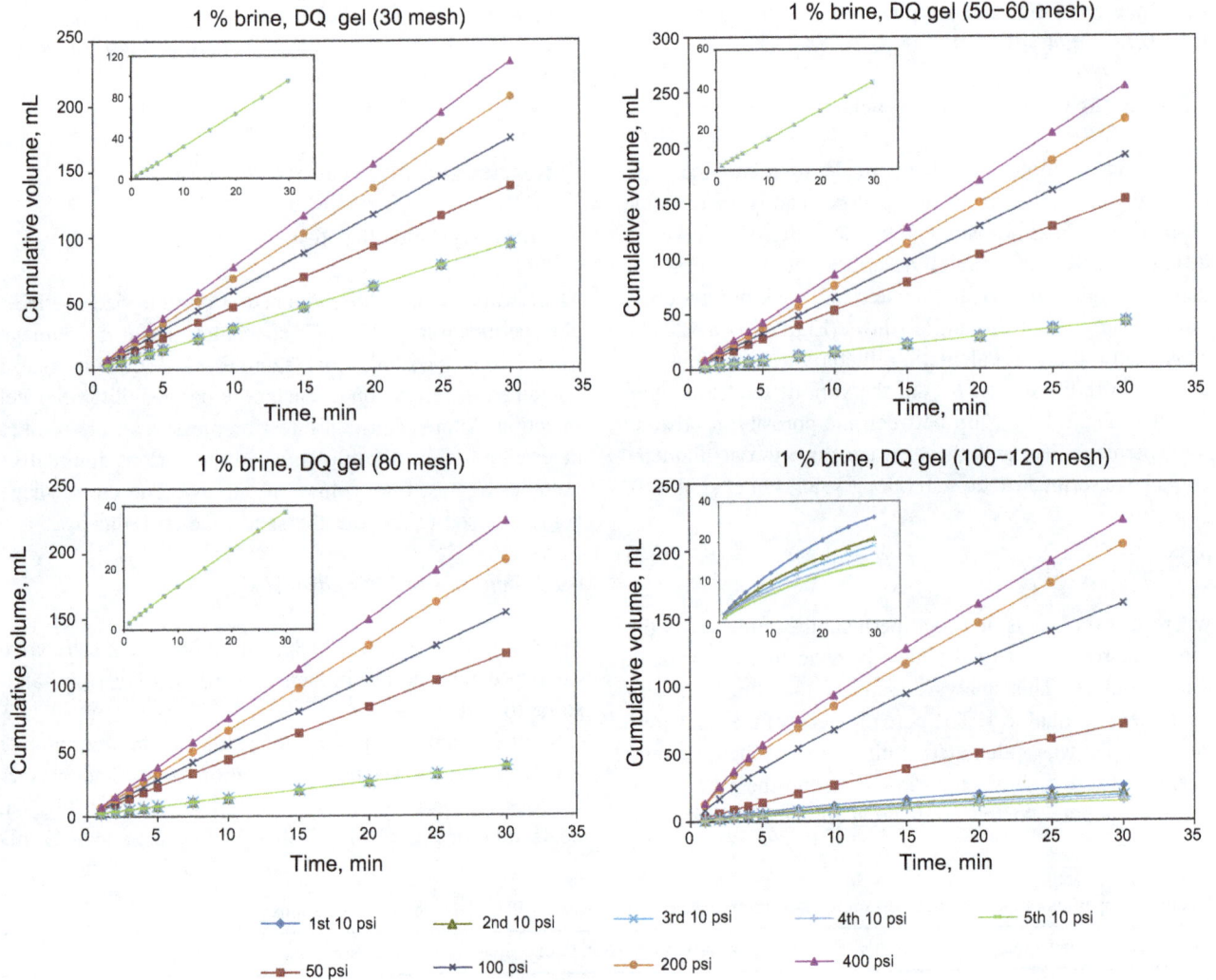

Fig. 6 Filtration test results at 1 % brine. **a** 30 mesh PPG, 10.65 mD core. **b** 50–60 mesh PPG, 20.45 mD core. **c** 80 mesh PPG, 12.35 mD core. **d** 100–120 mesh PPG, 9.75 mD core

for Fig. 6d. Figure 6a, b, and c respectively, show the influence of various PPG particle sizes (30, 50–60, and 80 mesh) on the core damage using different injection pressures (50, 100, 200, and 400 psi).

The relationships between the cumulative filtration volumes and the filtration test times (filtration curves) for each 10 psi, displayed in inserts Fig. 6a–c remain a straight line. There was a change in the y-scale (cumulative volume) for Fig. 6a–d because of different permeabilities of PPG packs. The PPG of larger particle sizes has higher PPG pack permeability than PPG of smaller particle sizes. As a result, the flow rate was affected by the PPG pack permeability. The flow rate decreased with a decrease in the particle size. The flow rate was constant with time, which indicates that the swollen particle gels of 30, 50–60, and 80 mesh did not damage the porous media. No PPG was observed to penetrate into the cores because the particle size of PPGs was larger than the pore throat of the

cores (no piston used). As a result, no cake was formed on the surface of the sandstone core. Other filtration curves in these figures remain straight lines. That means that the flow rates did not change with time and, therefore, no damage occurred at high pressures of 50, 100, 200, and 400 psi. The filtration model included a round tube withstanding a maximum pressure of 400 psi. The round tube will be broken if the pressure exceeds the limit pressure. The curve shapes of the different 10 psi pressures proved that when the injection pressure increased, no sandstone core damage occurred because all 10 psi pressure curves remain overlaid.

Figure 6d shows the influence of 100–120 mesh PPGs on the core damage. The curves are not similar for all 10 psi pressure. The filtration curves at 50, 100, 200, and 400 psi did not remain linear, implying that the core damage occurred at those pressures. DQ particle gel of small sizes of 100–120 mesh may damage the sandstone cores of

Table 6 Experimental results for the DQ gel with several particle sizes and NaCl solutions for filtration tests and load pressure models

No.	Type of sandstone	ϕ, %	k_b, mD	PPG particle size, mesh	NaCl concentration, %	k_a, mD	K_R, %	k_{ac}, mD	K_{Rac}, %
1	Missouri sandstone	14.0	10.65	30	1	10.65	0	5.65	46.94
2	Berea sandstone	15.0	20.45	50–60	1	20.45	0	8.50	58.43
3	Berea sandstone	15.0	12.35	80	1	12.35	0	5.05	59.10
4	Missouri sandstone	14.0	9.75	100–120	1	6.65	31.79	3.75	61.50
5	Roubidoux sandstone	16.0	114.20	30	1	113.90	0.26	7.07	93.75
6	Roubidoux sandstone	16.0	111.80	50–60	1	110.26	1.37	5.09	95.40
7	Roubidoux sandstone	16.0	110.00	100–120	1	30.55	72.27	3.90	96.45
8	Roubidoux sandstone	18.5	306.00	30	1	304.00	0.65	0.84	99.70
9	Roubidoux sandstone	18.5	300.28	50–60	1	295.60	1.56	0.77	99.74
10	Roubidoux sandstone	18.5	294.22	80	1	288.40	1.97	0.62	99.78
11	Roubidoux sandstone	18.5	293.20	100–120	1	20.70	92.93	0.36	99.87
12	Berea sandstone	14.0	10.80	30	0.05	10.80	0	5.05	53.24
13	Berea sandstone	14.0	10.65	30	1	10.65	0	5.65	46.94
14	Berea sandstone	14.0	10.35	30	10	10.35	0	5.75	44.44
15	Roubidoux sandstone	16.0	111.80	30	0.05	111.37	0.38	5.35	95.19
16	Roubidoux sandstone	16.0	113.23	30	1	112.90	0.29	7.07	93.73
17	Roubidoux sandstone	16.0	114.70	30	10	114.50	0.17	8.25	92.79
18	Roubidoux sandstone	18.5	300.50	30	0.05	298.00	0.84	0.64	99.78
19	Roubidoux sandstone	18.5	306.00	30	1	304.00	0.65	0.84	99.72
20	Roubidoux sandstone	18.5	305.00	30	10	303.50	0.49	0.90	99.70

permeability of 5–10 mD. PPGs were observed in the core because PPGs of 100–120 mesh have a lower gel strength.

4.1.2 Influence of core permeability

Twenty cores of permeabilities of 5–25, 110–115, and 290–310 mD were selected to investigate the influence of core permeability on sandstone core damage, and experimental results are listed in Table 6.

Table 6 provides the influence of permeability on the reduction in the core permeability (K_R). The permeability reduction means the decreasing of the original core permeability in percentage value after the core was damaged by PPGs for both static and load pressure models. k_b is the core permeability before PPG usage and k_a is the core permeability after PPG usage (without a piston). Table 6 also shows the permeability of each core (k_{ac}) and the core permeability reduction (K_{Rac}) after PPGs were compressed by a piston. The core permeability was determined at flow rates of 1, 2, and 3 mL/min after PPGs were poured out from the round tube.

For low permeability cores of 5–25 mD, the core permeability was not changed when PPGs of 30, 50–60, and 80 mesh were used. Less PPGs penetrated into the sandstone cores when PPGs of large particle sizes were used in low permeability cores. Figure 7 shows the influence of

core permeability on the permeability reduction of sandstone cores under the conditions of different PPG particle sizes and brine concentrations, respectively.

The core permeability reduced by 0.26 %–1.97 % for cores of 110–310 mD when PPGs of 30, 50–60, and 80 mesh were used. On the contrary, the core permeability increase caused an increase in the influence of 100–120 mesh PPGs on the rock damage. The core permeability reduction increased from 31.8 % to 92.9 % for cores of 5–310 mD.

4.1.3 Influence of NaCl concentration

Different brines (0.05 %, 1 %, and 10 % NaCl solutions) were chosen and used to investigate the influence of the NaCl concentration on the core damage. Several completely swollen PPGs were prepared from 30 mesh PPGs and different brines. As also shown in Fig. 7b, PPGs did not damage the low permeability cores (10–15 mD) at different NaCl concentrations. Figure 7b also shows the effect of different brine concentrations on the permeability reduction of higher permeability cores (110–115 mD and 300–310 mD). More core permeability reduction occurred when the NaCl concentration was lower for filtration tests before the gel was compressed by a piston. Table 3 shows the gel strength of the PPGs swollen with different brines

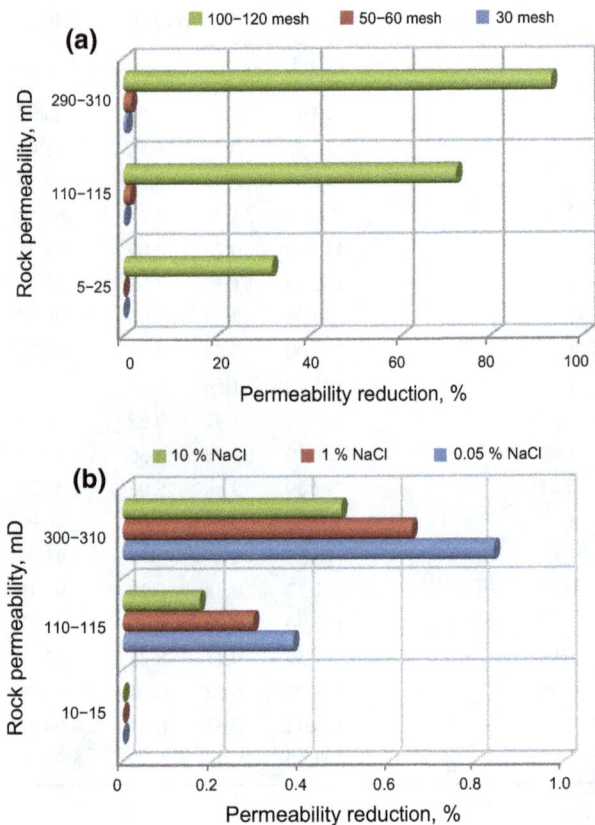

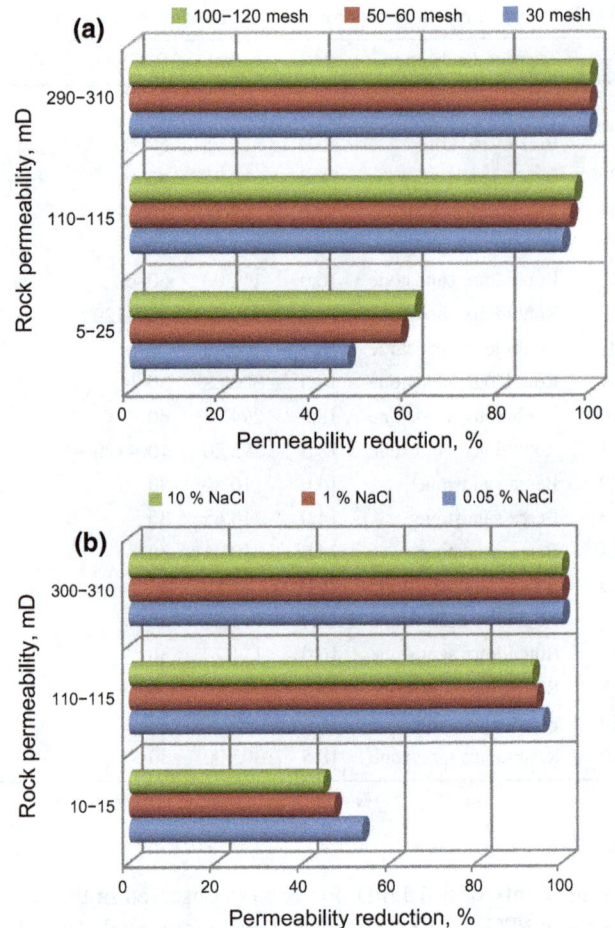

Fig. 7 Influence of core permeability on permeability reduction. **a** PPGs of different particle sizes (1 % brine). **b** Different NaCl concentrations (30 mesh PPGs)

Fig. 8 Influence of sandstone rock permeability on load pressure results. **a** PPGs of different particle sizes (1 % brine). **b** Different NaCl concentrations (30 mesh PPG)

before and after the gels were compressed by a piston. The PPGs with low NaCl concentrations have less gel strength. G_b' was the gel strength before PPGs were compacted by a round piston and G_a' was the gel strength after PPGs were compressed by a piston. After being compressed by a round piston, the PPG strength increased. The increase in the PPG strength was due to the water loss of swollen PPGs (El-sharafi and Bai 2012).

4.2 Results of load pressure tests

Sequences of experiments were done to study the influence of load pressure on rock damage as shown in Table 6. A round piston was placed at the top of the PPGs inside the round tube and the PPGs were compacted by the piston with 300 psi as a load pressure. After the gel was compressed, the permeability of each core was determined at the flow rates of 0.5, 0.75, and 1 mL/min. These lower flow rates were used because the stabilized pressure around the sandstone core increased so much after the gel was compressed and the core was damaged further. To confirm the level to which the injection pressure influenced the damage

to various cores, the load pressure tests were needed. Table 6 displays the DQ gel effect on the core damage as well as the core permeability reduction. The core permeability was determined before and after the PPGs were compressed. These measurements were taken to find out the influence of the load pressure on the damage to sandstone core samples.

Figure 8 shows the influence of the particle gel size, rock permeability, and the NaCl concentration on the damage to sandstone core samples after being compressed with a piston. The load pressure was 300 psi. Figure 8a shows the influence of the particle size of PPGs on the core permeability. Smaller-sized particles damaged the cores more than the larger-sized particles. This is because the smaller-sized particles may enter further into the porous media, particularly high permeability cores. Figure 8b shows the influence of NaCl concentration on the sandstone core permeability. The swollen PPGs prepared with solutions of higher NaCl concentrations decreased the core

(a)

DQ gel (100–120 mesh)

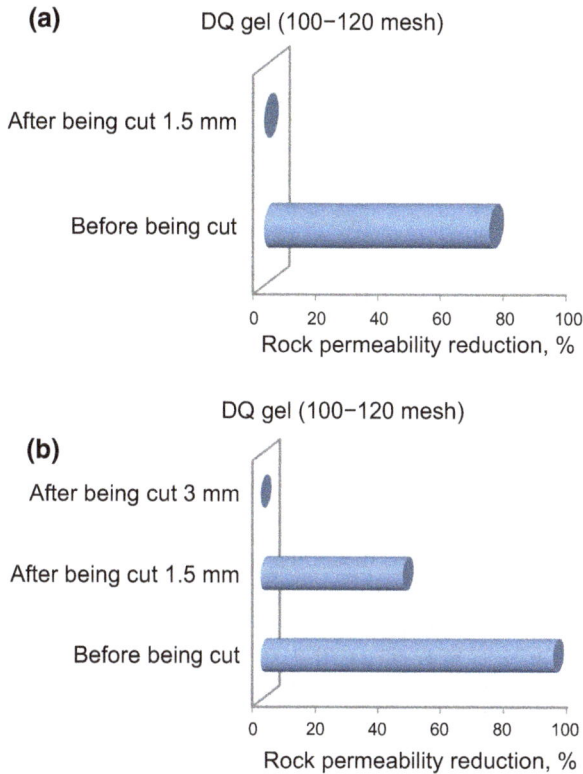

After being cut 1.5 mm

Before being cut

0   20   40   60   80   100
Rock permeability reduction, %

DQ gel (100–120 mesh)

(b)

After being cut 3 mm

After being cut 1.5 mm

Before being cut

0   20   40   60   80   100
Rock permeability reduction, %

Fig. 9 Permeability reduction in sandstone core samples before and after the rock surface was cut. **a** 110–120 mD core. **b** 290–310 mD core

permeability less than PPGs prepared with solutions of lower NaCl concentrations. The gel strength of the PPGs with higher NaCl concentrations was higher than that prepared with lower NaCl concentrations. Figure 8 shows that in high permeability cores, PPGs damaged the cores further.

4.3 Damage removed from the surface of sandstone rocks

After each load pressure experiment the damage was removed by cutting a slice or slices from the core surface first, 1.5 mm and then to 3 mm. This would remove the damage on the core surface and would not affect the core permeability of the non-damaged area since the core was fixed inside the round tube using two O-rings and a heavy duty glue. The core was cut with a sharp steel cutter which scratched the core surface many times until the core damage was removed. The purpose of cutting the core surface was to determine the penetration of the DQ gel into the core. This study included evaluating the effect of 100–120 mesh of DQ gel on the core permeability. The core permeability was determined after being cut. Figure 9 displays the reduction in the core permeability before and after each cut for the sandstone cores with permeabilities of

110–120 mD and 290– 310 mD, respectively. Zhang and Bai (2011) found that the swollen gel particles will propagate through porous media of super high permeability. Elsharafi (2013) used a quantitative analytical model to determine the formation damage in the low permeable zones/areas.

PPGs penetration into the core surface was proved after removing the core damage by cutting the surface of the cores. The gel penetration was only a few millimeters even if a PPG of small particle size (100–120 mesh) was used. After the core permeability was measured, it was found that the core permeability returned to its original value when the damage removed. Thus, the PPG could not propagate through cores of permeability < 310 mD. In these cases, as shown in Fig. 10, PPGs were found to form an internal filter cake or an external filter cake (Azizi et al. 1997). On the sandstone core surface, an external filter cake was created (Fig. 10a). When PPGs propagated a few millimeters into the rock surface, an internal filter cake was created (Fig. 10b). In this work, there was no deep penetration of PPGs from the surface of sandstone cores (Fig. 10c). Hence, PPGs cannot transmit from side to side of sandstone cores while the ratio of particle size of PPGs to pore throat size is greater than 17 for reservoir formations with rock permeability < 310 mD. Core damage was also dependent on the pore size. The damage increased if the pore throat size increased in high permeability rocks.

5 Comparison between weak (LiquiBlock™ 40K) and strong (Daqing) gels

Elsharafi and Bai (2012) investigated the influence of weak PPGs on low permeable formations. A comparison between the effect of a strong gel (DQ gel) and a weak gel (LiquiBlock™ 40K gel) on rocks is significant to select the best PPG type for use in a specific mature reservoir. The chosen PPG should improve sweep efficiency and minimize formation damage.

5.1 Filtration test results

A comparison of filtration test results indicate that the DQ gels of 30–80 mesh are a good choice to protect low permeable formation from gel penetration. This gel would not damage low permeability cores (less than 25 mD) when no piston was used as it can be seen in Fig. 11a. Additionally, the permeability reduction caused by gels of 30, 50–60, and 80 mesh was less than 2 % while the core permeability was 110–115 mD and 290–310 mD, respectively. In contrast, the particle gels of 100–120 mesh damaged the cores and reduced their permeability. Experimental results show that the weak gels with a low brine concentration are softer and

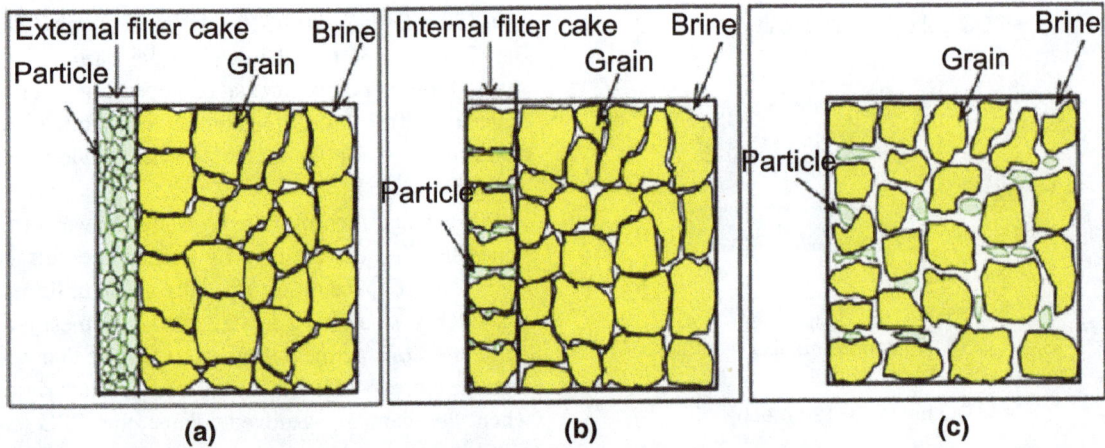

Fig. 10 Diagram of PPG damage. **a** PPGs form an external filter cake. **b** PPGs form an internal filter cake. **c** PPGs propagate in the core

Fig. 11 Photos of the PPG effect on the core damage. **a** Not damaged (strong gel). **b** Damaged (weak gel)

more deformable than those with a high brine concentration. Therefore, low brine concentration caused more core damage. The weak gel damaged the formation more than the strong gel because the weak gel had less strength and compressed further than the strong gel. Figure 11b shows a cake formed on the core surface when LiquiBlock™ 40K gel was used. Figure 12 illustrates the initial core

permeability (core permeability before the filtration test) versus the final rock permeability (after the filtration test) for both gels (LiquiBlock™ 40K gel, and DQ gel). LiquiBlock™ 40K gel of 30–120 mesh penetrated into the low permeable formations and decreased their permeabilities more than the DQ gel.

5.2 Load pressure results

The DQ gel, after being compressed by a piston, influenced core damage similar to the LiquiBlock™ 40K gel. The compressed DQ gel also formed a cake on the core surface and decreased the core permeability. The core damage under the load pressure was higher than that under the filtration test because the PPGs were compressed more under the load pressure. As a result the pressure around the PPG pack inside the round tube and the pressure on the top of the core surface increased. Therefore, more particle gels penetrated into the core surface and caused further damage. The change from the initial rock permeability to the final rock permeability for LiquiBlock™ 40K gel and DQ gels was depending on the particle size of gels, gel strength, and the original core permeability. Figure 13 shows the relationship between the initial core permeability and the final core permeability for LiquiBlockTM 40K and DQ gels. Figure 13 illustrates the final core permeability after PPGs were compacted by a piston for different permeability cores and PPGs of various particle sizes, respectively. A syringe pump with 300 psi injection pressure was used for LiquiBlock™ 40K and DQ gels. This load pressure reduced the core permeability more while using higher permeability cores for both gels. Both gels produced serious core damage and more permeability reduction when higher permeability cores were used. More damage occurred when higher permeability cores were used because higher permeability cores had larger pore throat

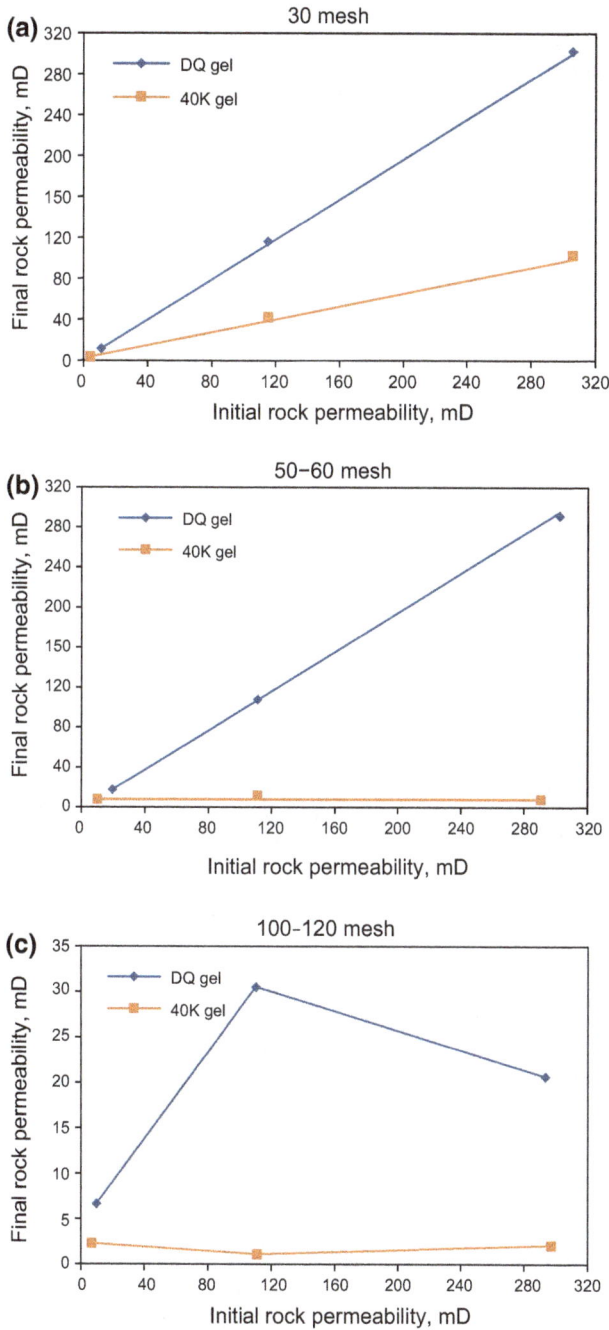

Fig. 12 Initial core permeability versus final rock permeability for both gels obtained from filtration tests. a 30 mesh PPGs. b 50–60 mesh PPGs. c 100–120 mesh PPGs

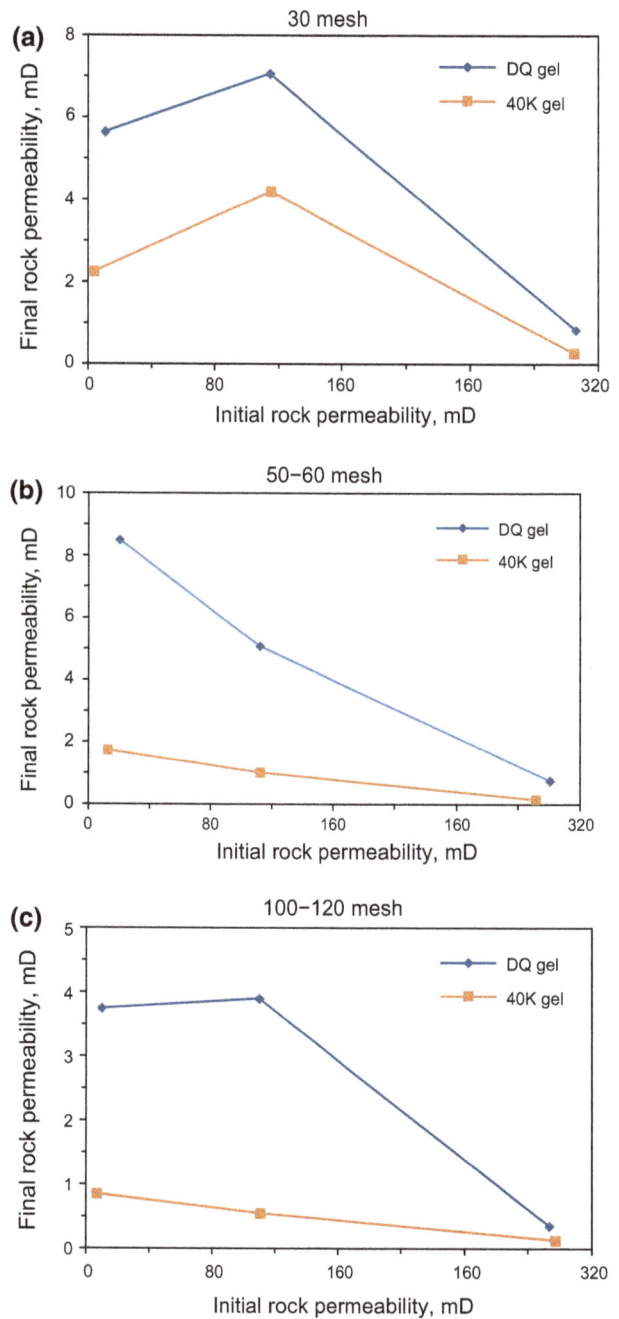

Fig. 13 Initial core permeability versus final rock permeability for both gels obtained from load pressure tests. a 30 mesh PPGs. b 50–60 mesh PPGs. c 100–120 mesh PPGs

sizes which allowed PPGs to penetrate into the core surface and form a cake easier than lower permeability cores. In field applications, the strong gel causes less formation damage to the unswept, low permeable zones/areas than the weak gel. In addition, the formation damage of unswept, low permeability, oil-rich zones could be controlled by controlling the strength, type, and particle size of PPGs, and the brine concentration.

6 Conclusions

(1) Filtration test results demonstrate that the strong DQ gel (30, 50–60, and 80 mesh) did not damage low permeability cores of 5–25 mD.

(2) The PPG did not propagate through sandstone cores and did not create a cake on the surface of sandstone cores.

(3) The PPG damage to cores was influenced by the PPG size and the NaCl concentration; 100–120 mesh DQ gel damaged low permeability cores (5–25 mD) and the core permeability reduced by up to 32 %.

(4) Load pressure test results demonstrate that the PPG damage to cores was affected by the load pressure; more damage occurred when higher load pressure was applied.

(5) The PPG damage to cores was influenced by the rock permeability; more damage occurred when using sandstone cores of high permeability of 290–310 mD.

(6) A comparison between weak and strong gels shows that a strong gel is a better selection when considering formation damage protection.

Acknowledgments The authors would like to thank the Research Partnership to Secure Energy for America (RPSEA) for its financial support for this work. The authors want also to thank McCoy School of Engineering at Midwestern State University in Wichita Falls, Texas, United States; Petroleum Engineering Department at Missouri University of Science and Technology in Rolla, Missouri, United States; Petroleum Engineering Department at Sirte University, Libya; and Baker-Hughes for their support.

References

Al-Abduwani FAH, Shirzadi A, Currie PK. Formation damage vs. solid particles deposition profile during laboratory simulated produced-water reinjection. SPE J. 2005;10(2):138–51. doi:10.2118/112509-PA.

Al-Anaza HA, Sharma MM. Use of a pH sensitive polymer for conformance control. In: SPE international symposium and exhibition on formation damage control, February 20–21, Lafayette, LA; 2002. doi:10.2118/73782-MS.

Ali M, Currie P, Salman M. The effect of residual oil on deep-bed filtration of particles in injection water. SPE Prod Oper. 2009;24(1):117–23. doi:10.2118/107619-PA.

Al-Muntasheri G, Nasr-El-Din HA, Al-Noaimi K, et al. A study of polyacrylamide-based gels crosslinked with polyethyleneimine. SPE J. 2009;14(2):245–51. doi:10.2118/105925-PA.

Al-Muntasheri G, Zitha PL. Gel under dynamic stress in porous media: new insights using computed tomography. In: SPE Saudi Arabia section technical symposium, 9–11 May, Al Khobar, Saudi Arabia; 2009. doi:10.2118/126068-MS.

Al-Muntasheri G, Zitha PL, Nasr-El-Din HA. A new organic gel system for water control: a computed tomography study. SPE J. 2010;15(1):197–207. doi:10.2118/129659-PA.

Azizi T, Jin W, Rahman SS. Management of formation damage by improved mud design. In: SPE Asia Pacific oil and gas conference and exhibition, 14–16 April, Kuala Lumpur, Malaysia; 1997. doi:10.2118/38039-MS.

Bailey B, Crabtree M, Elphick J, Kuchuk F, et al. Water control. Oilfield Rev. 2000;12:30–51.

Bai B, Liu Y, Coste JP, et al. Preformed particle gel for conformance control: transport mechanism through porous media. SPE Reservoir Eval Eng. 2007a;10(2):176–84. doi:10.2118/89468-PA.

Bai B, Li L, Liu Y, et al. Conformance control by preformed particle gel: factors affecting its properties and applications. SPE Reservoir Eval Eng. 2007b;10(4):415–21. doi:10.2118/89389-PA.

Bai B, Huang F, Liu Y, et al. Case study of preformed particle gel for in-depth fluid diversion. In: SPE/DOE symposium on improved oil recovery, 20–23 April, Tulsa, Oklahoma; 2008. doi:10.2118/113997-MS.

Bai B, Zhang H, Shuler P, et al. Preformed particle gel for conformance control. Final report, RPSEA, Contract Number 07123-2; 2012. www.netl.doe.gov.

Chauveteau G, Omari A, Tabary R, et al. Controlling gelation time and microgel size for water shutoff. In: SPE/DOE improved oil recovery symposium, 3–5 April, Tulsa, Oklahoma; 2000. doi:10.2118/59317-MS.

Chauveteau G, Omari A, Bordeaux U, et al. New size-controlled microgels for oil production. In: SPE international symposium on oilfield chemistry, 13–16 February, Houston, TX; 2001. doi:10.2118/64988-MS.

Coleman R, McLelland G. Produced water re-injection: how clean is clean?. In: SPE formation damage control symposium, 7–10 February, Lafayette, Louisiana; 1994. doi:10.2118/27394-MS.

Coste JP, Liu Y, Bai B, et al. In-depth fluid diversion by pre-gelled particles: laboratory study and pilot testing. In: SPE/DOE improved oil recovery symposium, 3–5 April, Tulsa, Oklahoma; 2000. doi:10.2118/59362-MS.

Dalrymple ED. Water control treatment design technology. In: SPE at the 15th world petroleum congress, 12–17 October, Beijing; 1997.

Elgmati M, Zhang H, Bai B, et al. Submicron-pore characterization of shale gas plays. In: SPE at North American unconventional gas conference and exhibition, 14–16 June, Woodlands, TX; 2011. doi:10.2118/144050-MS.

Elsharafi M, Bai B. Effect of weak preformed particle gel on unswept oil zones/areas during conformance control treatments. Ind Eng Chem Res. 2012;51(35):11547–54. doi:10.1021/ie3007227.

Elsharafi M. Minimizing formation damage for preformed particle gels treatment in mature reservoirs. Dissertation, Missouri University of Science and Technology; 2013.

Elsharafi M, Bai, B. Gel pack—a novel concept to optimize preformed particle gel conformance control treatment design. In: The 62nd southwestern petroleum conference, 22–23 April, Lubbock, TX; 2015. doi:10.2118/38039-MS.

Ershaghi I, Hashemi R, Caothien SC, et al. Injectivity losses under particle cake buildup and particle invasion. In: SPE California regional meeting, 2–4 April, Oakland, CA; 1986. doi:10.2118/15073-MS.

Frampton H, Morgan JC, Cheung SK, et al. Development of a novel waterflood conformance control system. In: SPE/DOE 14th symposium on improved oil recovery, 17–21 April, Tulsa, Oklahoma; 2004. doi:10.2118/89391-MS.

Gomaa AM, Mahmoud MA, Nasr-El-Din HA. Laboratory study of diversion using polymer-based in situ-gelled acids. SPE Prod Oper. 2011;26(3):278–90. doi:10.2118/132535-PA.

Gomaa AM, Nasr-El-Din HA. New insights into the viscosity of polymer-based in situ gelled acids. SPE Prod Oper. 2010;25(3):367–75. doi:10.2118/121728-PA.

Hong SD. Basics of reservoir petrophysics. Beijing: Petroleum Industry Publishing Company; 1985. p. 65–8 (in Chinese).

Hsi D, Dudzik S, Lane H, et al. Formation injectivity damage due to produced water reinjection. In: SPE formation damage control symposium, 7–10 February, Lafayette, Louisiana; 1994. doi:10.2118/27395-MS.

Huh C, Choi SK, Sharma MM. A rheological model for pH-sensitive ionic polymer solutions for optimal mobility-control applications. In: SPE annual technical conference and exhibition, 9–12 October, Dallas, TX; 2005. doi:10.2118/96914-MS.

Huh CU, Choi SK, Shrman MM. pH sensitive polymers for novel conformance control and polymer flood applications. In: SPE international symposium oilfields chemistry, 20–22 April, Woodlands, TX; 2009. doi:10.2118/121686-MS.

James JS. Modern chemical enhanced oil recovery theory and practice. Houston: Gulf Professional Publishing; 2011. p. 101–206. doi:10.1016/B978-1-85617-745-0.00001-2.

Jia H, Pu W, Zhao J, et al. Experimental investigation of the novel phenol–formaldehyde cross-linking HPAM gel system: based on the secondary cross-linking method of organic cross-linkers and its gelation performance study after flowing through porous media. Energy Fuels. 2011;25(1):727–36. doi:10.1021/ef101334y.

Juntail S, Abdoljalil V, Chun H, et al. Viscosity model of preformed microgels for conformance and mobility control. Energy Fuels. 2011a;25(1):5033–7. doi:10.1021/ef200408u.

Juntail S, Abdoljalil V, Chun H, et al. Transport model implementation and simulation of microgel processes for conformance and mobility control purposes. Energy Fuels. 2011b;25(1):5063–75. doi:10.1021/ef200835c.

Larkin R, Creel P. Methodologies and solutions to remediate inner-well communication problems on the SACROC CO_2 EOR project: a case study. In: SPE/DOE symposium on improved oil recovery, 20–23 April, Tulsa, Oklahoma; 2008. doi:10.2118/113305-MS.

Legemah M, Guerin M, Sun H, et al. Novel high-efficiency boron crosslinkers for low-polymer-loading fracturing fluids. SPE J. 2014;19(4):737–43. doi:10.2118/164118-PA.

Lei G, Li L, Nasr-El-Din H. New gel aggregates for water shut-off treatments. In: SPE improved oil recovery symposium, 24–28 April, Tulsa, Oklahoma; 2010. doi:10.2118/129960-MS.

Liu Y, Bai B, Wang Y. Applied technologies and prospects of conformance control treatments in China. Oil Gas Sci Technol. 2010;65(6):1–20. doi:10.2516/ogst/2009057.

Nemec R. Water, water everywhere in California oil production. NGI Shale Daily. 2014, Articles 99848. www.naturalgasintel.com.

Pritchett J, Frampton H, Brinkman J, et al. Field application of a new in-depth waterflood conformance improvement tool. In: SPE international improved oil recovery conference in Asia Pacific, 20–21 October, Kuala Lumpur, Malaysia; 2003. doi:10.2118/84897-MS.

Pyziak D, Smith D. Update on Anton Irish conformance effort. In: The 6th international conference on production optimization-reservoir-profile control-water and gas shutoff, 7–9 November, Houston, TX; 2007. doi:10.2118/103044-MS.

Rabie AI, Gomaa AM, Nasr-El-Din HA. Reaction of in situ gelled acids with calcite: reaction rate study. SPE J. 2011;16(4):981–92. doi:10.2118/133501-PA.

Rabie AI, Gomaa AM, Nasr-El-Din HA. HCl/formic in situ-gelled acids as diverting agents for carbonate acidizing. SPE Prod Oper. 2012;27(2):170–84. doi:10.2118/140138-PA.

Reddy BR. Laboratory characterization of gel filter cake and development of nonoxidizing gel breakers for zirconium-cross-linked fracturing fluids. SPE J. 2014;19(4):662–73. doi:10.2118/164116-PA.

Rousseau D, Chauveteau G, Renard M, et al. Rheology and transport in porous media of new water shutoff/conformance control microgels. In: SPE international symposium on oilfield chemistry, 2–4 February, Houston, TX; 2005. doi:10.2118/93254-MS.

Roussennac B, Toschi C. Brightwater trial in Salema Field (Campos Basin, Brazil). In: SPE EUROPEC/EAGE annual conference and exhibition, 14–17 June, Barcelona, Spain; 2010. doi:10.2118/131299-MS.

Seright RS. Washout of Cr(III)–acetate–HPAM gels from fractures. In: SPE international symposium on oilfield chemistry, 5–7 February, Houston, TX; 2003. doi:10.2118/80200-MS.

Smith D, Giraud M, Kemp C, et al. The successful evolution of Anton Irish conformance efforts. In: SPE annual technical conference and exhibition, 24–27 September, San Antonio, TX; 2006. doi:10.2118/103044-MS.

Sun Y, Wu Q, Wei M. Experimental study of friction reducer flows in microfracture. Fuel. 2014;131(1):28–35. doi:10.1016/j.fuel.2014.04.050.

Sydansk RD, Xiong Y, Al-Dhafeeri AM. Characterization of partially formed polymer gels for application to fractured production wells for water-shutoff purposes. SPE Prod Facil. 2005;20(3):240–9. doi:10.2118/89401-PA.

Thomas FB, Bennion DB, Anderson GE. Water shut-off treatments reduce water and accelerate oil production. J Can Pet Technol. 2000;39(4):25–9. doi:10.2118/00-04-TN.

Veil JA, Puder MG, Elcock D. A white paper describing produced water from production of crude oil, natural gas and coal bed methane. Technical report prepared for National Energy Technology Laboratory (U.S DOE, under Contract No. W-31-109-Eng-38), Argonne National Laboratory, Argonne, IL; 2004.

Vetter O J, Kandarpa V, Stratton M, et al. Particle invasion into porous medium and related injectivity problems. In: SPE international symposium on oilfield chemistry, 4–6 February, San Antonio, TX; 1987. doi:10.2118/16255-MS.

Wang Y, Bai B, Gao H, et al. Enhanced oil production through a combined application of gel treatment and surfactant huff'n' puff technology. In: SPE international symposium and exhibition on formation damage control, 13–15 February, Lafayette, Louisiana; 2008. doi:10.2118/112495-MS.

Wiedeman A. Regulation of produced water by the U.S. Environmental Protection Agency. In: Reed M, Johnsen S, editors. Produced water 2: environmental issues and mitigation technologies. New York: Plenum Press; 1996.

Wu Y, Tang T, Bai B, et al. An experimental study of interaction between surfactant and particle hydrogels. Polymer. 2011;52(2):452–60. doi:10.1016/j.polymer.2010.12.003.

Zaitoun A, Tabary R, Rousseau D, et al. Using microgels to shutoff water in gas storage wells. In: SPE international symposium on oilfield chemistry, February 28–March 2, Houston, TX; 2007. doi:10.2118/106042-MS.

Zhang H, Challa R, Bai B, et al. Using screening test results to predict the effective viscosity of swollen superabsorbent polymer particles extrusion through an open fracture. Ind Eng Chem Res. 2010;49(23):12284–93. doi:10.1021/ie100917m.

Zhang H, Bai B. Preformed particle gel transport through open fractures and its effect on water flow. SPE J. 2011;16(2):388–400. doi:10.2118/129908-PA.

Zitha PL, Darwish M. Effect of bridging adsorption on the placement of gels for water control. In: SPE Asia Pacific improved oil recovery conference, 25–26 October, Kuala Lumpur, Malaysia; 1999. doi:10.2118/57269-MS.

8

A review of down-hole tubular string buckling in well engineering

De-Li Gao[1] · Wen-Jun Huang[1]

Abstract Down-hole tubular string buckling is the most classic and complex part of tubular string mechanics in well engineering. Studies of down-hole tubular string buckling not only have theoretical significance in revealing the buckling mechanism but also have prominent practical value in design and control of tubular strings. In this review, the basic principles and applicable scope of three classic research methods (the beam-column model, buckling differential equation, and energy method) are introduced. The critical buckling loads and the post-buckling behavior under different buckling modes in vertical, inclined, horizontal, and curved wellbores from different researchers are presented and compared. The current understanding of the effects of torque, boundary conditions, friction force, and connectors on down-hole tubular string buckling is illustrated. Meanwhile, some unsolved problems and controversial conclusions are discussed. Future research should be focused on sophisticated description of buckling behavior and the coupling effect of multiple factors. In addition, active control of down-hole tubular string buckling behavior needs some attention urgently.

Keywords Down-hole tubular mechanics · Tubular string buckling · Wellbore configuration · Boundary condition · Friction force

✉ De-Li Gao
  gaodeli@cast.org.cn; huangwenjun1986@126.com

[1] MOE Key Laboratory of Petroleum Engineering, China University of Petroleum, Beijing 102249, China

Edited by Yan-Hua Sun

1 Introduction

Tubular string buckling is an important issue in well engineering. Buckling makes the initially straight tubular string buckle into curved shapes, which is an important reason for the well deviation problem. Buckling can also increase both bending stress on the tubular string and the contact force between the tubular string and the wellbore, which may further lead to serious down-hole problems such as tubular string failure, casing wear, hard slack off, or even "lock up." Down-hole tubular string buckling is usually taken as analogous to the Euler buckling problem for a rod with axial compressive forces on both ends. The rod remains straight until the axial force exceeds a certain value, namely the critical load. When the axial force is larger than the critical value, the initial configuration becomes unstable and the rod buckles into a laterally deformed configuration. However, unlike the free post-buckling deflection of the Euler rod, various external factors, such as the constraint of wellbore, tubular string weight, torque, friction force, etc., make tubular string buckling behavior more complex.

The first systematic research on tubular string buckling was conducted by Lubinski (1950). His pioneering work revealed the buckling mechanism of rotary drill strings in vertical wellbores and gave the critical buckling condition and post-buckling behavior of the drill string. Since then, a lot of improvement has been made in theoretical models and experiments. Many tubular string buckling models in vertical, horizontal, inclined, and curved wellbores under the action of torque, boundary condition, friction force, connectors, etc. have been proposed, some of which have been verified in later experiments and actual engineering operations. Despite all these achievements, some problems remain and need to be solved. For example, there is no accurate model so far to depict the transition process from two-dimensional lateral buckling to three-dimensional buckling in vertical wellbores. Different researchers derive inconsistent results of critical helical buckling loads for slightly inclined wellbores, for the transition process from sinusoidal buckling to helical buckling is a rather complex phenomenon and different researchers have adopted different assumptions. The studies of the effects of friction force, connectors, etc. are not mature and lack in-depth and systematic work.

In this paper, we review the progress in down-hole tubular string buckling. Classical research methods about down-hole tubular string buckling are presented. The effects of wellbore configuration, torque, boundary condition, friction force, connectors on tubular string buckling are discussed. Meanwhile, comments on some unsolved problems and controversial conclusions are presented.

2 Research methods

2.1 Beam-column model

Because the lateral displacement of the tubular string in wellbores is much smaller than the axial length, the linear elastic theory based on a small displacement is satisfied. The governing equation of tubular string buckling can be expressed by linear differential equations with respect to lateral displacements. When the tubular string does not contact the borehole wall or is in contact with the well wall at several single points, the distributed force on the down-hole tubular string is equal to tubular string weight and the governing equation for a inclined straight wellbore is expressed as follows (Gao 2006; Gao and Liu 2013):

$$\begin{cases} \dfrac{d^4u}{dz^4} + \dfrac{M_T}{EI}\dfrac{d^3v}{dz^3} + \dfrac{d}{dz}\left(\dfrac{F - qz\cos\alpha}{EI}\dfrac{du}{dz}\right) - \dfrac{q\sin\alpha}{EI} = 0 \\ \dfrac{d^4v}{dz^4} - \dfrac{M_T}{EI}\dfrac{d^3u}{dz^3} + \dfrac{d}{dz}\left(\dfrac{F - qz\cos\alpha}{EI}\dfrac{dv}{dz}\right) = 0 \end{cases}, \quad (1)$$

where u and v are the lateral displacements along x and y coordinates, respectively; z is the axial distance; F is the axial compressive force at the bottom end; M_T is the torque; EI is the bending stiffness; q is the weight per unit length of the down-hole tubular string; α is the hole angle.

The general solution of Eq. (1) can be expressed as the linear combination of certain linearly independent functions, namely

$$w = \mathbf{G}^T\mathbf{X} + w_g, \quad (2)$$

where w is the lateral displacement (u or v); $\mathbf{X}$ is the vector of undetermined constants and $\mathbf{G}$ is the vector of linearly independent power series with respect to the variable; w_g is

the lateral displacement caused by lateral tubular string weight (namely $q\sin\alpha$). With the general solution Eq. (2), the differential equation problem for down-hole tubular string buckling can be converted to an algebraic equation problem, which significantly reduces the complexity of the tubular string buckling problem.

The beam-column model has been applied to lateral buckling of a drill string under a zero-torque condition in vertical wellbores ($\alpha = 0$ and $M_T = 0$) (Lubinski 1950). According to Lubinski's analysis, a system of linear equations in the form of $\mathbf{MX} = 0$ is obtained by substituting the general solution Eq. (2) into boundary conditions at the top and bottom ends of the drill string. The critical lateral buckling force is obtained when $\mathbf{X}$ is not a zero vector, namely the determinant of the coefficient matrix $\mathbf{M}$ is equal to zero.

Generally, the beam-column model is used to depict the suspended section for the down-hole tubular string with multiple connectors distributed discretely (Mitchell 1982, 2000, 2003a; Huang and Gao 2014a,b, 2015). The suspended section between two adjacent connectors is usually very short, so the axial force on every suspended section can be approximately taken as a constant. Hence, the vector $\mathbf{G}$ in Eq. (2) can be written in a simpler form (Timoshenko and Gere 1963):

$$\mathbf{G} = \begin{bmatrix} 1 & z & \sin\left(\sqrt{F/EI} \cdot z\right) & \cos\left(\sqrt{F/EI} \cdot z\right) \end{bmatrix}^T. \quad (3)$$

Every suspended section is depicted by the above general solution and the terms including the lateral displacement, tangent direction, bending moment, and tangent shear force are all continuous at the connecting points between every two adjacent suspended sections. Introducing the boundary conditions at the ends of the integral tubular string, a system of nonlinear algebraic equations can be obtained. In this way, the down-hole tubular string buckling problem is transformed into the properties of the solutions of the nonlinear algebraic equations.

2.2 Buckling differential equation

For most cases, the down-hole tubular string is in continuous contact with the wellbore. The distributed force on the tubular string is equal to the sum of tubular string weight and the contact force between the tubular string and the wellbore. By introducing the wellbore constraint equations $u = r_c \cos\theta$ and $v = r_c \sin\theta$, the governing equation is transformed into the following differential equation form (Gao 2006):

$$\begin{aligned} &\frac{d^4\theta}{dz^4} - 6\left(\frac{d\theta}{dz}\right)^2\frac{d^2\theta}{dz^2} + 3\frac{M_T}{EI}\frac{d\theta}{dz}\frac{d^2\theta}{dz^2} + \\ &\frac{d}{dz}\left(\frac{F}{EI}\frac{d\theta}{dz}\right) + \frac{q\sin\alpha}{EIr_c}\sin\theta = 0 \end{aligned} \quad (4)$$

and the contact force on the tubular string is calculated by the the the following equation:

$$
N = EIr_c \left[4\frac{d^3\theta}{dz^3}\frac{d\theta}{dz} + 3\left(\frac{d^2\theta}{dz^2}\right)^2 - \left(\frac{d\theta}{dz}\right)^4 \right] +
$$

$$
M_T r_c \left[\left(\frac{d\theta}{dz}\right)^3 - \frac{d^3\theta}{dz^3} \right] + Fr_c \left(\frac{d\theta}{dz}\right)^2 + q\sin\alpha\cos\theta, \tag{5}
$$

where θ is the angular displacement shown in Fig. 1; r_c is the radial clearance between the tubular string and the wellbore; N is the compressive contact force on the tubular string per unit length.

Different from the beam-column model expressed by two variables u and v, there is only one variable θ in the buckling differential equation. However, it is difficult to solve the buckling differential equation due to the existence of nonlinear terms. If the friction force is introduced, the axial force F is related to the contact force N, and the complexity of the buckling differential equation is increased a lot. Up to now, general analytical solutions for the buckling differential equation Eq. (4) have not been found. Sinusoidal buckling and helical buckling are considered to be two representative solutions for Eq. (4) at present. It is generally accepted that a long tubular string constrained in a wellbore goes through an initial straight configuration, then sinusoidal buckling and later helical buckling with an increase in the axial compressive force from zero. These two buckling modes have been observed in a lot of experiments.

Sinusoidal buckling means that the tubular string behaves like a snaking curve along the lower side of the inclined wellbore. The sinusoidal buckling solution is usually expressed by

$$
\theta = A\sin(\omega \cdot z), \tag{6}
$$

where A is the amplitude and ω is the angular velocity of the angular displacement fluctuation. The critical load F_{crs} for the sinusoidal buckling can be obtained by analyzing the stability of the approximate linear form of Eq. (4) (Gao et al. 1998; Gao and Miska 2009). The relationship between the amplitude A and the axial force F is calculated by solving Eq. (4) with a perturbation method (Gao and Miska 2009).

Helical buckling means that the down-hole tubular string becomes a helix which spirals around the inner surface of the wellbore. The helical buckling solution can be expressed as follows:

$$
\theta = \frac{2\pi}{p}z \quad \text{or} \quad \theta = \frac{2\pi}{p}z + A\sin\left(\frac{2\pi}{p}z\right), \tag{7}
$$

where p is the helix pitch; A is the fluctuation amplitude caused by the tubular string weight. The analytical solution for the parameter $p = 2\pi\sqrt{2EI/F}$ is deduced from Eq. (4) for a weightless tubular string without torque (Mitchell 1988; Gao 2006). The parameter A is approximately solved with the perturbation method by assuming A to be a small term (Liu 1999; Gao and Miska 2010a). The critical load F_{crh} which converts the sinusoidal buckling to helical buckling is obtained when the contact force N between the tubular string and the high side of the inclined wellbore is equal to zero (Liu 1999).

2.3 Energy method

The energy method is another effective tool for us to study down-hole tubular string buckling problems. Compared to approximate solutions Eqs. (6) and (7) directly from the buckling differential equation, the buckling solutions from the energy method can be assumed more freely to depict the buckling configuration. Substituting approximate solutions such as Eqs. (6) or (7) into the total potential energy expression and calculating its minimum value, the buckling solutions can be determined. Meanwhile, the energy method is better used to calculate the critical buckling load and to analyze the stability of the post-buckling configuration.

For the suspended section on which the tubular string is not in contact with the wellbore, the total potential energy of the tubular string in inclined wellbores is expressed as the function of lateral displacements (Gao 2006):

$$
\Pi = U_b - \Omega_F - \Omega_M - \Omega_q
$$

$$
= \frac{1}{2}\int\limits_{z=0}^{L} \left\{ EI\left[\left(\frac{\partial^2 u}{\partial z^2}\right)^2 + \left(\frac{\partial^2 v}{\partial z^2}\right)^2\right] - F\left[\left(\frac{\partial u}{\partial z}\right)^2 + \left(\frac{\partial v}{\partial z}\right)^2\right] \right.
$$

$$
\left. - M_T\left[\frac{\partial u}{\partial z}\frac{\partial^2 v}{\partial z^2} + \frac{\partial v}{\partial z}\frac{\partial^2 u}{\partial z^2}\right] + 2q\sin\alpha(r_c - u) \right\}dz. \tag{8}
$$

For the continuous contact section on which the the tubular string is in continuous contact with the wellbore,

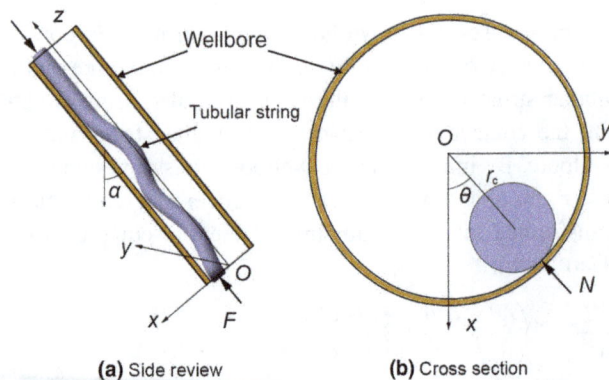

Fig. 1 Down-hole tubular string buckling in a inclined straight wellbore

(a) Side review **(b)** Cross section

the total potential energy is expressed as the function of angular displacement:

$$\Pi = \frac{1}{2} r_c^2 \int_{z=0}^{L} \left\{ EI \left[\left(\frac{\mathrm{d}\theta}{\mathrm{d}z} \right)^4 + \left(\frac{\mathrm{d}^2\theta}{\mathrm{d}z^2} \right)^2 \right] - F \left(\frac{\mathrm{d}\theta}{\mathrm{d}z} \right)^2 \right.$$
$$\left. - M_T \left(\frac{\mathrm{d}\theta}{\mathrm{d}z} \right)^3 + \frac{2q \sin\alpha(1 - \cos\theta)}{r_c} \right\} \mathrm{d}z, \tag{9}$$

where U_b is the elastic bending energy; Ω_F is the virtual work for axial force; Ω_M is the virtual work for torque; and Ω_q is the virtual work for tubular string weight.

To be specific, the amplitude A for sinusoidal buckling is calculated by introducing $\theta = A \sin(\omega \cdot z)$ into the total potential energy equation for the continuous section and letting the total energy reach the minimum value $\frac{\partial \Pi}{\partial A} = 0$. With the critical stability condition $\frac{\partial^2 \Pi}{\partial A^2} = 0$, the critical load F_{crs} is obtained (Liu 1999; Gao and Miska 2009), which converts the initial straight configuration into sinusoidal buckling. Similar to the analysis in sinusoidal buckling, the helical buckling pitch p for a weightless tubular string is deduced by introducing $\theta = \frac{2\pi}{p} z$ into Eq. (9) and letting $\frac{\partial \Pi}{\partial p} = 0$ (Lubinski and Althouse 1962). For a tubular string with weight, the helical buckling solution is $\theta = \frac{2\pi}{p} z + A \sin\left(\frac{2\pi}{p} z \right)$, which is more complicated than that for the weightless tubular string, whereas the solution process for the helical buckling fluctuation amplitude A is similar. Different from the stability criterion for sinusoidal buckling, the critical force F_{crs} which converts sinusoidal buckling to helical buckling is deduced by letting $\Delta \Pi = 0$, where $\Delta \Pi$ means the difference of total potential energy at the helical buckling and the initial configuration stages (Chen et al. 1990; Wu 1992; Cunha 1995).

However, the above three research methods are usually not isolated from each other. The combination of the three methods can give a sophisticated description of the tubular string deflection. For example, in the derivation of the deflection curve on the transition section near the boundary, the beam-column model is used to depict the suspended section while the buckling differential equation to depict the perturbed buckling section (Sorenson and Cheatham 1986; Liu et al. 1999). Taking a down-hole tubular string with two connectors on its two ends as another example, the two portions of the tubular string near the connectors are suspended, while the middle portion is in continuous contact with the wellbore. Similar to the analysis about boundary conditions, the suspended section and the continuous contact section are, respectively, depicted by the beam-column model and the bucking differential equation. In addition, the relative angular

displacement between the two connectors is obtained when the total potential energy of the down-hole tubular string achieves the minimum value (Huang and Gao 2014b). However, it is not an easy task for the simultaneous applications of the three classic methods because of the complicated solution process of the nonlinear algebraic equation systems.

3 Effect of wellbore configuration

3.1 Vertical wellbore

The critical buckling load for a tubular string in a vertical wellbore is an important issue. Lubinski (1950) studied the two-dimensional lateral buckling problem with the beam-column model and gave the critical buckling forces for the lowest two modes of buckling solutions shown in Table 1. When the axial force on the bottom of the tubular string reaches the first critical force, the initially straight tubular string becomes unstable and buckles into a two-dimensional curve with only a first order vibration. The curved configuration of the tubular string is believed to be one important reason for well deviation. When the axial force increases further and exceeds the second critical value, the tubular string deflection curve is seen as a second order vibration function and the tubular string tends to touch the inner surface of the wellbore on both sides of the wellbore axis. However, the tubular string does not sequentially buckle in a higher order of two-dimensional lateral buckling but becomes a three-dimensional curve with the increase in the axial force. Lubinski and Althouse (1962) assumed the three-dimensional curve to be a helix and deduced the pitch–force relationship with the energy method:

$$F = \frac{8\pi^2 EI}{p^2}. \tag{10}$$

Although Eq. (10) is derived for a weightless tubular string, it is proved to be an effective approximation for a tubular string with weight (Gao et al. 1996). The contact force between the helically buckling tubular string and the wellbore is equal to (Mitchell 1988)

$$N = \frac{F r_c^2}{4EI}. \tag{11}$$

At first, the critical load which initiates helical buckling in vertical wellbores is approximately considered to be 0 ($F = 0$) for a tubular string with rather small bending stiffness, such as tubing. As a result, the down-hole tubular string deflection is divided into two parts with the neutral point ($F = 0$) as the dividing point: the initial straight configuration above the dividing point and a full helix

Table 1 The critical buckling loads for vertical wellbores

Buckling mode	Lateral buckling	Sinusoidal buckling	Helical buckling
Dimensionless force ($\sqrt[3]{EIq^2}$)	1.94 and 3.75 (Lubinski 1950)	2.55 (Wu 1992)	0.00 (Lubinski and Althouse 1962)
			5.55 (Wu 1992)
			5.62 (Gao 2006)

depicted by Eq. (10) below the dividing point. Later, it was realized that it is too conservative to take the neutral point as the critical helical buckling force for the drill string. Wu (1992) improved the calculation method for the critical force between the three-dimensional sinusoidal buckling and helical buckling in vertical wellbores with the energy method shown in Table 1. In Wu's analysis, the tubular string buckles into a half-sine wave under the critical sinusoidal buckling condition and into a pitch of helix under the critical helical buckling condition.

Gao et al. (2002) and Gao (2006) pointed out that the sinusoidal buckling is unstable but helical buckling is stable in vertical wellbores with energy stability analysis and deduced the critical helical buckling with the buckling differential equation by letting the contact compressive force be positive for a period of the helix. Although different methods are employed in Wu's and Gao's research, their results for the critical helical buckling force are close to each other.

Mitchell (1988) studied helical buckling by solving the buckling differential equation numerically and proved that Lubinski's helical buckling model (Lubinski and Althouse 1962) was just an approximate result. Mitchell's results show that the pitch–force relationship expressed in Eq. (10) becomes invalid to depict the tubular buckling behavior near the neutral point because the tubular string may be not in contact with the wellbore.

Previous studies indicate that a key but tough problem is the depiction of the transition between the top suspended section and the bottom helically buckled section. A comprehensive model should consider the two sections as a whole: the top suspended section is depicted by the beam-column model and the bottom continuous contact section is depicted by the buckling differential equation. As a result, the dividing point can be determined with continuity conditions of axial displacement, slopes, bending moments and shear contact force of the two sections. In addition, how the two-dimensional lateral buckling turns to three-dimensional helical buckling with an increase in axial force has not been accurately described until now. A model for depicting the whole transition process of buckling state from initial vertical configuration, two-dimensional lateral buckling to the final helical buckling should be proposed.

3.2 Straight inclined and horizontal wellbores

Different from the vertical wellbores, the tubular string bucking behavior is greatly affected by the component of the tubular string weight perpendicular to the wellbore axis. If there is no axial force applied, the straight tubular string lies on the lower side of the wellbore. With the axial compression increasing to a certain value, the tubular string moves up from the wellbore bottom and buckles into a certain configuration. During this process, the axial force, which makes the tubular string become unstable, contends with other stable factors such as the wellbore constraint, perpendicular weight component, and the bending stiffness of the tubular string. That is to say, the buckling state is a comprehensive function of the stable and unstable factors.

Paslay and Bogy (1964) first introduced a trigonometric series to represent the buckling shape of the tubular string and obtained the critical sinusoidal buckling load for a long tubular string in an inclined wellbore with the energy method:

$$F_{crs} = \frac{EIn^2\pi^2}{L^2} + \frac{L^2 q \sin\alpha}{n^2\pi^2 r_c}, \tag{12}$$

where L is the tubular length, n is the number of half-period sinusoidal curves, q is the tubular string weight, and α is the inclination angle of the wellbore, F_{crs} is the critical sinusoidal buckling load. The minimum value of Eq. (12) is obtained by letting $\frac{n\pi}{L} = \left(\frac{q\sin\alpha}{EIr_c}\right)^{\frac{1}{4}}$ (Dawson 1984),

$$F_{crs} = 2\sqrt{\frac{EIq\sin\alpha}{r_c}}. \tag{13}$$

Equation (13) quantitatively depicts the critical state under the combined effects of stable and unstable factors. To simplify Paslay's derivation, a sine function buckling shape is assumed and then Eq. (13) can be directly deduced with the energy method (Chen et al. 1990; Miska et al. 1996; Liu 1999), Gao et al. (1998) obtained an identical solution with stability analyses on the approximate linear form of the buckling equation. The theoretical results are close to the Dellinger's experiments (Dellinger et al. 1983), and the fitting formulas for horizontal wellbores from the experimental data are $F_{crs} = 2.93EI^{0.479}q^{0.522}r^{-0.436}$.

After the axial force exceeds the critical value, the deflection curve of the tubular string can be approximately expressed by Eq. (6), where A is the amplitude and ω is the angular velocity of the angular displacement fluctuation. Different solving methods may lead to different solutions. For example, Gao and Miska (2010a) obtained $\omega = \left(\frac{q\sin\alpha}{EIr_c}\right)^{\frac{1}{4}}$, $A = 4\sqrt{\frac{\beta-1}{11}}$ and $\beta = \frac{F}{F_{crs}}$ by solving the buckling differential equation with the perturbation method, while Gao (2006) obtained $\omega = \sqrt{\frac{F}{2EI}}$ and $A = \sqrt{\frac{8(\beta^2-1)}{12\beta^2-1}}$ with the energy method. In fact, these two results are rather close to each other when the axial force approaches the critical sinusoidal buckling force ($\beta \approx 1$).

With an increase in the axial force, the sinusoidal buckling configuration becomes unstable and then buckles into a helical configuration. Unlike the distinct critical point between the initial configuration and sinusoidal buckling, there is no theoretical model to accurately represent the transition between these two completely different buckling modes—sinusoidal buckling and helical buckling. Different researchers proposed various assumptions about the transition process and derived different forms for the critical buckling loads with the energy stability principle. The main transition processes are illustrated in Fig. 2 and the values of critical buckling loads are given in Table 2.

Chen et al. (1990) first deduced the critical force to cause helical buckling with the energy method and stated that critical helical buckling load was $\sqrt{2}$ times the critical sinusoidal buckling load. In Chen's research, the axial force on the tubular string was assumed to be constant during the process from the starting of sinusoidal buckling to helical buckling. By assuming that the axial force increases linearly from the start of sinusoidal buckling to helical buckling, Wu (1992) refined Chen's method and pointed out that Chen's result was an average value of the critical sinusoidal and helical buckling loads. According to Wu, the critical helical buckling load is $2\sqrt{2}-1$ times the critical sinusoidal buckling load. Miska et al. (1996) assumed that the axial force increased linearly in the entire loading process and found that the critical helical buckling load was $2\sqrt{2}$ times the critical sinusoidal buckling load. Miska et al. (1996) also proved that the sinusoidal buckling configuration became unstable ($\beta = 1.875$) before the tubular string changes to the helical buckling configuration. Therefore, the tubular string may be in a sinusoidal or helical buckling state when the axial force satisfies $1.875 < \beta < 2\sqrt{2}$. Cunha (1995) pointed out that there are two critical helical buckling loads, where the lower limit value is equal to Chen's result and the upper limit is equal to Miska's

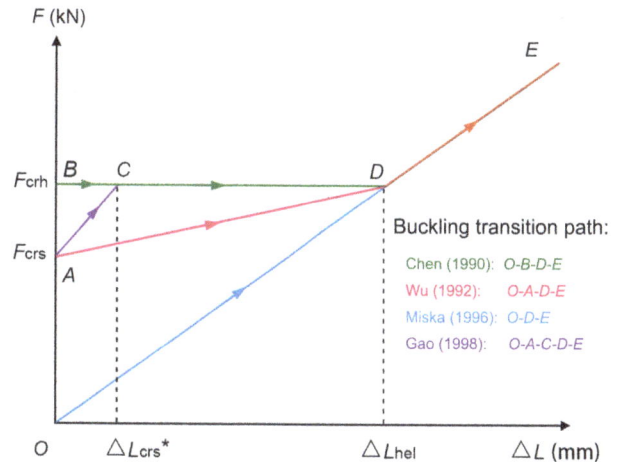

Fig. 2 The relationships between the axial force (F) and axial compressive displacement (ΔL) in the loading process (F_{crs} is the critical sinusoidal buckling load, F_{crh} is the critical helical buckling load, ΔL_{crs}^* is the maximum axial compressive displacement at the end of the sinusoidal buckling stage, and ΔL_{crh} is the minimum axial compressive displacement at the start of the helical buckling stage)

result. The upper limit means the minimum axial force causing helical buckling in the loading process, while the lower limit refers to the minimum axial force to keep the tubular string in the helical buckling stage for the unloading process. Mitchell (1997) obtained similar conclusions with the buckling differential equation. On the basis of previous studies, Gao et al. (2002) summarized the previous assumptions (Fig. 2) and considered that the axial force increases linearly in the sinusoidal buckling stage and then remains constant until the start of helical buckling based on experimental results.

In addition, the tubular string is always pushed on the inner surface of the wellbore whether in the sinusoidal or helical buckling stage. That is to say, the upper limit of the sinusoidal buckling is achieved when the contact force at the wellbore bottom is zero and the lower limit of the helical buckling is achieved when the contact force at the wellbore top is zero. Gao et al. (1998) calculated the critical buckling loads and found that these two limits are rather close to each other.

The differences among the results from different researchers mainly arise from two aspects. Firstly, previous buckling solutions are only approximate results with simplified assumptions. The accurate buckling solutions under sinusoidal buckling and helical buckling stages have not been obtained with the buckling differential equation and energy method. Secondly, a comprehensive description of the transition from the sinusoidal buckling to helical buckling has not been proposed due to the significant difference between these two buckling modes. The critical point between the sinusoidal buckling and helical buckling stages is usually assumed in theoretical analyses, but no

Table 2 The critical axial forces for different buckling modes

Researchers	Sinusoidal buckling (F/F_{crs})	Sinusoidal or helical buckling (F/F_{crs})	Helical buckling (F/F_{crs})
Chen et al. (1990)	$[1, \sqrt{2}]$	/	$[\sqrt{2}, \infty]$
Wu (1992)	$[1, 2\sqrt{2}-1]$	/	$[2\sqrt{2}-1, \infty]$
Miska et al. (1996)	$[1, 1.875]$	$[1.875, 2\sqrt{2}]$	$[2\sqrt{2}, \infty]$
Cunha (1995), Mitchell (1997)	$[1, \sqrt{2}]$	$[\sqrt{2}, 2\sqrt{2}]$	$[2\sqrt{2}, \infty]$
Gao et al. (1998)	$[1, 1.401]$	/	$[1.401, \infty]$

distinct dividing point is found in experiments. Therefore, sophisticated description of the buckling state transition and accurate buckling solutions under these two buckling modes still needs an in-depth research.

Previous studies assumed the radial deflection of the tubular string was fixed on the borehole wall and derived the pitch–force relationship (Eq. (10)) with the energy method (Chen et al. 1990; Miska et al. 1996). Cheatham (1984) removed the radial constraint from the wellbore in the energy method and proposed another new pitch–force relationship in the helical buckling stage:

$$F = \frac{4\pi^2 EI}{p^2}. \tag{14}$$

Cheatham pointed out that Eqs. (10) and (14) are, respectively, applicable for the loading and unloading processes. In the loading process, the pitch of the helix is variable but the tubular string is constrained by the wellbore. However, in the unloading process, the tubular string tends to lose contact with the wellbore but the pitch of the helix remains constant. Huang et al. (2015a) verified Cheatham's results from the view of the buckling differential equation and further pointed out that the contact force reaches its maximum value when Eq. (10) is satisfied and its minimum value when Eq. (14) is satisfied.

Gao et al. (1998) pointed out that the tubular string weight has an turbulent effect on the helix configuration, and a more accurate buckling solution should be expressed by

$$\theta = \frac{2\pi}{p} z + A \sin\left(\frac{2\pi}{p} z\right). \tag{15}$$

The perturbation magnitude $A = -\frac{4}{5\beta^2}$ is obtained by solving the buckling differential equation (Gao et al. 1998) and using the energy method (Gao 2006). Mitchell (2002a, b) used the Jacobi elliptic functions to solve the buckling differential equation and obtained the exact analytical helical buckling solution as follows:

$$\theta = 4 \sin^{-1}\left[\text{sn}\left(\sqrt{\frac{25+\sqrt{51}\beta}{50\sqrt{51}\beta}} \cdot \frac{2\pi z}{p}, \frac{50}{25+\sqrt{51}\beta} \right) \right]. \tag{16}$$

Meanwhile, Mitchell (2002a,b) found another new buckling solution in which the tubular string buckles into a periodically reversing curve which oscillates with a large angular amplitude about the top of the wellbore rather than the bottom of the wellbore. Huang et al. (2015a) further proved that sinusoidal buckling and helical buckling are just two special periodical solutions of the buckling differential equation.

The effect of tubular string buckling on bending moment, axial compressive displacement, and contact force is also an important issue. Table 3 lists the values of these three factors in the sinusoidal buckling and helical buckling stages. The bending moment from sinusoidal buckling can be neglected, while bending moment from helical buckling is significant especially under a large axial force with a big radial clearance. Because the axial compressive displacement in the helical buckling stage is far larger than that in the sinusoidal buckling stage, the transition from sinusoidal buckling to helical buckling includes a section where the tubular string is continuously shortened with almost no increase in the axial compression. This phenomenon has been observed in buckling experiments (Salies 1994; Zou 2002) and now it has been introduced as an important assumption to calculate the critical helical buckling loads.

The effect of the additional contact force from sinusoidal buckling is usually small for the length of the tubular string in the sinusoidal buckling stage is quite limited ($F_{crs} \leq F < F_{crh}$). However, the contact force increases significantly in the helical buckling stage ($F \geq F_{crh}$) due to the quadric relationship between the contact force and axial compressive load and it can seriously restrict the axial force transfer when the friction force effect is taken into consideration.

3.3 Curved wellbore

Experiments (McCann and Suryanarayana 1994; Salies 1994) have shown that the build wellbore curvature has a stabilizing effect on tubular string buckling. The tubular string may have been in "lock-up" or yielded while the tubular string does not enter the buckling state in the build section (Kyllingstad 1995).

Table 3 Relevant parameters in the post sinusoidal and helical buckling stages

Buckling mode	Maximum bending moment (M)	Axial compressive displacement ($\Delta L/L$)	Contact force (N)
Sinusoidal	$0.6302F^{0.08}(F-F_{crs})^{0.92}r_c$ (Mitchell 1999a)	$-0.7285\frac{r_c^2}{4EI}F^{0.08}(F-F_{crs})^{0.92}$ (Mitchell 1999a)	$q\sin\alpha$ (Mitchell 1999a)
			$q\sin\alpha+\frac{rF^2}{8EI}$ (Wu 1995)
		$-0.0843\frac{r_c^2F}{4EI}$ (Liu 1999)	
Helical	$0.5Fr_c$	$-\frac{r_c^2F}{4EI}$	$\frac{rF^2}{4EI}+q\sin\alpha$ (Mitchell 1999a)
			$\frac{rF^2}{4EI}$ (Liu 1999)

The first theoretical research into the critical buckling loads in a curved wellbore was to convert the buckling problem in a curved wellbore into the equivalent buckling problem in an inclined wellbore. Considering that the tubular weight component perpendicular to the wellbore axis is equal to the contact force in the unbuckling state in an inclined wellbore, the critical buckling load can be expressed by (He and Kyllingstad 1995).

$$F_{cr}=\beta\sqrt{\frac{4EI\cdot N}{r_c}},\tag{17}$$

where $\beta=1$ for Paslay's sinusoidal buckling and $\beta=\sqrt{2}$ for the Chen's helical buckling. Here, Eq. (17) is extended to the curved wellbore case, and the buckling load can be obtained by substituting the following contact force into a build section:

$$N=\frac{F_{cr}}{R}+q\sin\alpha,\tag{18}$$

where R is the curvature radius of the wellbore, q is the weight per unit length of the tubular string, and α is the inclination angle.

Wu and Juvkam-Wold (1995a) deduced the critical buckling loads in build and drop-off wellbore sections shown in Tables 4 and 5 with the energy method, in which the lateral component of tubular string weight and axial force was considered to do negative work. Wu's results show that the critical buckling loads in curved wellbores are usually larger than that in straight wellbores, except that the critical buckling loads become smaller in drop-off wellbores with small curvature. Later, Qui et al. (1998) obtained the critical buckling loads in which only the lateral component of tubular string weight was considered to do negative work and found that the sinusoidal buckling has become unstable before the tubular string enters the helical buckling stage. Kyllingstad (1995) pointed out that there is an overlapping section for two buckling states which acts as a barrier for buckling mode conversion. Mitchell (1999b) obtained Miska's critical sinusoidal buckling loads by both solving the buckling differential equation and using energy stability analysis, and resolved the conflict between the Miska and Wu's results. Mitchell

(1999b) pointed out that the lateral tubular string weight which appears to be part of the contact force can be neglected in Wu's studies as no work is done by the contact force. Liu (1999) solved the buckling differential equation with the Galerkin method and derived the critical helical buckling loads in build and drop-off wellbore sections by letting the minimum contact force on the tubular string be 0. Liu's results indicate that the up-limit value of the sinusoidal buckling is rather close to the critical helical buckling load.

Similar to the inclined wellbore case, the sinusoidal and helical buckling solutions of the tubular string in a curved wellbore can be expressed by $\theta=A\sin(\omega\cdot z)$ and $\theta=\frac{2\pi}{p}z+B\sin\left(\frac{2\pi}{p}z\right)$, in which the variable z is referred to the arc-length of the wellbore axis. The relevant parameters are calculated by $A=\sqrt{\frac{8(1-\lambda)}{12-\lambda}}$, $p=\sqrt{\frac{F}{2EI}}$ and $B=-\frac{1}{5}\lambda$, where $\lambda=\left(\frac{F}{q\sin\alpha R}+1\right)\cdot\frac{4EIq\sin\alpha}{r_c}\Big/F^2$ (Liu 1999).

The previous studies are mainly focused on build and drop-off wellbore sections in the vertical inclination plane. However, previous bucking models may not be applicable for three-dimensional wellbores of which both the inclination and azimuth angles change simultaneously in complex-structure wells. Therefore, a tubular string buckling model for arbitrary wellbores is needed.

4 Effect of other factors

4.1 Torque

Miska and Cunha (1995) proposed the critical helical buckling torque for a tubular string without axial force,

$$T_{crh}=2.087\sqrt[4]{\frac{(EI)^3q\sin\alpha}{r}},\tag{19}$$

and then the pitch of the helix is equal to

$$p=\frac{8EI\pi}{3T}.\tag{20}$$

Table 4 Critical buckling loads for build-up wellbores

Researchers	Critical sinusoidal buckling (F_{crs})	Critical helical buckling (F_{crh})
He et al. (1995)	$\frac{2EI}{Rr_c}\sqrt{1+\sqrt{1+\frac{R^2r_cq\sin\alpha}{EI}}}$	$\frac{4EI}{Rr_c}\sqrt{1+\sqrt{1+\frac{R^2r_cq\sin\alpha}{2EI}}}$
Wu and Juvkam-Wold (1995a)	$\frac{4EI}{Rr_c}\sqrt{1+\sqrt{1+\frac{R^2r_cq\sin\alpha}{4EI}}}$	$\frac{12EI}{Rr_c}\sqrt{1+\sqrt{1+\frac{R^2r_cq\sin\alpha}{8EI}}}$
Qui et al. (1998)	$\frac{2EI}{Rr_c}\sqrt{1+\sqrt{1+\frac{R^2r_cq\sin\alpha}{EI}}}$, $\frac{2.532EI}{Rr_c}\sqrt{1+\sqrt{1+\frac{R^2r_cq\sin\alpha}{3.52EI}}}$	$\frac{8EI}{Rr_c}\sqrt{1+\sqrt{1+\frac{R^2r_cq\sin\alpha}{2EI}}}$
Liu (1999)	$\frac{2EI}{Rr_c}\sqrt{1+\sqrt{1+\frac{R^2r_cq\sin\alpha}{EI}}}$	$\frac{3.77EI}{Rr_c}\sqrt{1+\sqrt{1+\frac{0.53R^2r_cq\sin\alpha}{EI}}}$

Table 5 Critical buckling loads for drop-off wellbores

Researchers	Critical sinusoidal buckling (F_{crs})	Critical helical buckling (F_{crh})
Wu and Juvkam-Wold (1995a)	$\frac{4EI}{Rr_c}\sqrt{1+\sqrt{1-\frac{R^2r_cq\sin\alpha}{4EI}}}$ (large curvature)	$\frac{12EI}{Rr_c}\sqrt{1+\sqrt{1-\frac{R^2r_cq\sin\alpha}{8EI}}}$ (large curvature)
	$\frac{4EI}{Rr_c}\sqrt{-1+\sqrt{1+\frac{R^2r_cq\sin\alpha}{4EI}}}$ (small curvature)	$\frac{12EI}{Rr_c}\sqrt{-1+\sqrt{1+\frac{R^2r_cq\sin\alpha}{8EI}}}$ (small curvature)
Liu (1999)	$\frac{2EI}{Rr_c}\sqrt{1+\sqrt{1-\frac{R^2r_cq\sin\alpha}{EI}}}$	$\frac{3.77EI}{Rr_c}\sqrt{1+\sqrt{1-\frac{0.53R^2r_cq\sin\alpha}{EI}}}$

However, the tubular string usually has yielded before torque reaches the critical value Eq. (19).

The critical helical buckling conditions while considering the effects of axial force, torque, and tubular string weight are shown in Table 6. Miska and Cunha (1995) derived the critical helical buckling load and pitch of the helix with the energy method. Later, He et al. (1995) treated the torque as a perturbation and obtained the approximate results for Miska's model. Wu (1997) assumed that there is only one period of helix when the axial force reaches its critical value and derived the critical buckling loads with the energy method. Their studies simultaneously show that the torque decreases the critical helical buckling load and pitch but increases the contact force. However, the pitch of the helix may be increased if the direction of the helix is opposite to that of the applied torque and torque has been proved to have no effect on sinusoidal buckling (Gao 2006). Paslay's experimental studies (Paslay 1994) indicated that the effect of torque on the critical helical buckling load was limited usually no more than 10 %.

Not only does torque affect the helical buckling configuration, but also helical buckling can also induce torque in turn. If a tubular string without torque at initial state is compressed axially and the two ends of the tubular string are constrained with no relative rotation, the induced torque due to helical buckling is equal to (Mitchell 2004)

$$T_{ind} = -\frac{Fr_c^2}{2}\sqrt{\frac{F}{2EI}}. \tag{21}$$

Mitchell (2004) pointed out that the induced torque may exceed the makeup torque for large radial clearance. Gao (2006) supplemented Mitchell's theory and referred to the induced torque as an incentive for the helix direction reversal observed in experiments (Salies 1994).

4.2 Boundary conditions

For a long pinned–pinned or clamped–clamped tubular string, the buckling configuration is divided into two parts: the transition section adjacent to the boundary condition and the full buckling section in the middle shown in Fig. 3 (Huang et al. 2015b). The transition section is further divided into suspended and perturbed buckling sections. For the suspended section, the tubular string loses contact with the wellbore due to the support of the boundary condition. For the full buckling section, the tubular string buckles into a sinusoidal or helical configuration. The perturbed buckling section, on which the tubular string is in continuous contact with the wellbore, is seen as the transition from no contact to full buckling section.

Wu and Juvkam-Wold (1995a) pointed out that if a tubular string has 3.5 or more pitches of helix, the transition section can be neglected. Gao and Miska (2009,

Table 6 Relevant parameters for the helical buckling with torque

Researchers	Critical helical buckling loads (F_{crh})	Pitch of helix (p)	Contact force (N)
Miska and Cunha (1995)	$F_{crh} = \frac{4\pi^2 EI}{p^2} + \frac{q\sin\alpha}{2\pi^2 r}p^2 - \frac{2\pi T}{p}$	$p^2 = \frac{8\pi^2 EI}{F}\frac{1}{(1+\frac{3\pi T}{Fp})}$	/
He et al. (1995)	$F_{crh} = 2\sqrt{\frac{2EIq\sin\alpha}{r}}\left(1 - 0.42T\sqrt[4]{\frac{r}{(EI)^3 q\sin\alpha}}\right)$	/	$\frac{rF^2}{4EI}\left(1 + \frac{T}{\sqrt{EI\cdot F/2}}\right)$
Wu (1997)	$F_{crh} = 2\sqrt{\frac{2EIq\sin\alpha}{r}}\left(1 - \frac{Tp}{4EI\pi}\right)$	$p = \sqrt[4]{\frac{8\pi^4 rEI}{q}} \cdot \sqrt[4]{1 - \frac{Tp}{4\pi EI}}$	/
Gao (2006)	/	$p^2 = \frac{8\pi^2 EI}{F}\left[\sqrt{1 + \frac{9T^2}{32EI\cdot F}} \pm \frac{3\sqrt{2}T}{8\sqrt{EI\cdot F}}\right]^2$	/

2010a) proved that the effect of the boundary condition on the full buckling section becomes negligible for a long pipe with the dimensionless length $\left(\frac{q}{EIr}\right)^{\frac{1}{4}}\cdot L$ larger than 5π.

The length of the transition section for fixed and pinned ends from different researchers is shown in Table 7. Mitchell (1982) assumed the transition section only included one suspended section. In Mitchell's analysis, the suspended section is described by the beam-column solutions and the full buckling section is expressed as a helix. On the basis of Mitchell's work, Liu et al. (1999) considered that the perturbed buckling section which connects the suspended section and the full buckling section cannot be ignored. In Liu's analysis, the perturbed buckling section is depicted described by the buckling differential equation, and the results show that the angular displacement rate decreases in an exponential manner in the perturbed buckling section. It is proved that the shear contact force on the contact point is not 0 in Mitchell's model and the contact force on the perturbed helix section is negative in Liu's model. Sorenson and Cheatham (1986) assumed that the suspended section is depicted by two suspended beams between which the adjacent point is in contact with the wellbore. The end of the perturbed buckling section is defined where the angular displacement change rate is 0.999 of that in the full buckling section. Later, Mitchell (2005) made some improvements on the calculation efficiency of Sorenson's model. All studies indicate that the

transition section is smaller than one pitch of helix, namely $\eta_1 + \eta_2 + \eta_3 < 2\pi$.

Huang et al. (2015a) pointed out that previous studies mainly considered the effect of axial force but neglected the bending moment and lateral force on the boundary constraints. Huang et al. proposed a novel classification method for the boundary conditions: if the virtual work of the bending moment and lateral force is 0, it is called the first category; otherwise the second category. It is shown that the boundary condition can also affect the full helical buckling section of a long tubular string under the second category case. Huang et al. (2015b) further verified that the boundary condition is closely related to the buckling configuration stability and found that the helical buckling direction may reverse abruptly when the boundary condition goes across some critical values.

4.3 Friction force

Friction force is considered to be a stability factor which inhibits the deviation of the tubular string away from its initial state. Experimental results also have indicated that friction force on the tubular string can delay the onset of the sinusoidal buckling (McCann and Suryanarayana 1994).

Mitchell (2007) pointed out that the initial buckling with friction is in the form of pipe rolling but not sliding. After the lateral rolling buckling increases to a certain amplitude,

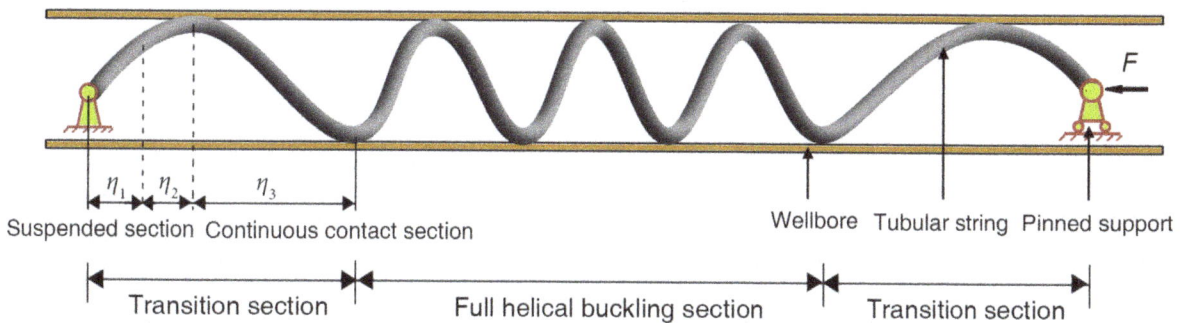

Fig. 3 Buckling configuration of a pinned–pinned tubular string in a horizontal wellbore. (η_1, η_2, and η_3, respectively, represent the first suspended section, the second suspended section, and the continuous contact section)

Table 7 Dimensionless lengths of the transition sections

Researchers	Boundary condition	η_1	η_2	$\eta_1 + \eta_2$	η_3	$\eta_1 + \eta_2 + \eta_3$
Mitchell (1982)	Fixed end	3.178	0	3.178	0	3.178
Liu et al. (1999)	Fixed end	2.718	0	2.718	1.211	3.929
Sorenson and Cheatham ((1986), Mitchell (2005), Huang et al. (2015b)	Pinned end	1.747	1.347	3.094	1.677	4.771
	Fixed end	2.700	0.903	3.603	1.846	5.449

η the dimensionless tubular string length $\sqrt{\frac{F}{2EI}} \cdot L$

it converts to the sliding buckling form. Then the critical buckling load is calculated by

$$\beta_{crs} = 1 + \frac{GJ}{r_p^2} \sqrt{\frac{r_c}{4EIq}}, \tag{22}$$

where GJ is the torsional stiffness of the tubular string. Gao and Miska (2010a) deduced the critical axial forces for sinusoidal buckling and helical buckling in a horizontal wellbore considering the sliding friction force with the energy method:

$$\beta_{crs} \approx 1 + 1.233\mu^{\frac{2}{3}}$$
$$\beta_{crh} \approx \sqrt{2}\left(1 + \frac{(30 + 7\pi^2)\mu}{30\pi}\right). \tag{23}$$

Note that only the first order approximation of μ is retained in Eq. (23). Gao's theoretical results are proved to be consistent with the results from later experiments.

Su et al. (2013) solve the tubular string buckling equation with the friction effect using Fourier series. The results show that the tubular string can tolerate substantial perturbation in a no-buckling state.

After the buckling is initiated, the axial friction force becomes dominant and then causes a rapid decrease in axial force transfer or even "lock up" (Kuru et al. 1999). Mitchell (1986) obtained the axial force distribution in the tubular string slack off process in vertical wellbores shown in Table 8, in which the tubular string below the neutral point is considered to stay in helical buckling. Later, Mitchell (1996) applied the friction force in the axial displacement-based governing equation and solved the axial force transfer with a finite element method. Wu and Juvkam-Wold (1995b) studied the axial force transfer on the sinusoidal and helical buckling sections in vertical and horizontal wellbores. The results show that the axial force on the helical buckling section in vertical wellbores increases in a hyperbolic tangent function with respect to the well depth and approaches the limit value of $\sqrt{\frac{4EIq}{\mu r_c}}$ with an increase in well depth. The axial force on the helical buckling section in horizontal wellbores decreases in a negative tangent function with respect to the well length and the maximum extending length on the helical buckling

section in a horizontal wellbore is $L = \frac{\pi}{\mu}\sqrt{\frac{EI}{qr_c}}$. Gao (2006) pointed out that Wu's tubular string buckling analysis was based on the non-friction case and further proposed a coupled model of buckling, contact force, and axial force. The coupled model is given in the form of the buckling differential equation with the sliding friction force. In Gao's model, the lateral component of friction force is dominant at the instant of lateral buckling, while the axial component of the friction force becomes dominant at the helical buckling stage. The differential equations are solved with the Galerkin method and the multiple-scale method, and the results show that buckling initiation is delayed a lot due to the friction force. Gao and Miska (2010a) re-built Liu's buckling differential equation with the energy method and calculated the axial force transfer in horizontal wellbores on the basis of the critical buckling loads Eq. (23). Gao's results indicate that the maximum extending length on the helical buckling section in a horizontal wellbore is $\frac{4EI}{\mu r_c F_{crh}}$.

One important method for improving the axial force transfer is rotating the tubular string. With rotation, the axial friction force can be dramatically decreased by converting the direction of the friction force from the axial direction to the rotational direction. Meanwhile, rotation also affects the tubular string buckling behavior. Menand et al. (2008, 2009) studied the effect of friction force and rotation on tubular string buckling with ABIS software and experiments, which indicates that rotation can reduce about 50 % of the critical helical buckling load from a non-rotating case. Gao and Miska (2010b) deduced the dynamic buckling equation of a rotating tubular string without friction and solved it with a perturbation method. The results show that there are two kinds of snaking motion: the first one is that the tubular string moves up and down about the static buckling configuration, while the other is the tubular string moves from one side to the other side of the wellbore periodically. Both theoretical and experimental results indicate that rotation does not affect the critical buckling load. Hydraulic vibration is another way to improve the axial force transfer and the "lock up" phenomenon caused by the combined effects of friction force

Table 8 Axial force transfer in vertical and horizontal wellbores

	Sinusoidal bucking	Helical bucking
Vertical wellbore	/	$F = \sqrt{\frac{4EIq}{\mu r_c}}\tanh\left(z\sqrt{\frac{\mu q r_c}{4EI}} + c\right)$ (Wu and Juvkam-Wold 1995b; Mitchell 1986)
Horizontal wellbore	$F = 2\sqrt{\frac{2EIq}{r_c}}\tan\left(-\mu z\sqrt{\frac{q r_c}{8EI}} + c\right)$ (Wu 1995) $F = \sqrt{\frac{4EIq}{r_c}}\left[-b_3 + \frac{1}{b_2}\tan\left(-\mu b_1 b_2\sqrt{\frac{q r_c}{4EI}}\cdot z + c\right)\right]$ (Gao and Miska 2010a) $b_1 = 0.7895 + 0.0428\mu^{2/3}$ $b_2 = 0.4460 - 0.1702\mu^{2/3}$ $b_3 = 1.1579 - 0.093\mu^{2/3}$	$F = \sqrt{\frac{4EIq}{r_c}}\tan\left(-\mu z\sqrt{\frac{q r_c}{4EI}} + c\right)$ (Wu and Juvkam-Wold 1995b) $F = F_{\mathrm{crh}}\frac{1}{1 - \frac{\mu r_c F_{\mathrm{crh}}}{4EI}z}$ (Gao and Miska 2010a)

and helical buckling can be overcome through tubular string vibration (Barakat et al. 2007). Some researchers also believed that the connectors on the tubular string can inhibit the buckling initiation and reduce the friction force on the tubular string. The effects of connectors on tubular string buckling are discussed below.

4.4 Connectors

Connectors distribute discretely along the tubular string and the diameters of connectors are larger than that of the tubular string body. Therefore, there exist three contact cases between the tubular string and the wellbore: no contact, point contact, and wrap contact. No contact means that the tubular string suspends between connectors and does not touch the inner surface of the wellbore, point contact means that the tubular string touches the wellbore at a single point, and wrap contact means that a segment of the tubular string is in continuous contact with the wellbore. In addition, there are three buckling states: non-buckling, sinusoidal buckling, and helical buckling. Huang et al. (2015c) pointed out that there are 9 deflection states and 12 transition conditions by the combination of three contact cases and three buckling states shown in Fig. 4. Generally speaking, a tubular string constrained in a straight wellbore goes through no contact, point contact to wrap contact for the contact states and non-buckling, sinusoidal buckling to helical buckling for the buckling states. However, the combination of connectors, tubular string buckling and other factors, such as wellbore configuration and tubular string gravity, is a complex problem.

In Fig. 4, "C" represents critical condition, "N," "P," and "W," respectively, denote no contact, point contact, and wrap contact, "I," "S," and "H," respectively, represent initial non-buckling, sinusoidal buckling, and helical buckling. For example, "Non-buckling & No contact" means the initial non-buckling state under a no contact case, "C_{NP_I}" means the critical condition between no contact and point contact in the non-buckling state, and

"C_{IS_N}" means the critical condition from initial non-buckling to sinusoidal buckling under no contact state.

Lubinski (1977) studied the two-dimensional deflection of a weightless tubular string with axial tension in a curved wellbore and obtained the bending moment magnification due to the existence of connectors in the no contact, point contact, and wrap contact cases. Later, Paslay and Cernocky (1991) extended Lubinski's work to the axial compression case. The results show that the local bending curvature with connectors is larger than the wellbore curvature. Therefore, the bending moment magnification should be taken into consideration for the tubular string design. On the basis of Paslay's work, Huang et al. (2015c) further considered the effect of tubular string weight and assumed that the wellbore curvature is equivalent to an additional tubular string weight. The results show that the tubular string weight and wellbore curvature affect the critical transition conditions between different contact cases a lot. All these studies focus only on the two-dimensional lateral deflection situations.

Mitchell (2003a) studied the three-dimensional sinusoidal buckling problem of a tubular string constrained in a horizontal wellbore and gave the buckling solutions for the no contact case. According to Mitchell, the connector/wellbore radial clearance r_c should be used in the critical sinusoidal buckling expression instead of the tubular string body/wellbore radial clearance r_b:

$$F_{\mathrm{crs}} = 1.991\sqrt{\frac{EIq}{r_c}} \approx 2\sqrt{\frac{EIq}{r_c}}. \tag{24}$$

From Eq. (24), we can see that the critical buckling load can be improved to $\sqrt{r_b/r_c}$ times of that under no connector case. Later, Mitchell (2003b) extended his work to the sinusoidal buckling problem in a curved wellbore. Mitchell (2000) and Mitchell and Stefan (2006) further studied the helical buckling of a tubular string in a vertical wellbore for the no contact case. The results show that the tubular string approximately buckles helically when the axial compression is low while the effect of connectors on bending stress

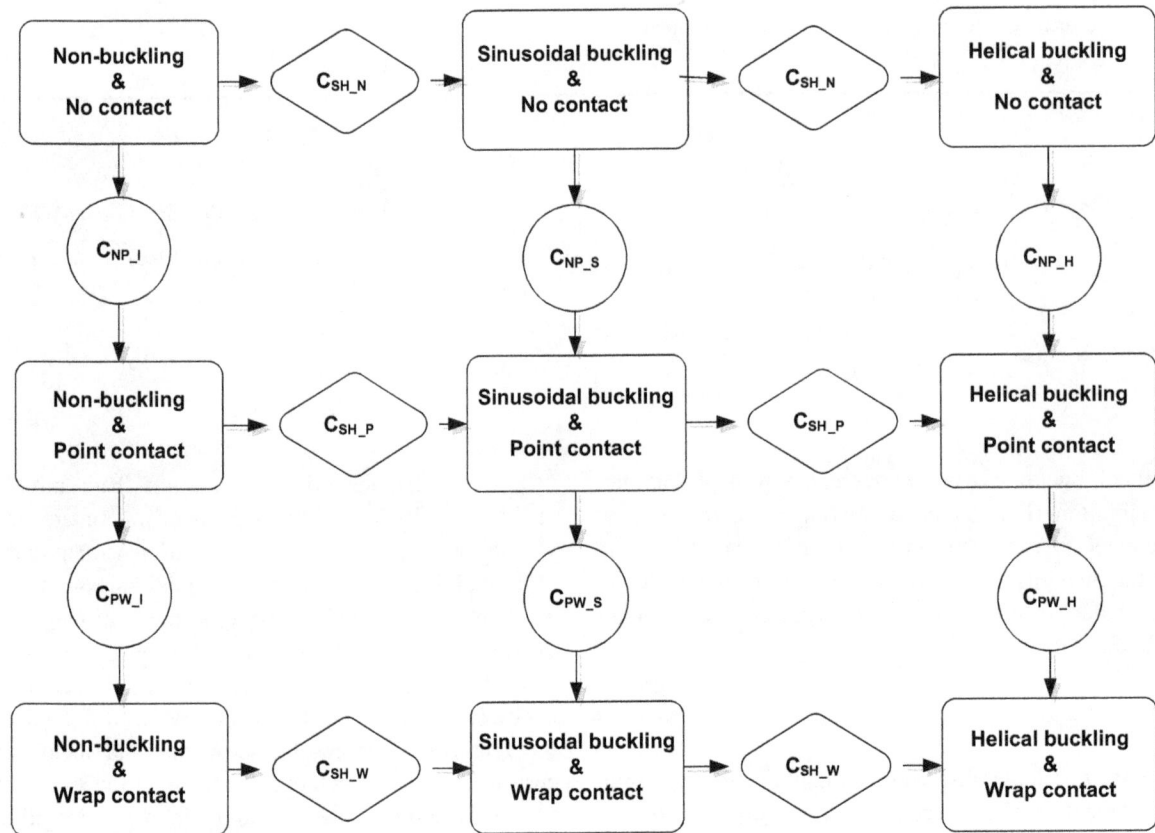

Fig. 4 Phase diagram of deflection states and critical conditions

becomes significant when the axial compression is high. Mitchell's results are verified by Duman's experiments (Duman et al. 2001, 2003) which show that connectors have no effect on the critical sinusoidal buckling load but increase the critical helical buckling load by 20 %.

Gao et al. (2012) studied the critical sinusoidal buckling load for a initially straight tubular string in no contact, point contact, and wrap contact cases. The results show that both the length between two adjacent connectors and the radial clearance difference between the tubular string body and connector affect the critical buckling loads a lot. With a decrease in the length between two adjacent connectors, the effective radial clearance approaches the connector radial clearance. When the length between two adjacent connectors reaches the critical value where the tubular string is just in point contact with the wellbore, the critical buckling load achieves its minimum value. Gao's results were proved to be in good agreement with numerical calculation (Daily et al. 2013; Hajianmaleki and Daily 2014).

On the basis of Mitchell's and Gao's studies, Huang and Gao (2014a, b, 2015) presented the complete phase diagram of the tubular string deflection states and studied the sinusoidal buckling and helical buckling problems for a

tubular string constrained in horizontal and curved wellbores. Huang's studies indicate that the effect of connectors on the tubular string buckling behavior is significant and reaches a maximum value under the critical condition from no contact to point contact cases.

Previous results indicate that connectors can delay the buckling initiation, reduce the contact force between tubular string and wellbore, and improve the axial force transfer on the tubular string. Therefore, a reasonable optimization of connector parameters may greatly improve the tubular string extension limit in long horizontal-section and extended-reach wells. For practical application, theoretical results about the effect of connectors on tubular string buckling and contact force should be obtained first.

5 Conclusions

In this article, our aim is to introduce three main research methods in down-hole tubular string buckling, summarize the effects of relevant factors on the critical buckling loads and post-buckling behavior, and draw a picture for future research in this field. Previous studies built the basic knowledge framework of down-hole tubular string buckling theory in which three issues are studied in detail: (1) the description of the transition between different buckling modes and the corresponding critical buckling loads; (2) the buckling behavior including the deflection curve, bending moment, and contact force under different bucking modes; (3) the effects of relevant factors such as wellbore configuration, torque, and boundary conditions. It is indicated that down-hole tubular string buckling is a complex problem because of the involvement of instability, nonlinearity, multiple factors etc. Future research on downhole tubular string buckling should still be conducted from two aspects: (1) a more sophisticated and more accurate description of the critical buckling conditions and postbuckling behavior, such as the quantitative description of the full path for the buckling mode transition in a vertical wellbore and an arbitrary three-dimensional wellbore; (2) the combination of as many factors as possible, such as the buckling behavior of a tubular string with connectors under the action of friction force in a three-dimensional wellbore. To achieve the above objects, the three research methods should be combined to solve the down-hole tubular string buckling problems and an efficient calculation program to solve the generated nonlinear algebraic equations should be proposed. In addition, the previous studies mainly focused on the mechanism of the down-hole tubular string buckling but neglected the active control of the tubular string buckling behaviors, which provides an important guidance for improving the extending limit of the tubular string in well engineering. Therefore, the achievement of active control is also a prominent research direction in the downhole tubular buckling.

Acknowledgments The authors gratefully acknowledge the financial support from the Natural Science Foundation of China (NSFC, 51221003, U1262201) and the Science Foundation of China University of Petroleum, Beijing (No. 00000). This research is also supported by other projects (Grant Numbers: 2014A-4214, 2013AA064803, 2011ZX05009-005).

References

Barakat ER, Miska SZ, Yu M, et al. The effect of hydraulic vibrations on initiation of buckling and axial force transfer for helically buckled pipes at simulated horizontal wellbore conditions. SPE/IADC drilling conference, 20–22 February Amsterdam, 2007. http://dx.doi.org/10.2118/105123-MS.

Cheatham JB Jr. Helical postbuckling configuration of a weightless column under the action of m axial load. Soc Pet Eng J. 1984; 24(4):467–72.

Chen Y, Lin Y, Cheatham JB. Tubing and casing buckling in horizontal wells (includes associated papers 21257 and 21308). SPE J Pet Technol. 1990;42(2):140–91.

Cunha JCDS. Buckling behavior of tubulars in oil and gas wells: a theoretical and experimental study with emphasis on the torque effect. Ph.D. dissertation. The University of Tulsa. 1995.

Daily JS, Ring L, Hajianmaleki M, et al. Critical buckling load assessment of drill strings in different wellbores using the explicit finite element method. SPE offshore europe oil and gas conference and exhibition, Aberdeen, 2013. http://dx.doi.org/10.2118/166592-MS.

Dawson R. Drill pipe buckling in inclined holes. J Pet Technol. 1984;36(10):1734–8.

Dellinger T, Gravley W, Walraven JE. Preventing buckling in drill string. US patent: 4384483. 1983.

Duman OB, Miska S, Kuru E. Effect of tool joints on contact force and axial force transfer in horizontal wellbores. SPE/IADC middle east drilling technology conference, 2001. http://dx.doi.org/10.2118/72278-MS.

Duman OB, Miska S, Kuru E. Effect of tool joints on contact force and axial–force transfer in horizontal wellbores. SPE Drill Complet. 2003;18(03):267–74.

Gao GH, Li Q, Zhang J. Buckling analysis of pipe string in a vertical borehole. J Xi'an Pet Inst. 1996;11(1):33–5 (in Chinese).

Gao GH, Miska SZ. Effects of boundary conditions and friction on static buckling of pipe in a horizontal well. SPE J. 2009;14(4): 782–96.

Gao GH, Miska SZ. Effects of friction on post–buckling behavior and axial load transfer in a horizontal well. SPE J. 2010a;15(4): 1104–18.

Gao GH, Miska SZ. Dynamic buckling and snaking motion of rotating drilling pipe in a horizontal well. SPE J. 2010b;15(3): 867–77.

Gao GH, Di Q, Miska SZ, et al. Stability analysis of pipe with connectors in horizontal wells. SPE J. 2012;17(3):931–41.

Gao DL. Down-hole tubular mechanics and its applications. Dongying: China University of Petroleum Press; 2006 (in Chinese).

Gao DL, Liu FW, Xu BY. An analysis of helical buckling of long tubulars in horizontal wells. SPE international oil and gas conference and exhibition in China, 2–6 November Beijing, 1998. http://dx.doi.org/10.2118/50931-MS

Gao DL, Lui FW, Xu BY. Buckling behavior of pipes in oil & gas wells. Prog Nat Sci. 2002;12(2):126–30 (in Chinese).

Gao DL, Liu FW. The post–buckling behavior of a tubular string in an inclined wellbore. Comput Model Eng Sci. 2013;90(1):17–36.

Hajianmaleki M, Daily JS. Critical-buckling-load assessment of drillstrings in different wellbores by use of the explicit finite-element method. SPE Drill Complet. 2014;29(2):256–64.

He X, Kyllingstad A. Helical buckling and lock–up conditions for coiled tubing in curved wells. SPE Drill Complet. 1995;10(1): 10–5.

He X, Halsey GW, Kyllingstad A. Interactions between torque and helical buckling in drilling. SPE annual technical conference and exhibition, 22–25 October, Dallas, 1995. http://dx.doi.org/10.2118/166592-MS.

Huang WJ, Gao DL. Sinusoidal buckling of a thin rod with connectors constrained in a cylinder. J Natl Gas Sci Eng. 2014a;18:237–46.

Huang WJ, Gao DL. Helical buckling of a thin rod with connectors constrained in a cylinder. Int J Mech Sci. 2014b;84:189–98.

Huang WJ, Gao DL. Helical buckling of a thin rod with connectors constrained in a torus. Int J Mech Sci. 2015;98:14–28.

Huang WJ, Gao DL, Liu FW. Buckling analysis of tubular strings in horizontal wells. SPE J. 2015a;20(2):405–16. http://dx.doi.org/10.2118/171551-PA.

Huang WJ, Gao DL, Wei SL. Boundary condition: a key factor in tubular string buckling. SPE J. 2015b. Preprint. http://dx.doi.org/10.2118/174087-PA.

Huang WJ, Gao DL, Wei SL. Local mechanical model of down-hole tubular strings constrained in curved wellbores. J Pet Sci Eng. 2015c;192:233–42.

Kuru E, Martinez A, Miska S, et al. The buckling behavior of pipes and its influence on the axial force transfer in directional wells. SPE/IADC drilling conference, 9–11 March, Amsterdam, 1999. http://dx.doi.org/10.2118/52840-MS.

Kyllingstad Å. Buckling of tubular strings in curved wells. J Pet Sci Eng. 1995;12(3):209–18.

Liu FW, Xu BY, Gao DL. Packer effect analysis of helical buckling of well tubing. J Tsinghua Univ. 1999;39(8):105–8 (in Chinese).

Liu FW. Post–buckling behaviors of tubulars within circular cylinders. Ph.D. dissertation. Beijing: Tsinghua University. 1999. (in Chinese).

Lubinski A. A study of the buckling of rotary drilling strings. Am Pet Inst. 1950;224(1):123–65.

Lubinski A. Fatigue of range 3 drill pipe. Revue de l'Inst Français du Pétrole. 1977;32(2):209–32.

Lubinski A, Althouse WS. Helical buckling of tubing sealed in packers. J Pet Technol. 1962;14(6):655–70.

McCann RC, Suryanarayana PVR. Experimental study of curvature and frictional effects on buckling. Offshore technology conference, 5/2/1994, Houston, 1994. http://dx.doi.org/10.4043/7568-MS.

Menand S, Sellami H, Akowanou J, et al. How drillstring rotation affects critical buckling load? IADC/SPE drilling conference, 4–6 March, Orlando, 2008. http://dx.doi.org/10.2118/112571-MS.

Menand S, Sellami H, Tijani M, et al. Buckling of tubulars in simulated field conditions. SPE Drill Complet. 2009;24(2):276–85.

Miska S, Cunha JC. An analysis of helical buckling of tubulars subjected to axial and torsional loading in inclined wellbores. SPE production operations symposium, 2–4 April, Oklahoma City, 1995. http://dx.doi.org/10.2118/29460-MS.

Miska S, Qiu W, Volk L, et al. An improved analysis of axial force along coiled tubing in inclined/horizontal wellbores. International conference on horizontal well technology, 18–20 November, Calgary, 1996. http://dx.doi.org/10.2118/37056-MS.

Mitchell RF. Buckling behavior of well tubing: the Packer effect. Soc Pet Eng J. 1982;22(5):616–24.

Mitchell RF. Simple frictional analysis of helical buckling of tubing. SPE Drill Eng. 1986;1(6):457–65.

Mitchell RF. New concepts for helical buckling. SPE Drill Eng. 1988;3(3):303–10.

Mitchell RF. Comprehensive analysis of buckling with friction. SPE Drill Complet. 1996;11(3):178–84.

Mitchell RF. Effects of well deviation on helical buckling. SPE Drill Complet. 1997;12(1):63–70.

Mitchell RF. Buckling analysis in deviated wells: a practical method. SPE Drill Complet. 1999a;14(1):11–20.

Mitchell RF. A buckling criterion for constant–curvature wellbores. SPE J. 1999b;4(4):349–52.

Mitchell RF. Helical buckling of pipe with connectors in vertical wells. SPE Dril Complet. 2000;15(3):162–6.

Mitchell RF. Exact analytic solutions for pipe buckling in vertical and horizontal wells. SPE J. 2002a;7(4):373–90.

Mitchell RF. New buckling solutions for extended reach wells. IADC/SPE drilling conference, 26–28 February, Dallas, 2002b. http://dx.doi.org/10.2118/74566-MS.

Mitchell RF. Lateral buckling of pipe with connectors in horizontal wells. SPE J. 2003a;8(2):124–37.

Mitchell RF. Lateral buckling of pipe with connectors in curved wellbores. SPE Drill Complet. 2003b;18(1):22–32.

Mitchell RF. The twist and shear of helically buckled pipe. SPE Drill Complet. 2004;19(1):20–8.

Mitchell RF. The pitch of helically buckled pipe. SPE/IADC drilling conference, 23–25 February, Amsterdam, 2005. http://dx.doi.org/10.2118/74566-MS.

Mitchell RF. The effect of friction on initial buckling of tubing and flowlines. SPE Drill Complet. 2007;22(2):112–8.

Mitchell RF, Stefan ZM. Helical buckling of pipe with connectors and torque. SPE Drill Complet. 2006;21(2):108–15.

Paslay PR. Stress analysis of drillstrings. University of Tulsa Centennial Petroleum engineering symposium, 29–31 August, Tulsa, 1994. http://dx.doi.org/10.2118/27976-MS.

Paslay PR, Bogy DB. The stability of a circular rod laterally constrained to be in contact with an inclined circular cylinder. J Appl Mech. 1964;31(4):605–10.

Paslay PR, Cernocky EP. Bending stress magnification in constant curvature doglegs with impact on drillstring and casing. SPE annual technical conference and exhibition, 6–9 October, Dallas, 1991. http://dx.doi.org/10.2118/22547-MS.

Qui W, Miska S, Volk L. Drill pipe/coiled tubing buckling analysis in a hole of constant curvature. SPE Permian Basin oil and gas recovery conference, 23–26 March, 1998. http://dx.doi.org/10.2118/39795-MS.

Salies JB. Experimental study and mathematical modeling of helical buckling of tubulars in inclined wellbores. Ph.D. dissertation. The University of Tulsa. 1994.

Sorenson KG, Cheatham JJB. Post-buckling behavior of a circular rod constrained within a circular cylinder. J Appl Mech. 1986;53(4):929–34.

Su T, Wicks N, Pabon J, et al. Mechanism by which a frictionally confined rod loses stability under initial velocity and position perturbations. Int J Solids Struct. 2013;50(14–15):2468–76.

Timoshenko SP, Gere JM. Theory of elastic stability. 2nd ed. New York: Tata McGraw-Hill Education; 1963.

Wu J. Buckling behavior of pipes in directional and horizontal wells. Ph.D. dissertation. Texas: Texas A&M University. 1992.

Wu J. Slack-off load transmission in horizontal and inclined wells. SPE production operations symposium, 2–4 April, Oklahoma City, 1995. http://dx.doi.org/10.2118/29496-MS.

Wu J, Juvkam-Wold HC. The effect of wellbore curvature on tubular buckling and lockup. J Energy Resour Technol. 1995a;117(3):214–8.

Wu J, Juvkam-Wold HC. Coiled tubing buckling implication in drilling and completing horizontal wells. SPE Drill Complet. 1995b;10(01):16–21.

Wu J. Torsional load effect on drill-string buckling. SPE production operations symposium, 9–11 March, Oklahoma City, 1997. http://dx.doi.org/10.2118/37477-MS.

Zou HH. Study of tubular string buckling in inclined straight wellbores. MS thesis. Beijing: China University of Petroleum. 2002. (in Chinese).

9

Experimental study of the azimuthal performance of 3D acoustic transmitter stations

Xiao-Hua Che[1,2] · Wen-Xiao Qiao[1,2] · Xiao-Dong Ju[1,2] · Jun-Qiang Lu[1,2] ·
Jin-Ping Wu[1,2] · Ming Cai[1,2]

Abstract Better well logging techniques for geologic investigations are urgently needed to identify and evaluate complex reservoirs. We describe a new type of 3D transmitter station with corresponding circuits and bodies. They can be used in a promising new technique of acoustic reflection well logging, that features better azimuthal detection capabilities, as well as better investigation depth. The transmitter stations consist of three-level subarrays that can radiate acoustic energy in any required azimuth of 3D space by circularly exciting various combinations at different levels. We tested the 3D acoustic transmitter stations and obtained laboratory directivity measurements with the 3D acoustic transmitter stations for the first time. The results show that the 3-dB beam width in the horizontal plane ranges from 59° to 67° as a result of phase-delayed excitation. The main beam is steered in the vertical plane at a deflection angle that ranges from 0° to 16° when the delay time of the excitation pulse between each pair of adjacent arc arrays is gradually adjusted. The 3-dB beam width is equal to 11°, whereas the deflection angle in the vertical plane is equal to 14°. Each of the four third-level subarrays in the same circumferential direction display consistent horizontal and vertical directivities, thus satisfying the requirements of azimuthal acoustic reflection logging.

Keywords Azimuthal performance · 3D · Acoustic transmitter stations · 3-dB beam width · Directivity

1 Introduction

In conventional monopole acoustic well logging, symmetrical acoustic sources facilitate shallow investigations, but they fail to detect fractures and small-scale geologic structures near boreholes, and they cannot evaluate the azimuthal properties of the formations around the boreholes (Haldorsen et al. 2006a). The acoustic waves reflected by the near-borehole interfaces with non-continuous acoustic impedance are obtained through acoustic reflection logging (Hornby 1989; Ellis et al. 1996; Esmersoy et al. 1997, 1998; Chang et al. 1998; Yamamoto et al. 1998; 1999). Migration imaging techniques that are similar to those used in seismic exploration are then employed to visualize small-scale geologic structures from a few meters to dozens of meters away from boreholes (Yamamoto et al. 2000; Tang 2004; Pistre et al. 2005; Li et al. 2008; Chai et al. 2009; Tang et al. 2007). Acoustic reflection logging generally facilitates more in-depth investigation than conventional acoustic logging, and it also yields images with higher resolution than those obtained through seismic exploration. Thus, this method is promising for future complex reservoir exploration.

The Borehole Acoustic Reflection Survey developed by Schlumberger employs monopole acoustic sources that radiate acoustic energy evenly in the circumferential direction. Monopole sources with single receivers cannot detect reflector azimuths, which is why Schlumberger used multi-receivers for their survey (Yamamoto et al. 2000; Al Rougha et al. 2005; Maia et al. 2006; Haldorsen et al. 2006b, 2010; Jervis et al. 2012). The acoustic reflection

✉ Xiao-Hua Che
aclab@cup.edu.cn

[1] State Key Laboratory of Petroleum Resources and Prospecting, China University of Petroleum, Beijing 102249, China

[2] Key Laboratory of Earth Prospecting and Information Technology, China University of Petroleum, Beijing 102249, China

Edited by Jie Hao

well logging tool (Zhao et al. 2004; Chai et al. 2009) developed by Bohai Drilling of the Dagang Oilfield Well Logging Branch employs a linear phased-array transmitter that is also a symmetrical acoustic source. This tool can identify high-angle fractures within 10 m of the well; however, reflector azimuth information cannot be extracted because of the axial symmetry of radiated acoustic fields.

In dipole remote acoustic reflection imaging, dipole acoustic sources are used to image small-scale, near-borehole geologic structures (Tang 2004; Patterson et al. 2008; Tang et al. 2007; Tang and Patterson 2009; Tang and Wei 2012a, b; Wei et al. 2013). Although low-frequency dipole acoustic sources provide for more thorough radial investigations, the directivities of the sources and the receivers limit the azimuthal resolution with 180° azimuthal ambiguity. Furthermore, the logging results are related to the positions of the logging tool in boreholes, and the sampling time is long.

Therefore, well logging tools with azimuthal resolution and remote-detecting functions are urgently required to invert the detailed formation information that is essential to geologic evaluations, reservoir characterizations, and oil-in-place assessments. To eliminate the azimuth ambiguity of single-well imaging, Zhang and Hu proposed a technique based on the pressure and displacement component, and they validated it by simulated examples (Zhang and Hu; 2014), whereas Gong et al. proposed a method using 3C reception data to eliminate the 180° azimuth ambiguity of dipole reflection imaging logging (Gong et al. 2015). However, data measured with current tools cannot be used with their method because only the vector receiver is capable of obtaining the displacement. Therefore, the method can only be verified after the development of a new acoustic logging tool. Qiao et al. proposed an acoustic phased-arc array transmitter with azimuthal directivity (Qiao et al. 2006, 2008, 2009); Che et al. numerically simulated the acoustic field in fluid-filled open holes, cased holes and formations generated by phased-arc array transmitters (Che and Qiao 2009; Che et al. 2010, 2014); and Wu et al. investigated the radiation characteristics of a phase-combined arc array transmitter that can be used in 3D acoustic well logging (Wu et al. 2013). However, the above studies all focused on the transmitter itself and were relatively simple because they did not consider other sections, such as corresponding circuits and tool bodies.

In the current study, we designed 3D acoustic transmitter stations with circuits and bodies. These stations are composed of a phased-arc combined array that consists of dozens of independent transducers. The 3D acoustic transmitter stations can be used directly downhole with other tool sections for azimuthal acoustic reflection well logging. We also tested and analyzed the azimuthal performance of the transmitter stations in the laboratory.

2 3D acoustic transmitter stations

Three-dimensional acoustic transmitter stations (Fig. 1) are the primary modules used by tools for azimuthal acoustic reflection well logging. The 3D structure consisting of four first-level subarrays that are evenly spaced along an axis is shown in Fig. 1a. A first-level subarray (Fig. 1b) is called a transmitter station, and it consists of eight elements that are distributed in a circle and numbered clockwise as TAx-1, TAx-2, TAx-3, TAx-4, TAx-5, TAx-6, TAx-7, and TAx-8, with x ranging from 1 to 4; thus, the four transmitter stations have a total of 32 independent transducers. The mandrel is made of steel and located inside the eight elements. The mandrel and the eight elements are sealed in a capsule and filled with silicone oil. The steel body near the elements is slotted to allow more acoustic energy to radiate into the borehole fluid. The four transmitter stations, which are numbered TA1, TA2, TA3, and TA4, are spaced 104 mm apart along the axis of the stations. Each of the four elements located along the axis in the same circumferential direction contains a phased linear array, which is defined as a second-level subarray. In each transmitter station, three adjacent elements constitute a third-level subarray. Therefore, one transmitter station can be recombined into eight three-element, third-level subarrays numbered clockwise as SUBx-1, SUBx-2, SUBx-3, SUBx-4, SUBx-5, SUBx-6, SUBx-7, and SUBx-8. The four third-level subarrays situated along the axis in the same circumferential direction consist of a combined arc array. The combined arc arrays are numbered clockwise as CAR-1, CAR-2, CAR-3, CAR-4, CAR-5, CAR-6, CAR-7, and CAR-8.

A 32-channel excitation circuit is integrated near the transmitter stations. Direct excitation with high-voltage pulses is applied to accurately control the pulse width, delay, polarity, and amplitude of each array element. When this circuit is controlled through a phased delay, the 3D acoustic transmitter stations can scan radiating acoustic energy with a circumferential stepping angle of 45° and an axial stepping angle of 1°. This pulse radiation mode can typically be used in azimuthal acoustic reflection logging tools. The main frequency is approximately 15 kHz. The excitation signal is a square wave with a signal width of one-half of the transducer main frequency reciprocal.

3 Experimental setup

The experimental setup for the 3D acoustic transmitter stations is depicted in Fig. 2. The transmitter stations are placed in a 5 m × 5 m × 4 m pool with a standard hydrophone. The hydrophone is moved and accurately

(a)

(b)

(c)

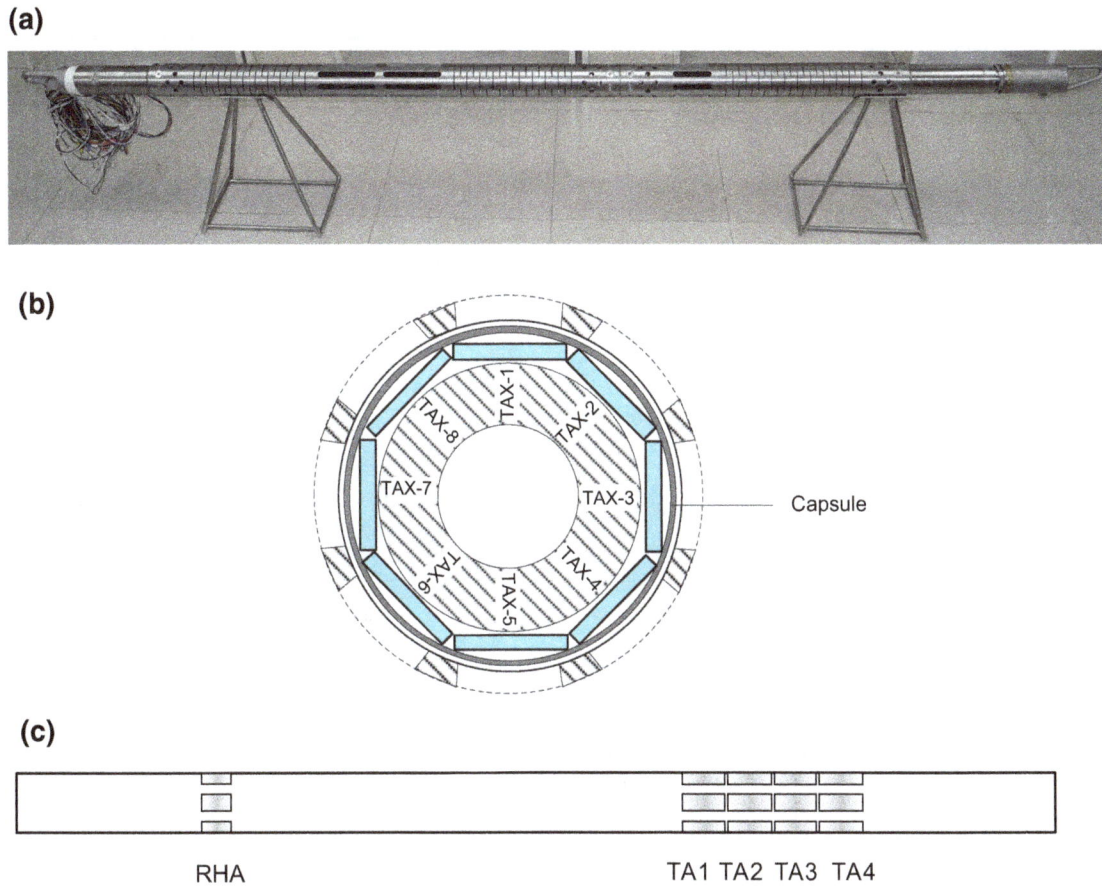

Fig. 1 **a** A photo of the 3D acoustic transmitter stations, **b** sketch of the element distribution of one transmitter station, and **c** 2D structure sketch of the 3D acoustic transmitter stations

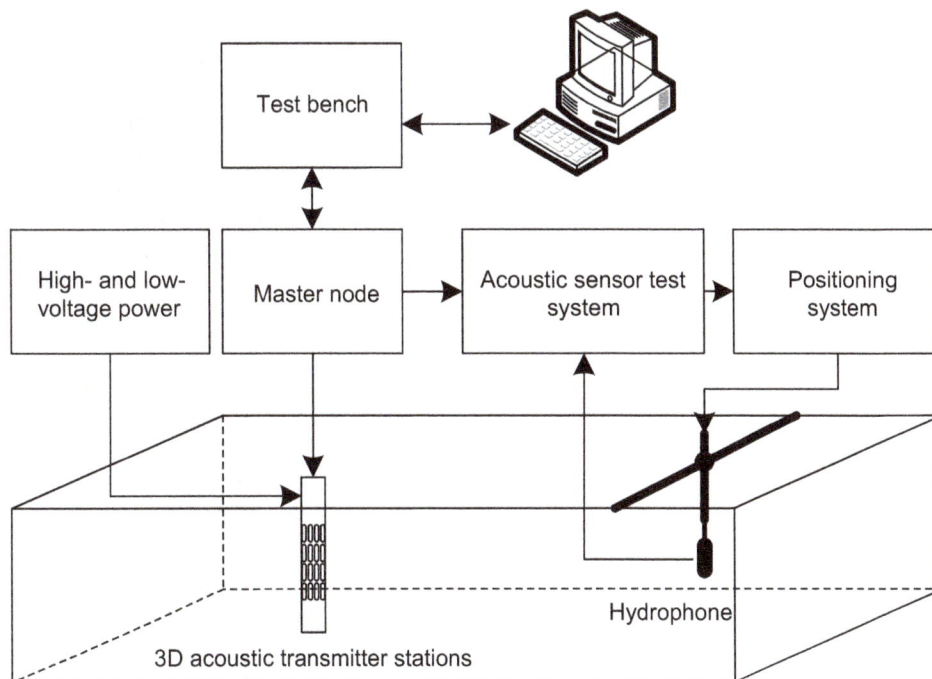

Fig. 2 Experimental setup for the 3D acoustic transmitter stations

positioned using a positioning system with four degrees of freedom. The test bench is an Ethernet-based embedded testing system that is specially designed to simulate logging ground controls and measurements. This system communicates with the 3D acoustic transmitter stations through a controller area network (CAN) bus. The master node is the main control circuit of the tool for azimuthal acoustic reflection logging. The programs are operated in the same mode used for actual well logging except that the codes utilized to control the acquisition nodes are blocked. For synchronicity with the acoustic sensor test system, an indicator signal SYN is introduced with the same synchronizing frame as the tool control bus. The test bench ensures the functionality of the 3D excitation circuits under the control of the host computer. The acoustic sensor test system initiates a synchronized collection in response to the synchronization signal. The hydrophone acquires data several times at each position and then moves along a predetermined trajectory under the control of the positioning system.

During experimental measurements, the 3D acoustic transmitter stations radiate acoustic waves that are then received by the hydrophone. The transmitter is maintained at a fixed height, whereas the hydrophone position is adjusted via the positioning system to situate the hydrophone at the same horizontal plane as the geometric center of the transmitter. The shortest distance to the water surface is 0.75 m, and the interval is 2.00 m. The transmitter stations are manually rotated clockwise to ensure that the normal exterior of each radiating surface of the eight elements is aligned with the geometric center of the hydrophone. The stepping angle α is 45°. The horizontal layout of the transmitter and the receiver during the experimental measurements is shown in Fig. 3. A coordinate system (xoy) is constructed with the axis center of the 3D acoustic transmitter stations as the origin. The positive direction of the y axis points to the geometric center of the hydrophone. Initially, the northern direction of the transmitter stations is parallel to the positive direction of the y axis. The excitation signal is a 600 V square wave with a pulse width of 30 μs. The time delay of the third-level subarray is 6 μs when the previous calculation method is applied (Che et al. 2010).

To measure the horizontal directivity of each combined arc array, which is shown in Fig. 4a, we first fix the position of the 3D acoustic transmitter stations (T) and excite these stations to radiate acoustic energy. The hydrophone moves along an arc with a radius of 2.00 m and central angle of 120° in the xy plane. The center of this plane is the geometric center (o) of the transmitter stations. The hydrophone receives acoustic waves generated by T at 61 evenly distributed positions along the arc, and the opening angle between each pair of adjacent receiver positions is 2° with respect to o. The hydrophone moves along an arc with

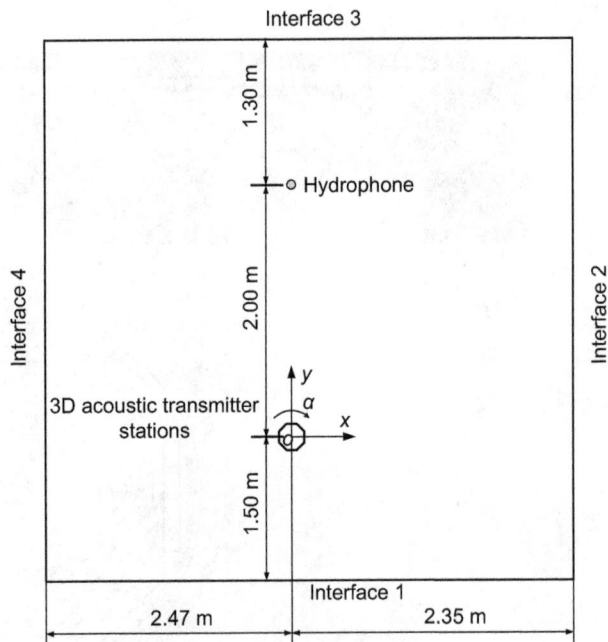

Fig. 3 Horizontal layout of the transmitter and receiver during experimental measurements for the 3D acoustic transmitter stations

a radius of 2.00 m and a central angle of 67° to measure vertical directivity. The geometric center (o) of the transmitter stations is the origin of the xoz plane as depicted in Fig. 4b. The hydrophone receives the acoustic waves generated by T at 68 evenly distributed positions along the arc, and the opening angle of each pair of adjacent receiver positions is 1° with respect to o.

4 Experimental results

First, we tested each of the elements in the four transmitter stations (i.e., TA1, TA2, TA3, and TA4). We then examined each of the third-level subarrays of the four transmitter stations and analyzed all of the measured waveforms as well as their corresponding spectra. Furthermore, we measured the horizontal directivity of the third-level subarrays as well as the horizontal and vertical directivities of the combined arc arrays.

4.1 Individual elements

When testing the individual elements, the height and angle of the 3D acoustic transmitter stations were adjusted to ensure that the tested element is on the same horizontal level as the hydrophone (the hydrophone is in the normal direction of the element radiating surface). Each of the elements, from TA1 to TA4, was excited to generate acoustic energy. The hydrophone was situated 2.00 m away from the tested element. We calculated the average

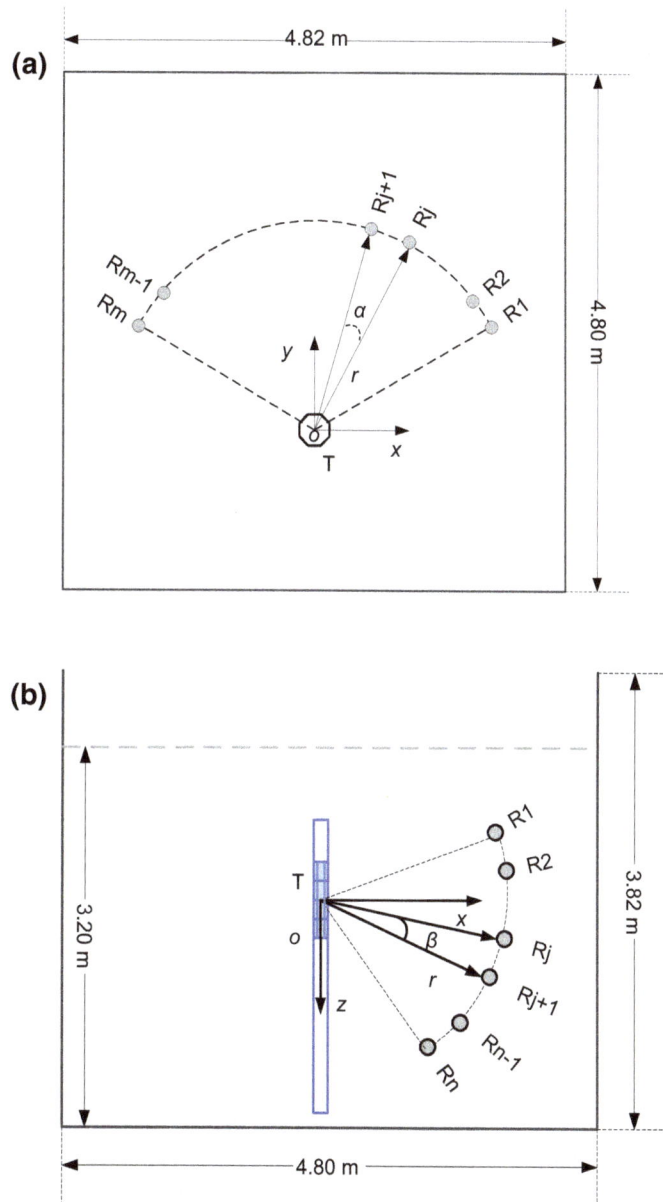

Fig. 4 Distribution of the transmitter and receiver for the **a** horizontal and **b** vertical directivity measurements

basic frequencies, average peak–peak voltages, and average peak–peak sound pressures of the direct waves for each transmitter station as shown in Table 1. Figures 5 and 6 display the direct waves received by the hydrophone and their corresponding spectra when transmitter stations TA1 and TA2 are excited, respectively. The acoustic waves generated by the elements of the transmitter stations exhibit almost identical waveform patterns with slightly different amplitudes. The transmission performance of the elements is consistent.

Table 1 Average basic frequencies, average peak–peak voltages, and average peak–peak sound pressures of the direct waves at 2.00 m intervals when testing each element of the four transmitter stations

Average value	TA1	TA2	TA3	TA4
Basic frequency, kHz	14.58	14.44	14.60	14.42
Peak–peak voltage, mV	21.45	22.24	23.75	21.30
Peak–peak sound pressure, Pa	1178	1222	1305	1170

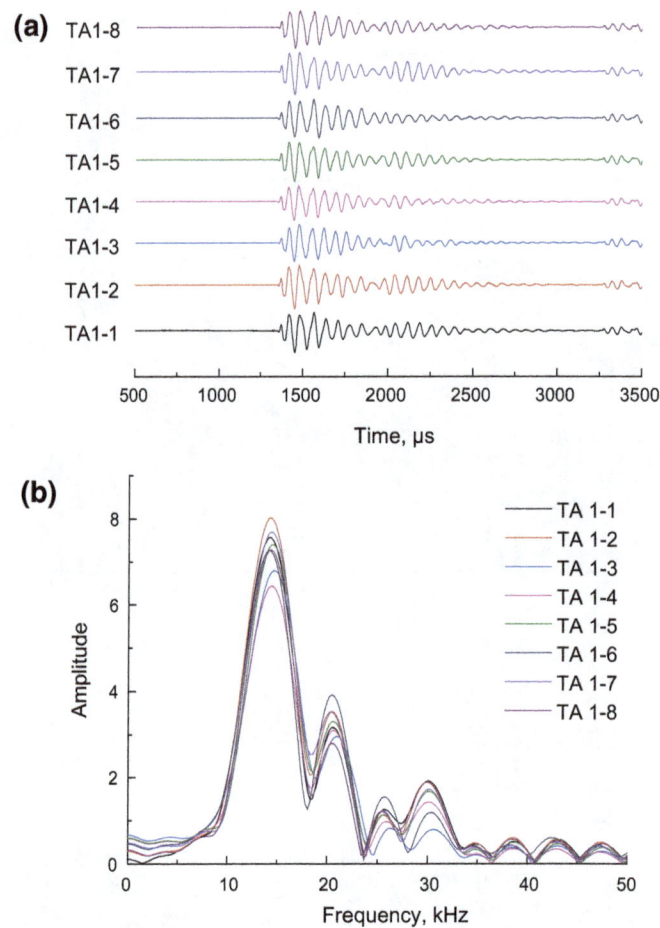

Fig. 5 **a** Experimental waveforms in the time-domain received by the hydrophone and **b** corresponding spectra when each of the TA1 elements radiates acoustic energy

4.2 Third-level subarrays

A transducer composed of the three adjacent elements of a transmitter station is considered to be a third-level subarray. To test the third-level subarrays, we adjusted the height and angle of the 3D acoustic transmitter stations to position the third-level subarray on the same horizontal plane as the hydrophone. In addition, the hydrophone was adjusted to the normal direction of the center element radiating surface of the third-level subarray. We separately excited all eight third-level subarrays of TA1, TA2, TA3, and TA4. The average basic frequencies, average peak–peak voltages, and average peak–peak sound pressures calculated from the waveforms in 2.00 m intervals are shown in Table 2. Figures 7 and 8 show the direct waves received by the hydrophone and their corresponding spectra when the third-level subarrays of TA1 and TA2 are excited, respectively. The waveforms generated by the third-level subarrays of the arc array exhibit almost identical patterns. Moreover, the spectra of the third-level

subarrays are identical with slightly different amplitudes. The radiation performance of the third-level subarrays is nearly consistent.

We also separately tested the horizontal directivities of the four third-level subarrays in the same circumferential direction. Elements 4, 5, and 6 were set as the centers of the third-level subarrays. Figure 9a–d shows the time-domain waveforms received by the hydrophone at different azimuth φ values when the four subarrays are in the same circumferential direction, and element 5 is the center. The direct wave amplitude is distributed symmetrically around the axis of the main lobe; this amplitude peaks at 0° and decreases gradually from 0° on both sides.

The amplitude of the time-domain waveforms was also analyzed to obtain the directivity curves (Fig. 10). This amplitude is shown in Fig. 9 during a time window of approximately 1320–1900 μs. The direction of the main radiated beam points consistently to 0° for all four third-level subarrays in the same circumferential direction when element 5 is the center. The radiated acoustic beams are

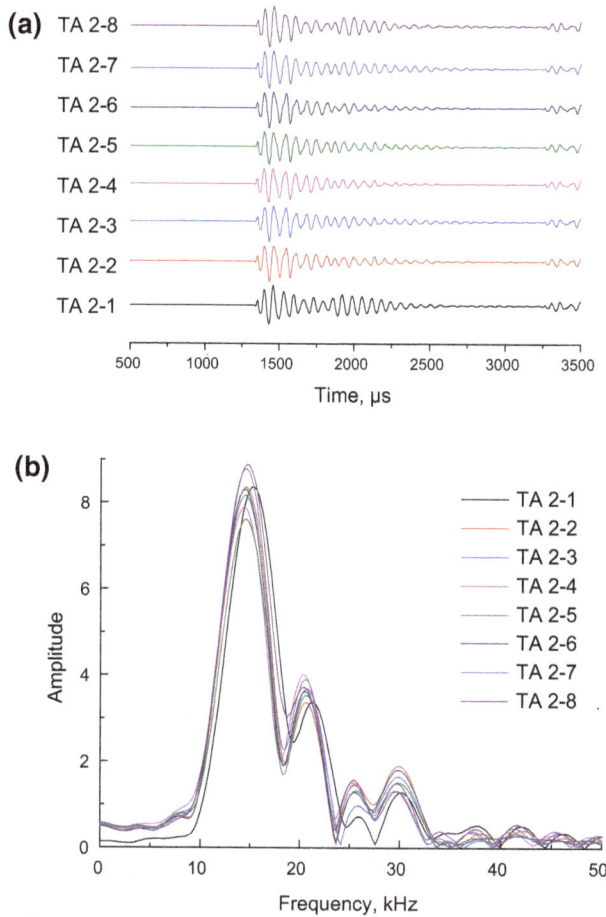

Fig. 6 **a** Experimental waveforms in the time-domain received by the hydrophone and **b** corresponding spectra when each of the TA2 elements radiates acoustic energy

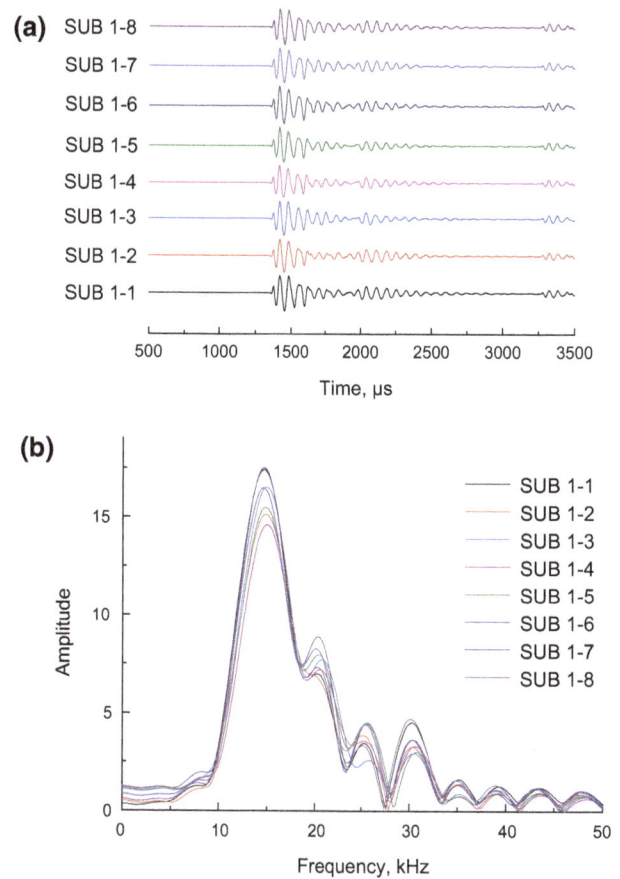

Fig. 7 **a** Experimental waveforms in the time-domain received by the hydrophone and **b** corresponding spectra when each of the TA1 subarrays radiates acoustic energy

distributed almost symmetrically around the main lobes. In addition, sound pressures and 3-dB beam widths are similar in the main lobe direction. Table 3 displays the peak–peak voltages, main frequencies, peak–peak sound pressures, and 3-dB beam widths of the direct waves generated by each of the third-level subarrays excited in the main lobe direction.

4.3 Combined arc arrays

When acoustic energy is radiated to the formations around a borehole during azimuthal acoustic reflection logging, the vertical deflection angle of the main acoustic beam radiated

by the transmitter is generally smaller than the first critical angle of the incident acoustic wave on the borehole wall from the borehole fluid. Thus, additional acoustic wave energy can enter the formation; this occurrence deepens the investigation and improves the signal-to-noise ratios of the useful signals. The main lobe of an acoustic beam radiated by a phased-combined arc array has a certain angular width. Therefore, the deflection angle of the main beam in the vertical plane is designed at approximately half of the first critical angle when a combined arc array is employed to radiate a 3D acoustic field to the formation. When measuring the directivity of the combined arc array, the method of Wu et al. was adopted to calculate the delay between different elements (Wu et al. 2013). The vertical

Table 2 Average basic frequencies, average peak–peak voltages, and average peak–peak sound pressures of the direct waves at 2.00 m intervals

Average value	TA1	TA2	TA3	TA4
Basic frequency, kHz	15.06	14.82	15.00	14.76
Peak–peak voltage, mV	52.18	56.27	59.94	52.47
Peak–peak sound pressure, Pa	2867	3092	3293	2883

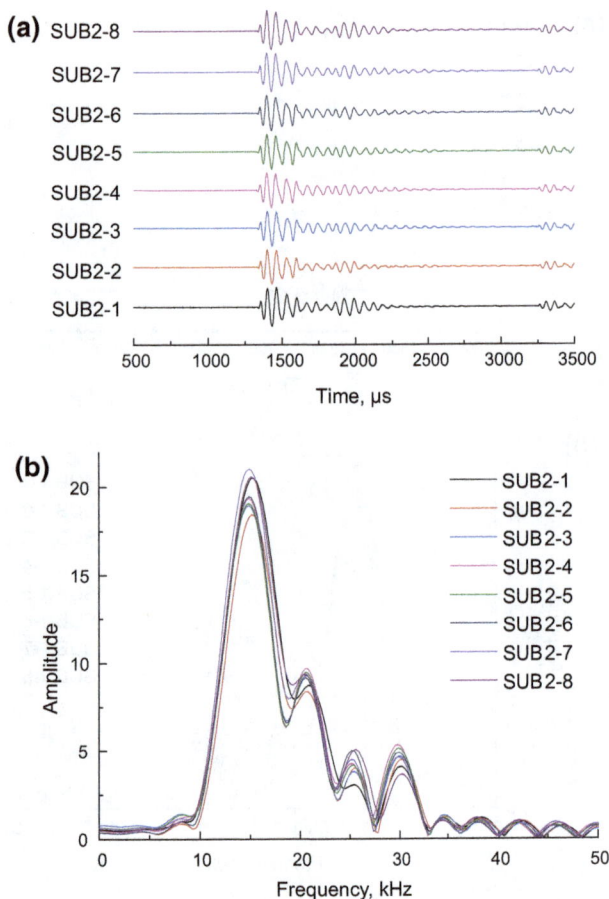

Fig. 8 **a** Experimental waveforms in the time-domain received by the hydrophone and **b** corresponding spectra when each of the TA2 subarrays radiates acoustic energy

deflection angle of the main acoustic beam is assumed to be $\theta_0 = 16°$. The horizontal directivity of each combined arc array in the *xoy* plane was first measured, and then the vertical directivity in the vertical plane $\varphi = 0°$ was measured; this plane is situated at the horizontal maximum direction of the main lobe. Figure 11 displays the time-domain waveforms received by the hydrophone at different azimuths when the combined arc array CAR-6 is excited. The distribution of the direct wave amplitude is symmetric in the *xoy* plane, with a peak at 0°. These results are consistent with the distribution of the direct wave amplitudes of the third-level subarrays. In the $\varphi = 0°$ plane, the direct wave amplitude changes with deflection angle variations. The amplitude peaks at a deflection angle of 14°; thus, the main lobe of radiation is steered at 14°.

Figures 12 and 13 show the experimental horizontal and vertical directivity curves of the combined arc arrays. The main lobe of the radiated acoustic beams for the three combined arc arrays points to approximately $\varphi = 0°$. The beams are distributed symmetrically around the direction of the main lobes, with a 3-dB beam width that ranges from

59° to 67°. The main lobe in the vertical plane deflects along the direction of $\theta = 14°$, which almost corresponds to the designed deflection angle of 16°. The 3-dB beam width is only 11°. Thus, the 3D acoustic transmitter stations exhibit acceptable azimuthal directivity. Because the phased-combined arc array transmitter radiates acoustic energy in a certain azimuth with a specific angle width, acoustic energy can then be radiated to a 3D space through scanning by the excitation of different combined arc arrays.

The time-domain waveforms received by the hydrophone at 2.00 m intervals were analyzed when each combined arc array radiates acoustic energy in the deflected direction. A statistical analysis was also performed on the time-domain waveforms and their corresponding spectra. Table 4 shows the peak–peak voltages, main frequencies, and peak–peak sound pressures when the three combined arc arrays are excited separately. Overall, the three combined arc arrays exhibit good transmission performance.

4.4 Discussion

As shown in Figs. 8b, 6b, and the corresponding tables, the third-level subarrays (three-element arc arrays) radiate acoustic waves with amplitudes that are approximately 2.5 times greater than those of the acoustic waves radiated by an individual element. The observed spectra are essentially identical. In addition, the 3-dB beam width of the vertical directivity that is radiated by the third-level subarray is approximately 60°, which reveals that three-element arc arrays can radiate focused acoustic energy in a certain azimuthal range while radiating weak acoustic energy in other directions.

The combined arc array radiates acoustic waves with similar spectra and amplitudes that are approximately 3.4 times greater than those of the acoustic waves radiated by a third-level subarray as shown in Figs. 11b and 8b as well as the corresponding tables. Figure 12 indicates that the horizontal directivity of the 3D acoustic transmitter stations displays a 3-dB beam width of less than 60°, which indicates that the system exhibits a high capacity for directional radiation. The results revealed that acoustic energy radiated in a certain azimuthal range was effectively increased by increasing the number of three-element arc arrays along the axial direction, and these results will help increase the amount of radiated acoustic energy that enters the formation when the transducer is used downhole.

Figure 13 shows that the 3-dB beam width in the vertical plane is only 11° when the radiated acoustic beam is deflected by 14° in the vertical plane. Such conditions maximize the amount of radiated acoustic wave energy that enters a formation surrounding a fluid-filled borehole. The results reveal that increasing the number of arc arrays along

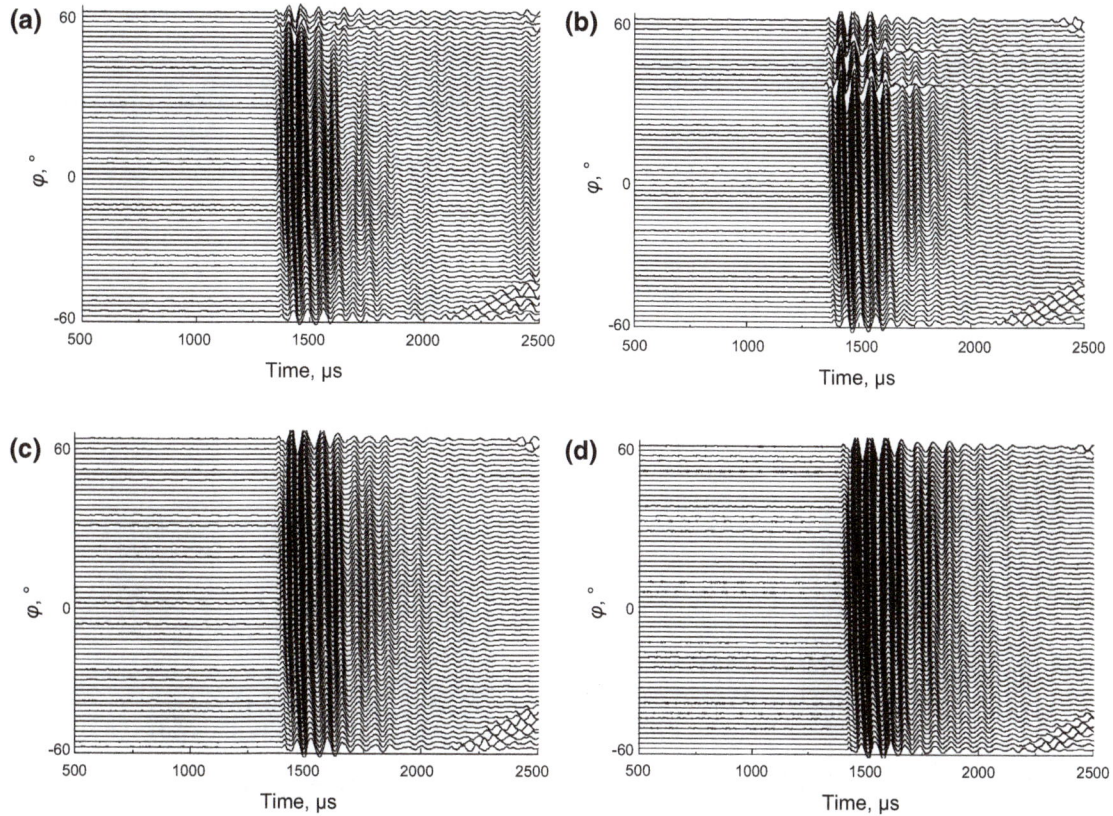

Fig. 9 Time-domain waveforms received by the hydrophone at different azimuth φ values when the four subarrays radiate in the same circumferential direction and element 5 is the center. **a** SUB1-5, **b** SUB2-5, **c** SUB3-5, and **d** SUB4-5

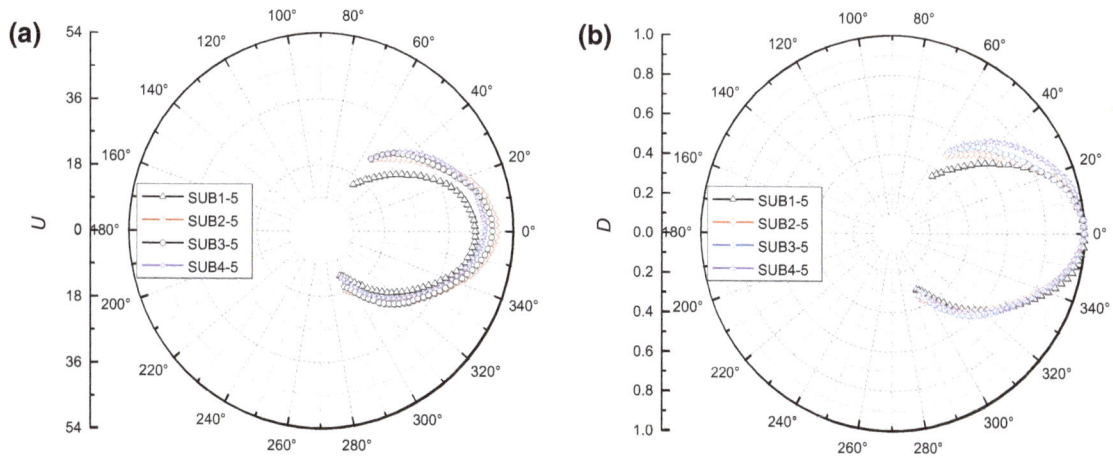

Fig. 10 Horizontal directivities of the third-level subarrays with element 5 as the center. **a** without normalization and **b** with normalization. D is a dimensionless variable

the axial direction can increase the acoustic energy radiated in a certain azimuthal range and also improve the vertical radiation directivity of the transducer. Thus, it is possible to impinge the radiated acoustic wave onto the borehole wall at an incident angle that is smaller than the first critical angle when using the downhole 3D acoustic transmitter stations. Therefore, the acoustic energy radiated by the transducer almost wholly converts to formation compressional energy.

The 3D acoustic transmitter stations can radiate acoustic waves in any desired direction by controlling the phase and amplitude of the 32-channel excitation pulse. Therefore, azimuthal acoustic reflection logging can be realized with the proposed transmitter stations.

Table 3 Peak–peak voltages, main frequencies, peak–peak sound pressures, and 3-dB beam width of the direct waves generated by each of the third-level subarrays

Numbering	Peak–peak voltage, mV	Main frequency, kHz	Peak–peak sound pressure, Pa	3-dB beam width
SUB1-4	31.32	14.89	1721	45°
SUB2-4	47.84	14.65	2629	65°
SUB3-4	48.04	14.40	2639	59°
SUB4-4	42.69	14.65	2346	71°
SUB1-5	43.40	14.65	2384	61°
SUB2-5	49.28	14.40	2708	63°
SUB3-5	48.26	14.65	2651	67°
SUB4-5	46.18	14.40	2537	73°
SUB1-6	42.50	14.89	2335	76°
SUB2-6	48.83	14.65	2683	65°
SUB3-6	50.47	14.65	2773	58°
SUB4-6	48.34	14.40	2656	57°

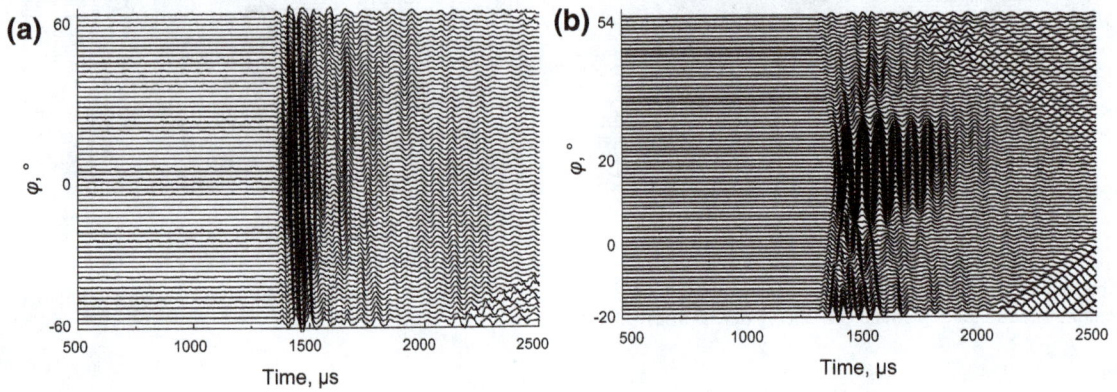

Fig. 11 Waveforms in the time-domain received by the hydrophone in different azimuths when the combined arc array CAR6 radiates acoustic energy; **a** *xoy* plane and **b** $\varphi = 0°$ plane

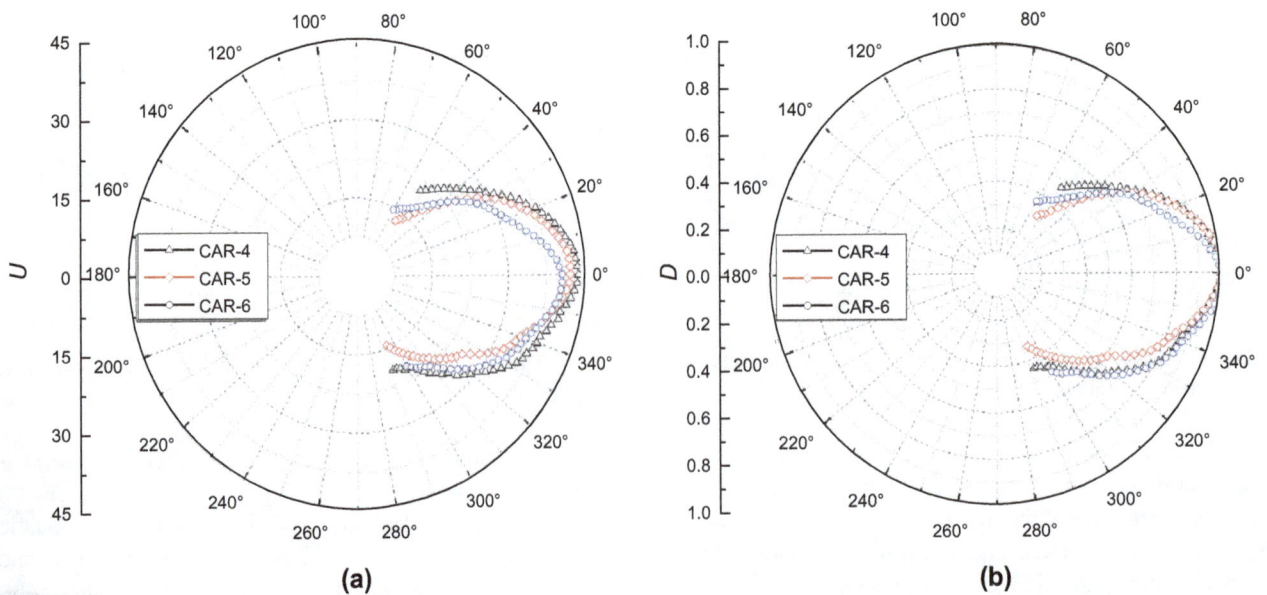

Fig. 12 Radiation directivity curves of the combined arc arrays in the 3D acoustic transmitter stations; **a** without normalization and **b** with normalization

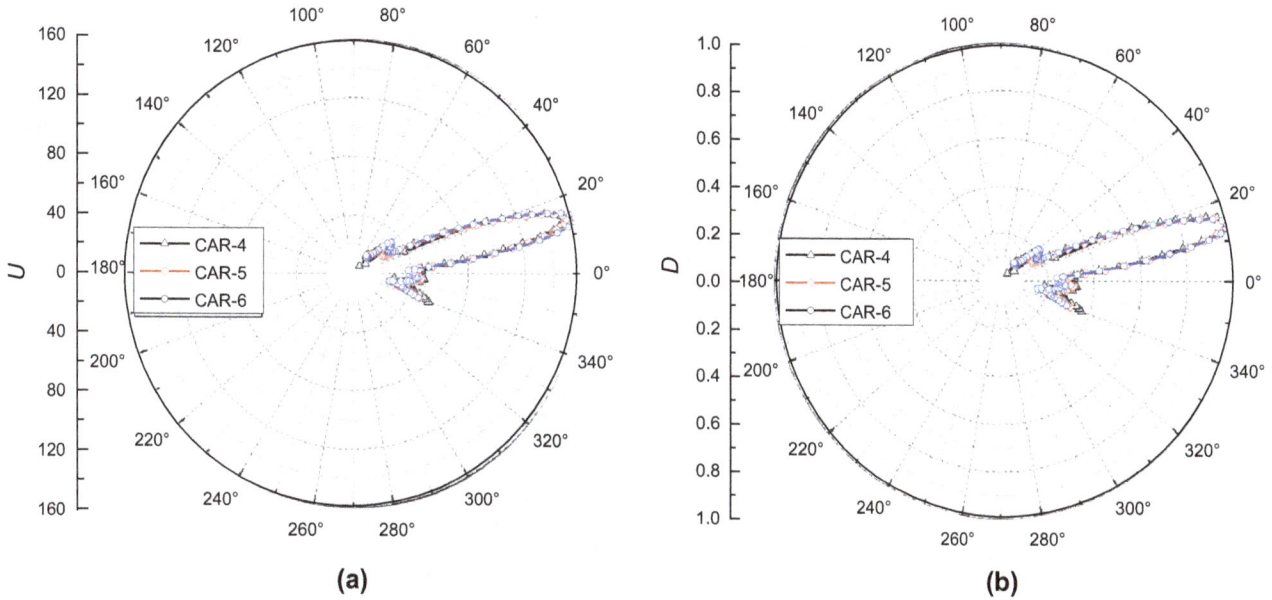

Fig. 13 Vertical directivities of the combined arc arrays in the 3D acoustic transmitter stations; **a** without normalization and **b** with normalization

Table 4 Peak–peak voltages, main frequencies, and peak–peak sound pressures of the combined arc arrays

Numbering	Peak–peak voltage, mV	Main frequency, kHz	Peak–peak sound pressure, Pa
CAR-4	151.29	14.40	8314
CAR-5	153.31	14.40	8424
CAR-6	159.59	14.40	8770

5 Conclusions

Here, we propose 3D acoustic transmitter stations with corresponding circuits and bodies for azimuthal acoustic reflection well logging. The transmitter stations consist of three-level subarrays that are evenly spaced along an axis, and each element is an individual transducer. We measured the azimuthal performance of the transmitter stations, including the vertical and horizontal directivity curves of the combined arc arrays.

The measured waveforms and their corresponding spectra show that the waveform pattern and the radiation performance levels of the individual elements and the third-level subarrays are almost consistent except for slight differences in amplitude. When the four third-level subarrays in the same circumferential direction are excited by a "quasi-square wave" pulse signal with a pulse width of 30 μs and a voltage of 600 V, the main lobes of the radiation acoustic beams point to 0° in the horizontal plane. The acoustic beams, which exhibit a 3-dB beam width that ranges from 59° to 67°, are distributed symmetrically around the main lobes. Furthermore, the horizontal directivities of the four third-level subarrays in the same circumferential direction are nearly consistent. The main lobe

of the acoustic beams for the combined arc arrays is steered by 14° in the vertical plane with a 3-dB beam width of 11° during individual operation. The combined arc arrays of the proposed 3D acoustic transmitter stations are also consistent, which indicates that the system can be used in azimuthal acoustic reflection logging.

Three-element arc arrays can radiate focused acoustic energy in a certain azimuthal range while radiating only weak acoustic energy in other directions. Increasing the number of arc arrays along the axial direction can increase the acoustic energy radiated in a certain azimuthal range and also improve the vertical radiation directivity of the transducer. The radiated acoustic wave can be made to impinge onto the borehole wall at an incident angle that is smaller than the first critical angle when using the 3D acoustic transmitter stations downhole. Therefore, the acoustic energy radiated by the transducer almost wholly converts to the formation compressional energy.

The exploration accuracy of the proposed 3D acoustic transmitter stations must be calibrated further for application in downhole tools for oil field tests. However, the large number of required transducers complicates the electronic circuits of the transmitter stations. Thus, accurate phase control is necessary to ensure good azimuthal performance.

Acknowledgments   The authors would like to thank the editors and reviewers for their valuable comments. This work is supported by the National Natural Science Foundation of China (Grant Nos. 11204380, 11374371, 61102102, 11134011), National Science and Technology Major Project (Grant No. 2011ZX05020-009), China National Petroleum Corporation (Grant Nos. 2014B-4011, 2014D-4105, 2014A-3912) and PetroChina Innovation Foundation (2014D-5006-0307). Che X. H. acknowledges the support from China Scholarship Council (No. 201306445018) and the Earth Resources Laboratory at Massachusetts Institute of Technology (MIT) for the opportunity as a visiting scientist in 2014–2015 and also thanks Prof. Toksöz M. N. for his support and help during this visit to MIT.

References

Al Rougha HAB, Sultan A, Haldorsen J, et al. Integration of microelectrical and sonic reflection imaging around the borehole offshore UAE. Paper IPTC 11021 presented at International Petroleum Technology Conference, Doha, Qatar; 21–23 Nov 2005.

Chai XY, Zhang WR, Wang GQ, et al. Application of remote exploration acoustic reflection imaging logging technique in fractured reservoirs. Well Logging Technol. 2009;33(6):539–43 (**in Chinese**).

Chang C, Hoyle D, Watanabe S, et al. Localized maps of the subsurface. Oilfield Rev. 1998;10:56–66.

Che XH, Qiao WX. Numerical simulation of an acoustic field generated by a phased arc array in a fluid-filled borehole. Pet Sci. 2009;6(3):225–9.

Che XH, Qiao WX, Ju XD. Characteristics of the acoustic field generated by an acoustic phased arc array transmitter in the borehole formation. Acta Pet Sin. 2010;31(2):343–6 (**in Chinese**).

Che XH, Qiao WX, Wang RJ, et al. Numerical simulation of an acoustic field generated by a phased arc array in a fluid-filled cased borehole. Pet Sci. 2014;11(3):385–90.

Ellis D, Engelman B, Fruchter J, et al. Environmental applications of oilfield technology. Oilfield Rev. 1996;8:44–57.

Esmersoy C, Chang C, Kane MR, et al. Sonic imaging: a tool for high-resolution reservoir description. 67th SEG Annual Meeting, Expanded Abstracts. 1997. pp. 278–281.

Esmersoy C, Chang C, Kane M, et al. Acoustic imaging of reservoir structure from a horizontal well. Lead Edge. 1998;17(7):940–6.

Gong H, Chen H, He X, et al. Eliminating the azimuth ambiguity in single-well imaging using 3C sonic data. Geophysics. 2015;80(1):A13–7.

Haldorsen JBU, Johnson DL, Plona T, et al. Borehole acoustic waves. Oilfield Rev. 2006;18:34–43.

Haldorsen J, Voskamp A, Thorsen R, et al. Borehole acoustic reflection survey for high resolution imaging. 76th SEG Annual Meeting, Expanded Abstracts. 2006b. pp. 314–18.

Haldorsen JBU, Zhu FP, Hirabyashi N, et al. Borehole acoustic reflection survey (BARS) using full waveform sonic data. First Break. 2010;28:33–8.

Hornby BE. Imaging of near-borehole structure using full-waveform sonic data. Geophysics. 1989;54(6):747–57.

Jervis M, Bakulin A, Tonellot TL, et al. High-resolution seismic imaging from a single borehole to detect a nearby well. Paper SEG 2012-0963.1 presented at 72th SEG Annual Meeting, Expanded Abstracts, Las Vegas; 4–9 Nov 2012.

Li LS, Wang LJ, Wang ZY, et al. Application of the data from distant detection acoustic imaging well logging. World Well Logging Technol. 2008;23(4):14–5 (**in Chinese**).

Maia W, Rubio R, Junior F, et al. First borehole acoustic reflection survey mapping a deepwater turbidite sand. 76th SEG Annual Meeting, Expanded Abstracts. 2006. pp. 1757–61.

Patterson D, Tang XM, Ratigan J. High-resolution borehole acoustic imaging through a salt dome. 78th SEG Annual Meeting, Expanded Abstracts. 2008. pp. 319–23.

Pistre V, Kinoshita K, Endo T, et al. A modular wireline sonic tool for measurements of 3D (azimuthal, radial, and axial) formation acoustic properties. SPWLA 46th Annual Logging Symposium, New Orleans, Louisiana; 26–29 Jun 2005.

Qiao WX, Ju XD, Chen XL, et al. Downhole acoustic arc array transmitter with controlled azimuthal directivity. China Patent ZL. 2006; 03137596. 0 (in Chinese).

Qiao WX, Che XH, Ju XD, et al. Acoustic logging phased arc array and its radiation directivity. Chin J Geophys. 2008;51(3):939–46 (**in Chinese**).

Qiao WX, Ju XD, Che XH. A kind of acoustic logging with phased arc array transducer. Well Logging Technol. 2009;33(1):22–5 (**in Chinese**).

Tang XM. Imaging near-borehole structure using directional acoustic-wave measurement. Geophysics. 2004;69(6):1378–86.

Tang XM, Zheng Y, Patterson D. Processing array acoustic logging data to image near-borehole geologic structures. Geophysics. 2007;72(2):87–97.

Tang XM, Wei ZT. Significant progress of acoustic logging technology: remote acoustic reflection imaging of a dipole acoustic system. Appl Acoust. 2012a;31(1):10–7 (**in Chinese**).

Tang XM, Wei ZT. Single-well acoustic reflection imaging using far-field radiation characteristics of a borehole dipole source. Chin J Geophys. 2012b;55(8):2798–807 (**in Chinese**).

Tang XM, Patterson D. Single-well S-wave imaging using multi-component dipole acoustic log data. Geophysics. 2009;74(6): 211–23.

Wei ZT, Tang XM, Zhuang CX. P-wave remote acoustic imaging of a borehole dipole source radiation in an acoustically slow formation. Acta Pet Sin. 2013;33(5):905–13 (**in Chinese**).

Wu JP, Qiao WX, Che XH, et al. Experimental study on the radiation characteristics of downhole acoustic phased combined arc array transmitter. Geophysics. 2013;78(1):D1–9.

Yamamoto H, Haldorsen JBU, Mikada H, et al. Fracture imaging from sonic reflections and mode conversion. 69th SEG Annual Meeting, Expanded Abstracts, Houston, Texas; 31 Oct–5 Nov 1999.

Yamamoto H, Watanabe S, Mikada H, et al. Fracture imaging using borehole acoustic reflection survey. Proceedings of the 4th SEGJ International Symposium. 1998. pp. 375–82.

Yamamoto H, Watanabe S, Koelman JMV, et al. Borehole Acoustic reflection survey experiments in horizontal wells for accurate well positioning. Paper SPE/PS-CIM 65538 presented at SPE/PS-CIM International Conference on Horizontal Well Technology, Calgary, Alberta, Canada; 6–8 Nov 2000.

Zhang YD, Hu HS. A technique to eliminate the azimuth ambiguity in single-well imaging. Geophysics. 2014;79(6):D409–16.

Zhao XD, Li GY, Liu BZ. On transmitting transducer of far detecting acoustic logging tool. Well Logging Technology. 2004;28(6):540–2 (**in Chinese**).

10

Liquid marbles containing petroleum and their properties

Edward Bormashenko[1,2] · Roman Pogreb[1] · Revital Balter[2,3] · Hadas Aharoni[2] ·
Doron Aurbach[3] · Vladimir Strelnikov[4]

Abstract Liquid marbles (non-stick droplets) containing crude petroleum are reported. Liquid marbles were obtained by use of fluorinated decyl polyhedral oligomeric silsequioxane (FD-POSS) powder. Marbles containing crude petroleum remained stable on a broad diversity of solid and liquid supports. The effective surface tension of marbles filled with petroleum was established. The mechanism of friction of the marbles is discussed. Actuation of liquid marbles containing crude petroleum with an electric field is presented.

Keywords Liquid marbles · Crude petroleum · Friction · Low energy oil transportation, effective surface tension · Electrical actuation

1 Introduction

Liquid marbles are non-stick droplets encapsulated with micro- or nano-scaled solid particles (Mahadevan 2001; Aussillous and Quéré 2001, 2004, 2006) Since liquid marbles were introduced in the pioneering works of

✉ Edward Bormashenko
edward@ariel.ac.il

[1] Physics Faculty, Ariel University, 40700 Ariel, Israel

[2] Department of Chemical Engineering and Biotechnology, Ariel University, 40700 Ariel, Israel

[3] Department of Chemistry, Bar-Ilan University, 52900 Ramat Gan, Israel

[4] Institute of Technical Chemistry of Ural Division of Russian Academy of Science, Perm, Russia

Edited by Xiu-Qin Zhu

Aussillous and Quèrè (2001), they have been exposed to intensive theoretical and experimental research (McHale et al. 2009; Arbatan and Shen 2011; Bormashenko et al. 2009a, 2011, 2013a; Planchette et al. 2012, 2013; Laborie et al. 2013). An interest in liquid marbles arises from both their very unusual physical properties and their promising applications. Liquid marbles present an alternate approach to superhydrophobicity, i.e., creating a non-stick situation for a liquid/solid pair. Usually, superhydrophobicity is achieved by a surface modification of a solid substrate. In the case of liquid marbles, the approach is opposite: the surface of a liquid is coated by particles which may be more or less hydrophobic (Zang et al. 2013). Marbles coated by graphite and carbon black, which are not strongly hydrophobic, were also reported (Dandan and Erbil 2009; Bormashenko et al. 2010a).

A variety of media, including ionic liquids and liquid metals, could be converted into liquid marbles (Gao and McCarthy 2007; Sivan et al. 2013). Liquid marbles were successfully exploited for microfluidics (Venkateswara Rao et al. 2005; Xue et al. 2010; Bormashenko et al. 2008, 2011, 2012), water pollution detection (Bormashenko and Musin 2009), gas sensing (Tian et al. 2010), electrowetting (Newton et al. 2007), blood typing (Arbatan et al. 2012), and optical probing (Zhao et al. 2012). Respirable liquid marbles for the cultivation of microorganisms and Daniel cells based on liquid marbles were reported recently by the group led by Shen (Tian et al. 2013; Li et al. 2013). Stimulus (pH, UV, and IR)-responsive liquid marbles were reported by several groups (Dupin et al. 2009; Nakai et al. 2013a, b). The stability of marbles is crucial for their microfluidics and sensing applications. Marbles possessing increased mechanical and time stability were prepared by Matsukuma et al. (2013). It is noteworthy that liquid marbles retain non-stick properties on a broad diversity of solid and liquid supports (Bormashenko et al. 2009b, c). Actually, liquid marbles are separated from the support by air cushions in a way similar to Leidenfrost droplets (Biance et al. 2003). The state-of-the-art in the study of properties and applications of liquid marbles is covered in recent reviews (McHale and Newton 2011; Bormashenko 2011, 2012). The majority of groups concentrated their efforts on the study of properties of marbles containing liquids possessing a high surface tension such as water, liquid metals, and ionic liquids, whereas reports devoted to marbles containing organic liquids are still scarce (Xue et al. 2010). Our study is devoted to manufacturing marbles containing crude petroleum, characterized by relatively low surface tension (γ) is in the range of 28–33 mJ/m^2 (Francis and Bennett 1922; Harvey 1925). We also demonstrate new possibilities in micro-manipulation of crude petroleum with the use of liquid marbles.

2 Experimental

Petroleum was supplied by Givot Olam Oil Ltd. Liquid marbles were manufactured with the use of fluorinated decyl polyhedral oligomeric silsequioxane (FD-POSS) powder. FD-POSS was synthesized according to the protocol described in detail by Mabry et al. (2008). FD-POSS powder is strongly hydrophobic and was successfully used by Xue et al. for manufacturing marbles of both aqueous solutions and organic liquids (Xue et al. 2010). FD-POSS powder particles were imaged with SEM (JEOL JSM 6510 LV, Japan). Highly developed topography of FD-POSS particles, displayed in Fig. 1, strengthening their pronounced inherent hydrophobicity is noteworthy.

FD-POSS powder was spread uniformly on a superoleophobic surface, manufactured according to the procedure described by Bormashenko et al. (2013b). Petroleum droplets with a volume of 5–400 µL were deposited using a precise microsyringe on the layer of FD-POSS powder. Rolling of petroleum droplets resulted in the formation of petroleum marbles enwrapped with the FD-POSS powder, as depicted in Fig. 2.

The kinematic viscosity of petroleum was established with a conventional Ostwald viscometer. The kinematic viscosity v at ambient conditions was established as 1.48×10^{-6} m^2/s. The relative density of the crude petroleum oil measured with a pycnometer at ambient conditions was 0.79 g/cm^3.

3 Results and discussion

3.1 Effective surface tension of marbles containing crude petroleum

Liquid marbles prepared according to the procedure described in the experimental section remained stable for a long

Fig. 2 **a** 10 µL liquid marble of crude petroleum coated with FD-POSS powder. **b** 20 µL liquid marble of crude petroleum coated by FD-POSS supported by 26 wt% NaCl water solution

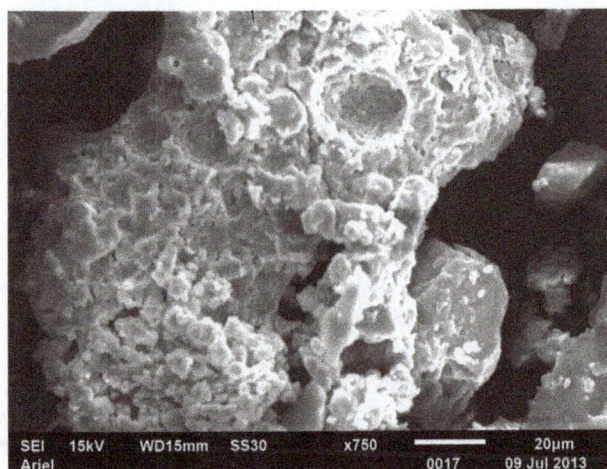

Fig. 1 SEM image of FD-POSS particle. The scale bar is 20 µm

time and demonstrated low rolling friction. Marbles remained stable on a diversity of solid and liquid substrates including water and aqueous NaCl solutions, as shown in Fig. 2a, b.

The effective surface tension of the marbles was measured according to the maximal puddle height method. We increased the volume of crude petroleum oil in a marble from 10 to 400 µL and measured its maximal height, H (see Fig. 3). The effective surface tension was calculated as described in detail by Aussillous and Newton (Aussillous and Quéré 2006; Newton et al. 2007):

$$r_{\text{eff}} = \frac{\rho g H^2}{4} \qquad (1)$$

The effective surface tension of crude petroleum marbles coated with the FD-POSS powder was established as 30.2 ± 0.2 mJ/m^2. The obtained value is close to the surface tension of the crude petroleum aforementioned in the Introduction Section (Francis and Bennett 1922; Harvey 1925). The reported value should be taken with a certain care due to the pronounced hysteretic nature of the effective surface tension of liquid marbles, which was demonstrated recently (Bormashenko et al. 2013a). Regrettably, we did not manage to establish the hysteresis of the surface tension of marbles containing petroleum, because the method of pumping of a "pendant marble" turned out to be unfeasible for the marbles containing low surface tension liquids.

3.2 Mechanisms of friction of crude petroleum liquid marbles

The mechanism of friction of rolling liquid marbles is rather complicated. The energy dissipation rate $\frac{dE}{dt}$ under liquid marbles rolling is given by

$$\frac{dE}{dt} = \frac{dE_{visc}}{dt} + \frac{dE_{CL}}{dt}, \qquad (2)$$

where $\frac{dE_{visc}}{dt}$ describes the viscous dissipation under rolling, and $\frac{dE_{CL}}{dt}$ is the energy dissipation rate due to the

disconnection (de-pinning) of the contact line, at which a marble contacts the solid support. The viscous dissipation may be calculated as follows (Landau and Lifschitz 1987):

$$\frac{dE_{visc}}{dt} = \eta \int\limits_{V_d} (\nabla \vec{u})^2 dV. \qquad (3)$$

Here V_d is the volume over which viscous dissipation occurs, η is the viscosity of the liquid, and $\vec{u}$ is the velocity field in the droplet. When a liquid marble contains a sufficiently viscous liquid (such as glycerol) it will stop rolling mainly by the viscous dissipation (Bormashenko et al. 2010b). Thus, the stopping distance of the marble S possessing an initial velocity of the center of mass U may be estimated as follows:

$$S \cong 1.5 \frac{\rho U R^5}{\eta a^3} \cong 1.5 \frac{U R^5}{v a^3}, \qquad (4)$$

where $R = \sqrt[3]{\frac{3V}{4\pi}}$ is the radius of the marble (V is its volume), a is the radius of the contact area (shown in Fig. 3), ρ is the density, and v is kinematic viscosity (Bormashenko et al. 2010b). We measured the stopping distance of marbles with a simple device depicted in Fig. 4: 12 µL crude petroleum marbles rolled downhill, and the velocity U was measured with a rapid camera at the origin of the horizontal portion of its pathway (see Fig. 4). Marbles rolled on the horizontal glass slides across the distance S necessary for their eventual deceleration, and the stopping distance of marbles S was measured. A series of 10 experiments were performed to establish the averaged value of the stopping distance.

For marbles possessing an initial velocity of the center of mass $U = 0.028$ m/s, the stopping distance S was established as $S = 7.7 \pm 0.2$ mm. Substituting $R = 1.42$ mm, $a = 1.16$ mm established for 12 µL crude petroleum marbles and $v = 1.48 \times 10^{-6}$ m^2/s (see the Sect. 2) into Exp. 4 yields the estimation $S \cong 100$ mm, which is an order of magnitude larger than the experimentally established $S = 7.7$ mm. This means that the condition:

$$\frac{dE_{visc}}{dt} << \frac{dE_{CL}}{dt} \qquad (5)$$

takes place for marbles filled by the crude petroleum. Thus, the energy dissipation under rolling is mainly due to the

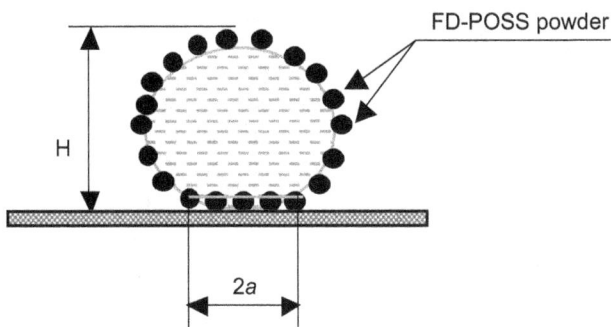

Fig. 3 Scheme illustrating the maximal high height method of the measurement of the effective surface tension of liquid marbles

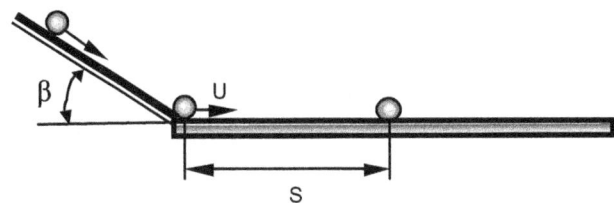

Fig. 4 Sketch of the experimental device used for the study of friction of liquid marbles. Marbles rolled from a height of 2 cm, $\beta = 15^0$.

disconnection (de-pinning) of the triple contact line, consuming the energy. This situation is typical for marbles containing low viscosity liquids (Bormashenko et al. 2010b). Low-energy transport of crude oil, exploiting non-stick lotus effect-inspired surfaces, was discussed recently (Wang et al. 2013). Non-stick liquid marbles filled with crude oil present a distinct alternative to lotus effect-based solutions.

3.3 Actuation of liquid marbles containing crude petroleum with an electric field

The effect of an electric field on crude petroleum oil marbles was studied. Crude petroleum oil marbles (20 μL) were placed on a glass slide located between two plain electrodes (as shown in Fig. 5); the electric field was increased from 0 to 10^6 V/m by a power supply Pasco (model SF-9586), as shown in Figs. 5, 6.

When the electric field reached $E \cong 1.2 \times 10^6$ V/m, the marble started to stretch until it touched the upper electrode, as depicted in Fig. 6. The value of the electric field necessary for electrical actuation of marbles filled with crude petroleum may be estimated as follows: The dimensionless constants ξ_1, ζ_2 describing the sensitivity of

the marble to the electric field could be introduced (Bormashenko et al. 2012). The constant ξ_1 describes interrelation of electrically induced effects and gravity:

$$\xi_1 = \frac{\varepsilon_0 \varepsilon}{\rho g R} E^2, \qquad (6)$$

where ε and ρ are the dielectric constant and density of the marble, respectively. The constant ξ_2 describes the interrelation of electrically induced and surface phenomena:

$$\xi_2 = \frac{\varepsilon_0 \varepsilon R}{\gamma_{\text{eff}}} E^2. \qquad (7)$$

When $\xi_1 \cong \xi_2 \cong 1$ takes place, the marble becomes sensitive to the electric field. Substituting $\varepsilon = 2.1$, $\rho \cong 800$ kg/m^3, and $\gamma_{\text{eff}} \cong 30$ mJ/m^2 yields the electric field E^* necessary for electrical actuation of petroleum-based marbles the estimation $E^* \cong 10^6$ V/m, which is in an excellent agreement with the experimental findings.

The established value of E^* supplies to both of the dimensionless constants ξ_1, ζ_2 the value close to unity, this is not surprising, because the radius of 20 μL marbles is close to the capillary length $l_{\text{ca}} = \sqrt{\frac{\gamma_{\text{eff}}}{\rho g}} \cong 2$ mm, thus the capillary and gravitational energies of marbles are comparable.

4 Conclusions

We conclude that the use of decyl polyhedral oligomeric silsequioxane powder allows manufacturing of liquid marbles containing crude petroleum. The effective surface tension of the marbles was established as 30.2 ± 0.2 mJ/m^2. The mechanisms of friction of the liquid marbles filled with crude petroleum were examined. The viscous dissipation is negligible for rolling marbles containing crude petroleum. The energy dissipation under rolling of marbles is mainly due to the disconnection of the triple contact line. Liquid marbles containing crude petroleum can be actuated by electric fields of 10^6 V/m. The mechanism of the electrical actuation of liquid marbles containing crude oil is

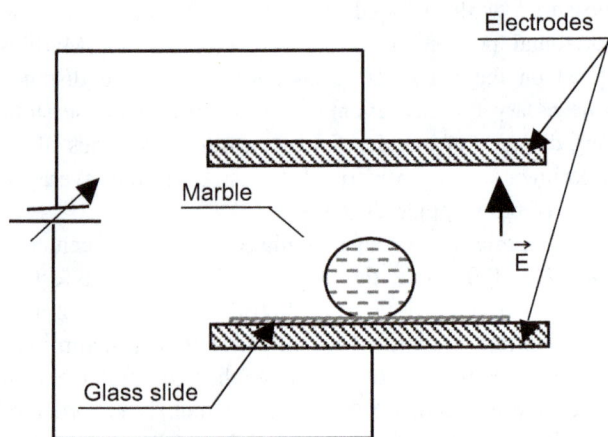

Fig. 5 Experimental setup used for investigation of marbles shape exposed to an electric field

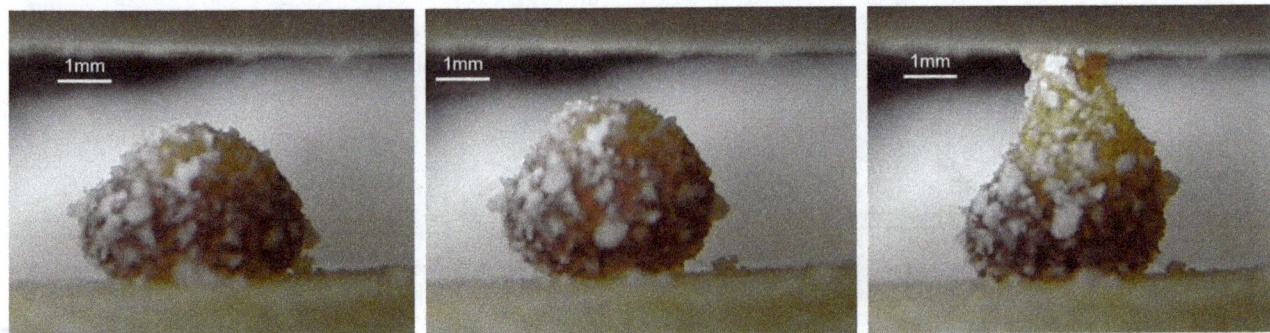

Fig. 6 The sequence of images illustrating the behavior of a 20 μL crude petroleum oil marble exposed to an electric field; the left image depicts the initial state of the marble and the right shows the marble exposed to the electric field of $E \cong 10^6 \frac{V}{m}$

discussed. Liquid marbles allow effective manipulation of micro-volumes of crude oil.

Acknowledgments Acknowledgment is made to the Donors of the American Chemical Society Petroleum Research Fund for support of this research (Grant 52043-UR5). The authors are grateful to Mrs. Yelena Bormashenko for her kind help in preparing this manuscript. We are thankful to Dr. R. Grynyov and Mrs. N. Litvak for SEM imaging.

References

Arbatan T, Li L, Tian J, et al. Liquid marbles as micro-bioreactors for rapid blood typing. Adv Healthc Mater. 2012;1:80–3.

Arbatan T, Shen W. Measurement of the surface tension of liquid marble. Langmuir. 2011;27:12923–9.

Aussillous P, Quéré D. Liquid marbles. Nature. 2001;411:924–7.

Aussillous P, Quéré D. Properties of liquid marbles. Proc R Soc A. 2006;462:973–99.

Aussillous P, Quéré D. Shapes of rolling liquid drops. J Fluid Mech. 2004;512:133–51.

Biance A-L, Clanet C, Quéré D. Leidenfrost drops. Phys Fluids. 2003;15:1632–7.

Bormashenko E. Liquid marbles: properties and applications. Curr Opin Colloid Interface Sci. 2011;16:266–71.

Bormashenko E. New insights into liquid marbles. Soft Matter. 2012;8:11018–21.

Bormashenko E, Musin A. Revealing of water surface pollution with liquid marbles. Appl Surf Sci. 2009;255:6429–31.

Bormashenko E, Pogreb R, Bormashenko Y, et al. New Investigations on ferrofluidics: ferrofluidic marbles and magnetic-field-driven drops on superhydrophobic surfaces. Langmuir. 2008;24:12119–22.

Bormashenko E, Pogreb R, Whyman G, et al. Shape, vibrations, and effective surface tension of water marbles. Langmuir. 2009a;25: 1893–6.

Bormashenko E, Bormashenko Y, Musin A. Water rolling and floating upon water: marbles supported by a water/marble interface. J Colloid Interface Sci. 2009b;333:419–21.

Bormashenko E, Bormashenko Y, Musin A, et al. On the mechanism of floating and sliding of liquid marbles. Chem Phys Chem. 2009c;10:654–6.

Bormashenko E, Bormashenko Y, Gendelman O. On the nature of the friction between nonstick droplets and solid substrates. Langmuir. 2010b;26:12479–82.

Bormashenko E, Pogreb R, Musin A, et al. Interfacial and conductive properties of liquid marbles coated with carbon black. Powder Technol. 2010a;203:529–33.

Bormashenko E, Pogreb R, Balter R, et al. Composite non-stick droplets and their actuation with electric field. Appl Phys Lett. 2012;100:151601.

Bormashenko E, Musin A, Whyman G, et al. Revisiting the surface tension of liquid marbles: measurement of the effective surface tension of liquid marbles with the pendant marble method. Colloids Surf A. 2013a;425:15–23.

Bormashenko E, Grynyov R, Chaniel G, et al. Robust technique allowing manufacturing superoleophobic surfaces. Appl Surf Sci. 2013b;270:98–103.

Bormashenko Y, Pogreb R, Gendelman O. Janus droplets: liquid marbles coated with dielectric/semiconductor particles. Langmuir. 2011;27:7–10.

Dandan M, Erbil HY. Evaporation rate of graphite liquid marbles: comparison with water droplets. Langmuir. 2009;25:8362–7.

Dupin D, Armes SP, Fujii S. Stimulus-responsive liquid marbles. J Am Chem Soc. 2009;131:5386–7.

Francis CK, Bennett HT. The surface tension of petroleum. Ind Eng Chem. 1922;14:626–7.

Gao L, McCarthy TJ. Ionic liquid marbles. Langmuir. 2007;23: 10445–7.

Harvey EH. The surface tension of crude oil. Ind Eng Chem. 1925;17(1):85.

Laborie B, Lachaussee F, Lorenceau E, et al. How coatings with hydrophobic particles may change the drying of water droplets: incompressible surface versus porous media effects. Soft Matter. 2013;9:4822–30.

Landau LD and Lifschitz TM. Fluid mechanics 2nd ed., Vol. 6 of the Course of theoretical physics. New York: Pergamon Press; 1987.

Li M, Tian J, Li L, et al. Charge transport between liquid marbles. Chem Eng Sci. 2013;97:337–43.

Mabry JM, Vij A, Iacono ST, et al. Fluorinated Polyhedral Oligomeric silsesquioxane (F-POSS). Angew Chem Int Ed. 2008;47:4137–40.

Mahadevan L. Non-stick water. Nature. 2001;411:895–6.

Matsukuma D, Watanabe H, Minn M, et al. Preparation of poly(lactic-acid)-particle stabilized liquid marble and the improvement of its stability by uniform shell formation through solvent vapor exposure. RSC Adv. 2013;3:7862–6.

McHale G, Elliott SJ, Newton MI, et al. Levitation-free vibrated droplets: resonant oscillations of liquid marbles. Langmuir. 2009;25:529–33.

McHale G, Newton MI. Liquid marbles: principles and applications. Soft Matter. 2011;7:5473–81.

Nakai K, Fujii S, Nakamura Y, et al. Ultraviolet-light-responsive liquid marble. Chem Lett. 2013a;42:586–8.

Nakai K, Nakagawa H, Kuroda K, et al. Near-infrared-responsive liquid marbles stabilized with carbon nanotubes. Chem Lett. 2013b;42:719–21.

Newton MI, Herbertson DL, Elliott SJ, et al. Electrowetting of liquid marbles. J Phys D Appl Phys. 2007;40:20–4.

Planchette C, Biance A-L, Lorenceau E. Transition of liquid marble impacts onto solid surfaces. Europhys Lett. 2012;97:14003.

Planchette C, Biance A-L, Pitois O, et al. Coalescence of armored interface under impact. Phys Fluids. 2013;25:042104.

Sivan V, Tang S-Y, O'Mullane AP, et al. Liquid metal marbles. Adv Funct Mater. 2013;23:144–52.

Tian J, Fu N, Chen XD, et al. Respirable liquid marble for the cultivation of microorganisms. Colloids Surf B. 2013;106:187–90.

Tian J, Arbatan T, Shen X, et al. Liquid marble for gas sensing. Chem Commun. 2010;46:4734–6.

Venkateswara Rao A, Kulkarni MM, Bhagat SD. Transport of liquids using superhydrophobic aerogels. J Colloid Interface Sci. 2005;285:413–8.

Wang Z, Zhu L, Liu H, et al. A conversion coating on carbon steel with good anti-wax performance in crude oil. J Petroleum Sci Eng. 2013;112:266–72.

Xue Y, Wang H, Zhao Y, et al. Magnetic liquid marbles: a "precise" miniature reactor. Adv Mater. 2010;22:4814–8.

Zang D, Chen Z, Zhang Y, et al. Effect of particle hydrophobicity on the properties of liquid water marbles. Soft Matter. 2013;9:5067–73.

Zhao Y, Hu Z, Parhizkar M, et al. Magnetic liquid marbles, their manipulation and application in optical probing. Microfluid Nanofluid. 2012;13:555–64.

11

Methods for seismic sedimentology research on continental basins

Ping-Sheng Wei · Ming-Jun Su

Abstract In contrast to marine deposits, continental deposits in China are characterized by diverse sedimentary types, rapid changes in sedimentary facies, complex lithology, and thin, small sand bodies. In seismic sedimentology studies on continental lacustrine basins, new thinking and more detailed and effective technical means are needed to generate lithological data cubes and conduct seismic geomorphologic analyses. Based on a series of tests and studies, this paper presents the concepts of time-equivalent seismic attributes and seismic sedimentary bodies and a "four-step approach" for the seismic sedimentologic study of continental basins: Step 1, build a time-equivalent stratigraphic framework based on vertical analysis and horizontal correlation of lithofacies, electrofacies, seismic facies, and paleontological combinations; Step 2, further build a sedimentary facies model based on the analysis of single-well facies with outcrop, coring, and lab test data; Step 3, convert the seismic data into a lithological data cube reflecting different lithologies by means of seismic techniques; and Step 4, perform a time-equivalent attribute analysis and convert the planar attribute into a sedimentary facies map under the guidance of the sedimentary facies model. The whole process, highlighting the verification and calibration of geological data, is an iteration and feedback procedure of geoseismic data. The key technologies include the following: (1) a seismic data-lithology conversion technique applicable to complex lithology, which can convert the seismic reflection from interface types to rock layers; and (2) time-equivalent seismic unit analysis and a time-equivalent seismic attribute extraction technique. Finally, this paper demonstrates the validity of the approach with an example from the Qikou Sag in the Bohai Bay Basin and subsequent drilling results.

Keywords Continental basin · Seismic sedimentology · Four-step approach · Time-equivalent seismic attribute · Seismic sedimentary body · Lithology conversion processing

1 Introduction

Since it was presented for the first time in the 1990s (Zeng et al. 1998a, b), seismic sedimentology has been improved in theory and methodology by many researchers and is playing an increasingly important role in oil and gas exploration and development (Posamentier 2002; Miall 2002; Zeng and Hentz 2004; Zhang et al. 2007; Zhu et al. 2009; Zeng 2010; Reijenstein et al. 2011; Hubbard et al. 2011). Early seismic sedimentology research mainly focused on discussions of the cases and workflows of overseas marine basins (Kolla et al. 2001; Hentz and Zeng 2003; Zeng and Hentz 2004; Loucks et al. 2011) and established two key techniques: 90° phase shifting of seismic traces and strata slicing (Zeng 2011). The concept of seismic sedimentology was introduced in China in approximately 2000, followed by the study of sedimentary facies in China's continental basins (Dong et al. 2006; Lin et al. 2007; Zhang et al. 2010; Zhu et al. 2011). However, few case studies have been reported; geologic researchers in China have mainly explored the applicability of seismic sedimentology for different types of continental basins in China and constructed seismic sedimentology models for different sedimentary types (Zhu et al. 2013). In practice, the

P.-S. Wei · M.-J. Su (✉)
Petrochina Research Institute of Petroleum Exploration & Development-Northwest, Lanzhou 730020, Gansu, China
e-mail: sumj@petrochina.com.cn

Edited by Jie Hao

existing workflows and key techniques were not fully suitable for the continental basins in China (Zeng 2011; Qian 2007). Therefore, we present a seismic sedimentology research method and workflow for the sedimentary features of the continental basins and demonstrate the efficacy of the method and its workflow using cases in the Qikou Sag in the Bohai Bay Basin and subsequent well drilling.

2 Sedimentary features of strata in continental basins and the difficulties of seismic sedimentology studies

It is well recognized by researchers in China that the seismic sedimentology method originally used for marine basins can provide useful guidance for studying the sedimentation of continental basins in this country (Zhao et al. 2011; Huang et al. 2011; Dong et al. 2011; Zeng et al. 2013). However, compared to marine basins, continental basins in China are quite different in their major control factors and petrophysical and seismic reflection features (Gu et al. 2005; Zeng et al. 2012) (Table 1), resulting in difficulties in seismic sedimentology interpretation due to the following characteristics: (1) complex depositional systems, various sedimentation types, and rapid changes in sedimentary facies within a small range; due to the filtering effect of wavelets, sands of different sedimentary types are similar in their geophysical responses, which is not only shown in seismic reflection but also on seismic slices, so a seismic sedimentology methodology for sands of different sedimentary types is still lacking; (2) thin and poor lateral continuity of sand bodies; seismic sedimentology investigates the sedimentary

features and evolutionary pattern of deposition, but due to the effects of different deposition velocities in different parts of the basin, seismic slices for thin sands are prone to diachroneity; and (3) complex lithology–wave impedance relations; in contrast to marine basins, continental basins in China are much more complex in their lithological distribution (in many cases, conglomerate, sandstone, mudstone, limestone, and coal coexist) and multi-polar in wave impedance, making it difficult to calibrate the lithology on the 90° phase seismic profile and resulting in multiple possible interpretations of the lithology based on seismic data. In view of these constraints, new thinking and more detailed and effective technical means are needed to generate lithological data cubes and conduct seismic geomorphological analyses for continental basins.

3 Seismic sedimentology research method for continental rifted basins

In line with the sedimentary features of continental basins, the basic research approach to seismic sedimentology is "geology → seismic → geology" (Wei et al. 2011). Expressed as "point → plane → cube", the approach stresses the calibration of geological data, which is also an effective way to resolve seismic ambiguity (Xu et al. 2010, 2011). The method can be summarized into four steps:

(1) Build a time-equivalent stratigraphic framework. First, through detailed vertical analysis and horizontal correlation of outcrop, core and logging data, identify the high-order discontinuity surfaces, abrupt

Table 1 Factors influencing seismic sedimentology research on marine and continental basins

Geological factor	Marine basin	Continental basin
Control factor	Change of global sea level, basin subsidence, sedimentation rate, and climate	Tectonic subsidence, lake level, source supply, paleoclimate, and paleotopography
Sedimentation range	Coastal zone, continental shelf, continental slope and deep sea, thousands of kilometers wide and tens to hundreds of kilometers thick or more	Alluvial fan, fluvial and lacustrine sedimentation zones, relatively small in deposition range and with rapid facies changes
Lateral continuity of sedimentary layer	Long lateral extension and good continuity	Short lateral extension and poor continuity
Thickness and change of sequence	Sequence is thick and stable	Sequence is thin and changeable
Sedimentation type	Relatively simple, large in area, stable in distribution, and includes mainly delta, coastal-shallow sea, and deep sea deposits	Relatively complex, small in area, unstable in distribution, and includes mainly fan delta, braided delta, delta, coastal-shallow lake and deep lake deposits
Change of sedimentary facies	Continuous and stable, gradual in transition	Exhibits rapid facies changes, sudden changes in sedimentary facies are common
Petrophysical features	Relatively simple lithology, simple lithology–wave impedance relationship	Complex lithology, lithology–wave impedance relationship commonly shows multi-polar distribution
Geophysical feature	Clear seismic reflection feature, easy to interpret	Interpretation is ambiguous
Difficulty of prediction	Large prediction area, easy to predict	Small prediction area, difficult to predict

boundary changes in sedimentary facies and marine flooding or lake extension boundary resulting from changes in sea (lake) level or sedimentary base level, and, combined with the analysis of 3D seismic data, build a high-resolution stratigraphy sequence framework. The precision of sequence division depends on the data quality. When the geological and geophysical data are high in resolution, third- and fourth-order (or even higher) sequences and depositional system tracts can be recognized. Then, when time-equivalent analysis is conducted on divided seismic reflectors, adjust the diachronous horizons according to the analysis results. Finally, build the well-seismic tied time-equivalent stratigraphic framework.

(2)   Analyze the sedimentary facies of coring intervals in a single well by using the surface outcrop data, coring data, and lab test results. Use the analysis results to calibrate the SP, GR, etc., logs of the coring intervals and represent the sedimentary facies in the form of logs. Then, use these logs to study the sedimentary facies of non-coring intervals, thus accomplishing the sedimentary facies analysis of the target interval (single-well facies division), and build the sedimentary facies model.

(3)   Convert the seismic data cube into a lithological data cube with wave impedance, SP, and GR via seismic data conversion (i.e., reservoir parameter inversion or 90° phase shifting).

(4)   Conduct time-equivalent stratal slicing or extract the time-equivalent unit's seismic attributes, and under the guidance of the sedimentary facies model, interpret these data cubes based on continuous vertical changes and relative spatial changes in the time-equivalent attributes. Obtain a relatively accurate sedimentary facies plan and profile, and finally build a 3D depositional system model based on modeling and 3D visualization (Fig. 1).

The method for seismic sedimentology research in continental basins involves two major techniques: (1) time-equivalent seismic unit analysis and a time-equivalent seismic attribute extraction technique; (2) a seismic data-lithology conversion processing technique.

3.1 Time-equivalent seismic unit analysis and time-equivalent seismic attribute extraction technique

3.1.1 Interpretation risks caused by seismic slice diachroneity

Time-equivalent stratigraphy is the foundation for sedimentary analysis. Among the several slicing techniques commonly used at present, stratal slicing is closer to time-

equivalent interfaces than time slicing and horizon slicing. However, even stratal slicing is prone to diachroneity.

Stratal slicing assumes that the vertically equal ratio division between two time-equivalent sedimentary interfaces is approximate to slicing along time-equivalent sedimentary interfaces. This assumption assumes that the deposition velocity in the vertical direction does not change with time. However, deposition velocity is affected by factors such as sediment supply rate and changes in accommodation, and these factors change with time due to the effects of tectonic movements and paleoclimate, among other variables. Thus, any geological process is a multivariate function including the time variable. For any point on a plane, the structure and sedimentary environment constantly change with time; thus, the deposition velocity at this point changes with time. Moreover, in many cases, the effects of this change on deposits in continental basins are not negligible (Lin and Zhang 2006), which is different from deposits in marine basins. In addition, the virtual 3D model built by Zeng et al. (1998a, b) reveals that in the 30 Hz Ricker wavelet synthetic data, the stratal slices controlled by two marker events have a left or right deviation of approximately 9 ms. Thus, diachroneity is likely to occur in stratal slices of thin continental interbeds.

Diachronous stratal slices complicate interpretation and are likely to lead to misinterpretations. Figure 2 shows the theoretical model of the fourth-period channel, in which the maximum thickness of the channel sands is 3 m, the maximum width of the channel is 50 m, and the velocities of the first-period, second-period, third-period, and fourth-period channel sands are 1,800 m/s, 1,700 m/s, 1,600 m/s, and 1,500 m/s, respectively. The mudstone is 4 m thick, and its velocities, from shallow to deep, are 1,000 m/s, 1,200 m/s, 1,400 m/s, 1,600 m/s, and 1,800 m/s. Figure 2a shows the model perpendicular to the channel. Figure 2b–e shows the planar geometry of the four periods of the channel (corresponding to slices 1, 2, 3, and 4, respectively, in Fig. 2a). Figure 2f is a 20 Hz Ricker wavelet synthetic section (corresponding to Fig. 2a). Slice A, as shown in Fig. 2g, picks the top of the wave crest corresponding to the location of the fourth-period channel. Affected by the seismic resolution and sand shale velocity difference, there is up to a 6 ms error between the geological time-equivalent surface of slice A and that corresponding to the fourth-period channel (slice B position). Compared with slice B (Fig. 2h), slice A can be interpreted in a variety of ways due to the interference of adjacent layers, but it is difficult to identify a real situation in which only one meandering river channel exists.

3.1.2 Time-equivalent slice processing technique

Local diachroneity on stratal slices can be removed by nonlinear processing. Figure 3a shows a stratal slice of the first member of the Nenjiang Formation in the Anda area of the

Fig. 1 Seismic sedimentology research approach and method for continental basins

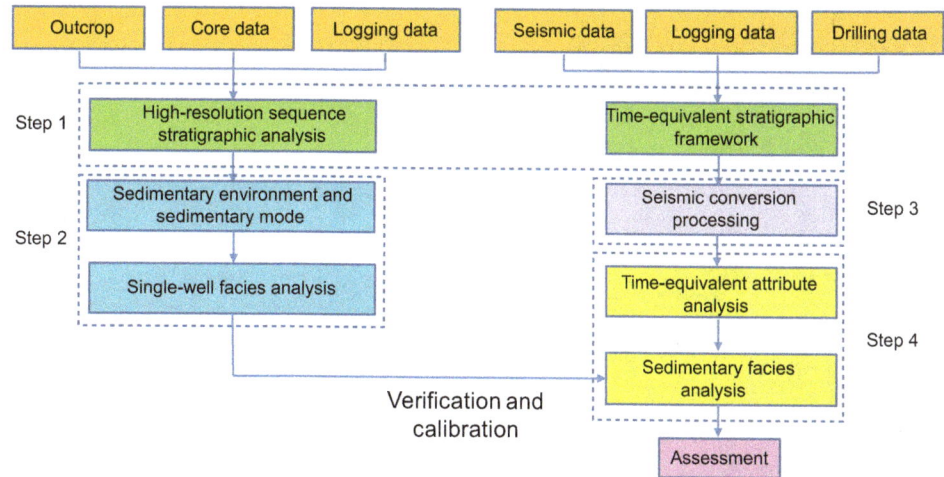

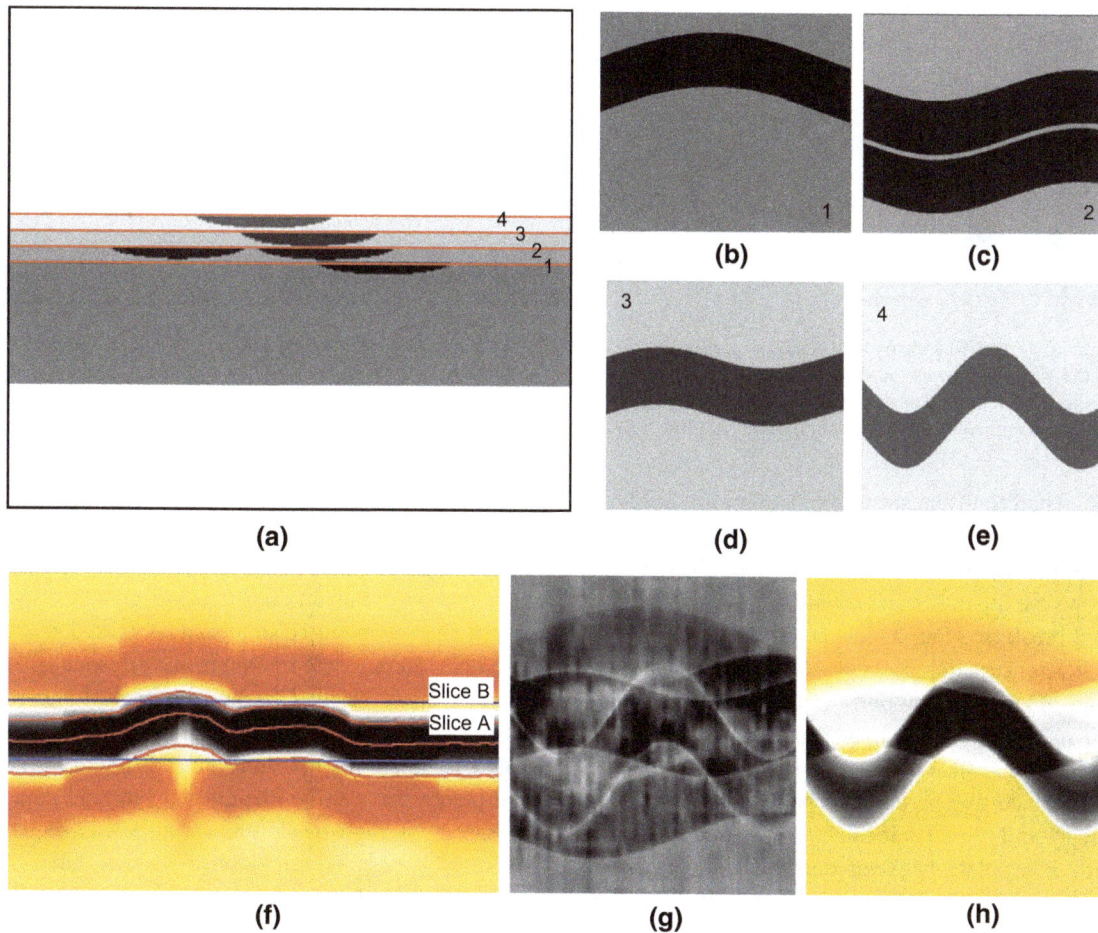

Fig. 2 Theoretical model of four periods of a channel. a The model perpendicular to the channel, b–e the planar geometry of the four periods of the channel (corresponding to slices 1, 2, 3, and 4), f a 20 Hz Ricker wavelet synthetic section (corresponding to (a)), g corresponding to slice A, and h corresponding to slice B

Songliao Basin, where two nearly NS channels in the study area developed and the channel in the middle-eastern area divided in the middle. To determine the reason for this phenomenon, we analyzed the isochronism of seismic reflection at the break in the channel. Zeng et al. (1998a, b)

asserted that the occurrence of time-equivalent reflection events did not change with seismic frequency. Generally, a series of filters are used for filtering in the seismic band width. If the occurrence of a seismic event is not related to its frequency, the event is time-equivalent; otherwise, it is

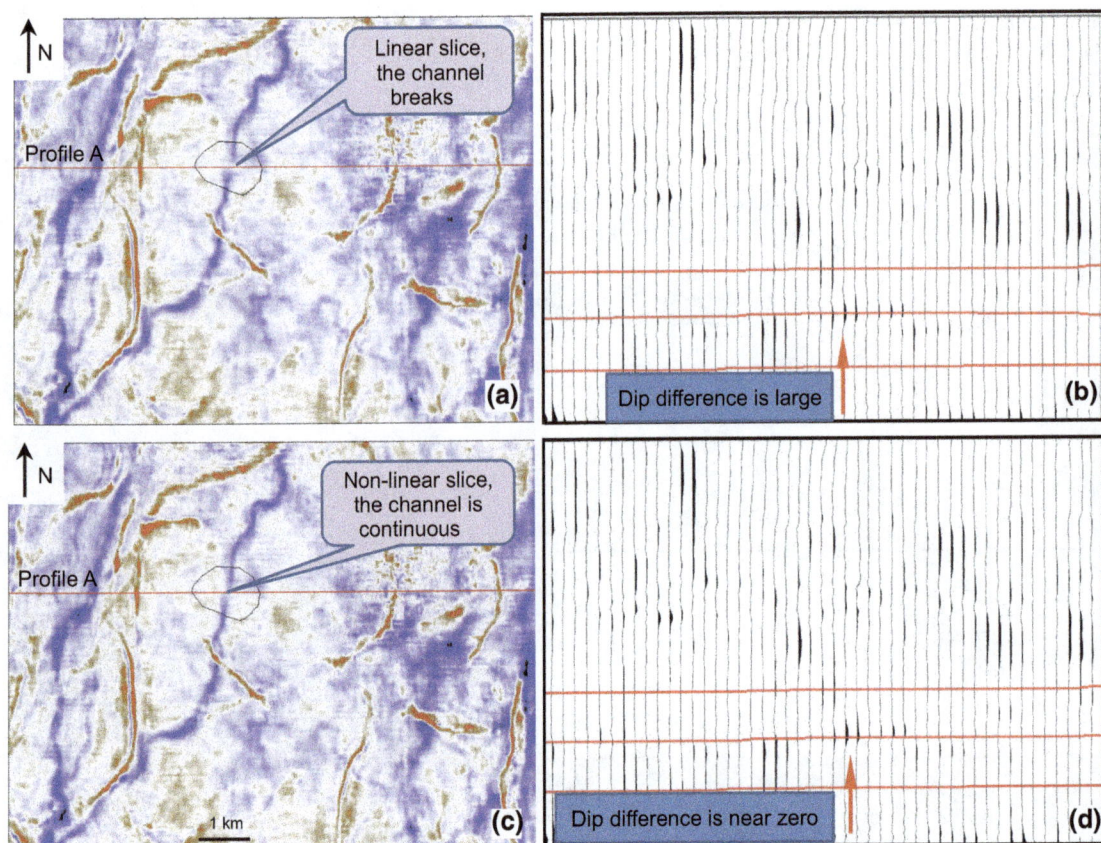

Fig. 3 Stratal slice and time-equivalent analysis profile for the 1st member of the Nenjiang Fm. in the Anda area of the Songliao Basin. **a** A stratal slice of the first member of the Nenjiang Formation in the Anda area of the Songliao Basin, **b** the dip difference profile of profile A crossing the channel break, **c** the amplitude slice after non-linear processing, and **d** the dip difference profile of profile A, showing the slice position after nonlinear processing, with a dip angle difference close to zero

diachronic. Therefore, the seismic reflection dip angles of different frequencies can be subtracted. If the difference is nearly zero, the seismic reflection is time-equivalent. Figure 3b shows the dip difference profile of profile A crossing the channel break (see Fig. 3a for the profile position). The dip angle difference at the channel break is approximately 6°, showing an obvious diachronous feature. To obtain time-equivalent stratal slices, non-linear processing can be conducted on the stratal slice at the channel break. The original slice was obtained by interpolation between the two time-equivalent layers T1 and T2. In this case, we used a Gaussian function to interpolate between the two time-equivalent layers at the break (area in the black circle in Fig. 3a). Figure 3d shows the slice position after non-linear processing, with a dip angle difference close to zero. Figure 3c is the amplitude slice after non-linear processing, which shows that the channel break has apparently improved.

3.1.3 Analysis of time-equivalent seismic unit

Because continental sands are thin and exhibit rapid facies changes, their stratal slices have universal diachroneity,

which makes it impossible to obtain time-equivalent stratal slices through non-linear processing. In this case, the vertical study scale should be adjusted to increase the time unit thickness. For this purpose, we proposed the concept of a time-equivalent seismic unit. A time-equivalent seismic unit refers to the comprehensive responses of a sedimentary body from a certain geological period on the seismic reflection data volume; the top and bottom of a time-equivalent seismic unit are sedimentary time-equivalent planes that can be identified by seismic data.

Based on high-resolution stratigraphic sequence analysis, it is possible to classify four or more orders of sequence and depositional system tracts. The sequence units correspond to time-equivalent seismic units. High-order sequence units correspond to the minimum time-equivalent seismic units that can be analyzed. The tops and bottoms of these sequences generally correspond to the lake expansion boundaries, which have relatively stable lateral distribution and feature strong reflection, good continuity on the seismic profile, and good isochronism of corresponding seismic slices. In contrast, the sand bodies inside the sequence are generally thin and unstable in distribution, appearing as discontinuous reflections on the seismic profile.

3.1.4 Extraction of time-equivalent seismic attributes

To analyze the sedimentation distribution features of the time-equivalent seismic unit, it is necessary to introduce the concepts of time-equivalent seismic attributes and seismic sedimentary bodies. The time-equivalent seismic attribute represents the geometric, kinematics, dynamic, and statistical features of seismic waves corresponding to the time-equivalent seismic unit. The seismic sedimentary body is a reflection of a sedimentary event on the seismic record. In addition to thin sands that cannot be analyzed by stratal slices, gravity flow events exhibit diachroneity in certain cases. Thus, conventional stratal slicing or inter-layer seismic attributes can hardly ensure the isochronism of attribute slices, and the time-equivalent seismic unit can ensure the study unit's unity with time. The extracted time-equivalent seismic attribute reflects the seismic reflection features of the predominant facies in this interval. The planar distribution feature of the time-equivalent seismic attribute provides the base map for sedimentation analysis and overcomes the diachroneity that commonly occurs in continental deposits. When the length of the time-equivalent unit time window approaches zero, the time-equivalent seismic attribute equals the time-equivalent stratal slice. Hence, the time-equivalent stratal slice can be regarded as a particular case of the time-equivalent seismic attribute.

Figure 4 shows a 90° phase shifting profile through Well X3 in the Qinan sub-sag of the Qikou Sag (see Fig. 5 for the position of the plane). The core data of Well X3 show that gravity flow sands in the SS1 sequence developed there due to their strong scouring and incision effects. The seismic sedimentary body corresponding to the sands exhibits up-warp at both ends and is concave in the middle. On the original seismic profile with relatively low predominant frequency (17 Hz) and the frequency-division profile of 40 Hz, the seismic sedimentary body is consistent in occurrence, indicating that the sand is syndeposited, but on the stratal slice, there is an obvious event-crossing phenomenon without time-equivalent meaning and thus without meaning for seismic sedimentology analysis. Therefore, a time-equivalent seismic attribute extraction was conducted.

Figure 5 shows the maximum peak amplitude of the SS1 sequence; the higher the amplitude, the better developed is the sand. The attribute shows there are two sand strips in a nearly NS direction in the western part of the work area, and there is a fan-shaped sand body in the eastern part of the work area. The attribute distribution matches fully with the sands revealed by drilling. Well X3 is in the middle strip, showing that these strips are gravity flow channels, and the fan body in the eastern part is a sub-lacustrine fan.

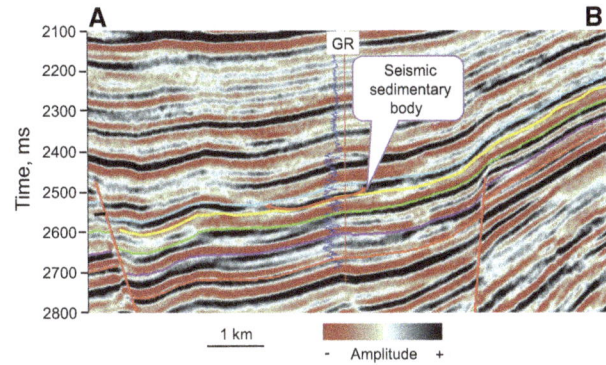

Fig. 4 90° phase shifting profile of the profile AB through Well X3 in the Qinan subsag (see Fig. 5 for the position of profile AB)

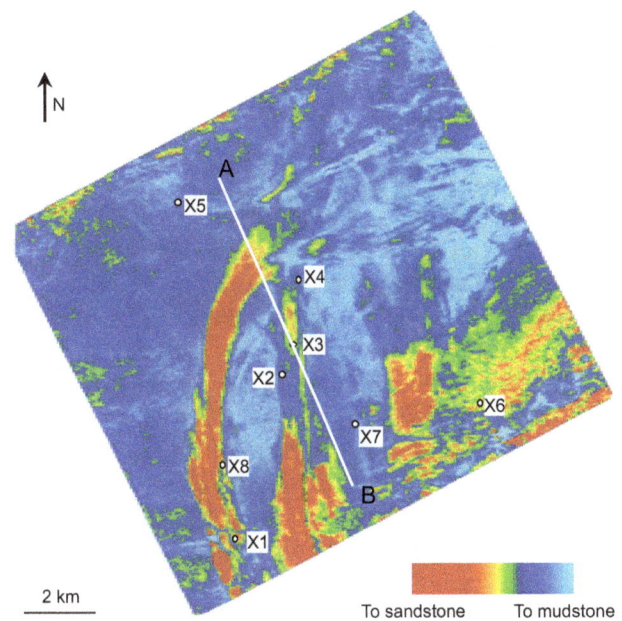

Fig. 5 Maximum peak amplitude of SS1 in the Shahejie Formation of the Qikou Sag

3.2 Seismic data lithological conversion technique

It is a prerequisite of seismic sedimentology research to convert the seismic data cube into a lithological data cube that can be directly correlated with logging data. Zeng and Backus (2005a, b) demonstrated the advantages of the 90° phase shifting of seismic data in the stratigraphic and lithological interpretation of thin beds by means of models and examples and concluded that the 90° phase wavelet shifted the main lobe (maximum amplitude) of the seismic response to the midpoint of the thin beds. Thus, the seismic response corresponded to the midpoint of the thin bed rather than its top and bottom boundaries, connecting the main seismic event to the reservoir unit defined in geology (e.g., sandstone reservoir). In this way, the seismic polarity can correspond to lithology within 0–1 wavelength ranges. Although the inaccuracy increases when the formation thickness is less than ¼ λ (λ is wavelength), the top and bottom boundaries of the formation can be determined at the position where the amplitude passes the zero point. When applying these improvements to real data, a one-to-one relationship will be established between the seismic event and the lithological unit of the thin beds, which will facilitate the seismic interpretation of sedimentary lithology (e.g., distinguishing sand from shale).

When the wave impedance can be used to effectively distinguish the lithology, the 90° phase shifting of seismic data is undoubtedly the most economic and effective seismic lithological method at present. However, in contrast to marine basins, the continental basins in China have multiple provenances, complex lithology, poor sorting, and a complicated lithology–wave impedance relationship, which make it difficult to interpret their lithology based on 90° phase shifting.

One possible solution is to convert the seismic data into a logging parameter data cube capable of reflecting the lithology (Su et al. 2013), which involves the following steps: (1) analyze the reservoir features and choose the logging data that best reflect the reservoir features in the work area; (2) perform detailed well-seismic calibration and convert the seismic data into a 90° phase data cube; and (3) take the 90° phase shifting data cube as the constraint cube and use the co-kriging and fractal interpolation reconstruction methods to accomplish the seismic trace logging parameter inversion. Constraint data cubes constrained by a 90° phase conversion data cube have two advantages: first, their amplitude fidelity, i.e., the amplitude of the original seismogram, is maintained; second, they have closely corresponding spatial positions with equivalent reservoirs (equivalent layers of a number of reservoirs).

The fractal theory provides an effective tool for the description of complex phenomena. Well logging and seismic signals have self-similarity, i.e., fractal features (Lu and Li 1996; Jiang et al. 2006). The fractional Brownian motion (FBM), a statistically self-similar and non-stationary random process, is currently one of the most common mathematical models used to study fractal signals and has been applied effectively in reservoir fracture prediction and high-resolution processing (Leary 1991; Cao et al. 2005; Li and Li 2008; Li et al. 2008; Chang and Liu 2002; Fen et al. 2011; Huang et al. 2009). In the present paper, FBM is applied in seismic lithological inversion.

In probability theory, a normalized FBM of Hurst index α ($0 < \alpha < 1$) is defined as X: $[0, \infty) \to$ R, a "Random Walk Process" within a certain probability space, which causes FBM to satisfy the following:

(1) with probability 1, $X(0) = 0$ and $X(t)$ is a continuous function of t,
(2) for $t \geq 0$ and $h > 0$, the increment $X(t + h) - X(t)$ is subject to normal distribution with a mean of 0 and variance h^{2a}; thus,

$$P(X(t+h) - X(t) \leq x) = (2\pi)^{-1/2}h^{-a}\int_{-\infty}^{x}\exp\left(\frac{-u^2}{2h^{2a}}\right)\mathrm{d}u \quad (1)$$

$B_a(t)$ is generally used to denote an FBM of Hurst index α.

Pentland (1984) extended FBM to a high-dimensional situation and defined the fractional Brownian random field (FBR): with X, $\Delta X \in R^2$, $0 \leq H \leq 1$, $F(y)$ is a Gaussian random function with a mean of 0, $P_r(\bullet)$ represents the probability measure and $\|\bullet\|$ represents the norm. If the random field $B_H(X)$ meets

$$P_r\left[\frac{B_H(X+\Delta X) - B_H(X)}{\|\Delta X\|^H} < y\right] = F(y) \quad (2)$$

the $B_H(X)$ is the FBR and is characterized by

$$E|B_H(X+\Delta X) - B_H(X)|^2 = E|B_H(X+1) - B_H(X)|^2\|\Delta X\|^{2H} \quad (3)$$

Then, the parameter H can be calculated conveniently using Eq. (3).

Essentially seismic lithological inversion is the joint application of the kriging technique and fractal interpolation. The co-kriging interpolation result is faithful to well point data and seismic lateral changes (Yang et al. 2012; Huang et al. 2012). The random fractal interpolation methods consist mainly of random midpoint shift, continuous random stacking, and spectrum synthesis. In this paper, the random midpoint shift interpolation method is used, which is in fact a recursive midpoint displacement method, with the following computing equations:

$$X(i) = X_1(i) + X_n(i) \quad (4)$$
$$X_l(i) = PA\{A, D_1, \cdots D_i, \cdots D_n\} \quad (5)$$
$$X_n(i) = K \cdot \sqrt{1 - 2^{2H-2}}\|\Delta X\| \cdot H \cdot \sigma \cdot G \quad (6)$$

where X_1 is a linear term representing large-scale feature information and is worked out using the co-kriging method; PA is a co-kriging interpolation operator; A is a data cube of 90° phase conversion; and D_i represents well data. When the geological condition is simple, X_1 can also be derived using the linear interpolation method.

X_n is a non-linear term, representing medium- and small-scale feature information; G is a Gaussian random variable and follows the distribution of N (0, 1); σ is a standard variance of the normal distribution of data; H is the Hurst index depicting the detail and roughness of the data information; $\|\Delta X\|$ is a sample interval; and K is the calibration coefficient of the non-linear term. We can see that the value of the interpolation point completely depends on the H index and σ characterizing the raw data. Therefore, the ultimate inversion results would closely match the well point data and seismic horizontal changes, and the vertical resolution is consistent with the well logging data.

4 Case study

4.1 Overview of the 3D study area

The study area, located in the Qibei sub-sag of the Qikou Sag, is a Cenozoic rifted basin (Zhou et al. 2011). The Sha2 member is the target zone for study and has a burial depth of 4,100–5,000 m, a reservoir (single layer) thickness of less than 12 m, and a lithology of sandstone, mudstone, calcareous mudstone, and dolomite interbeds with unequal thicknesses. In recent years, high-yielding oil and gas wells in the Sha2 member have been discovered in this area; thus, there is an urgent need to understand the reservoir distribution and sedimentary pattern. However, the predominant frequency of seismic data in this area is approximately 17 Hz; this resolution does not meet the demand for distinguishing thin interbed reservoirs less than 12 m thick. Moreover, due to the existence of calcareous mudstone, it is also difficult to distinguish sandstone from calcareous mudstone using the wave impedance absolute value. As a result, it is difficult to reflect the variation of the sedimentary reservoir by seismic slices. Therefore, the co-kriging and fractal theory-based lithology conversion technique and the seismic sedimentology research method for continental basins are utilized in this study to determine the spatial distribution and sedimentary features of the reservoirs.

4.2 Building the time-equivalent stratigraphic framework

Substantial research has been conducted on the Paleogene sequence division of the Qikou Sag (Wu et al. 2010; Huang et al. 2010), which concluded that the Sha2 member is a complete third-order sequence with an angular unconformity at the basin margin and a conformable contact in the central basin. The Sha2 member can be further divided into three fourth-order sequences, i.e., BinII (SB2), BinIII (SB3), and BinIV (SB4). The final sequence interpretation scheme is obtained based on the correlation and the verification of the seismic profile with the sequence interpretation scheme on well logs after a depth-time conversion.

With the AA′ profile that extends in the dip direction as an example (Fig. 6), the sequence boundary shows continuous seismic reflections. These reflections are mostly parallel to the maximum flooding surface, and their occurrence does not change with frequency, thereby exhibiting time-equivalent stratigraphy significance. The building of the sequence framework is not only a prerequisite for time-equivalent stratal slicing but also provides constraint conditions for seismic lithology conversion.

4.3 Single-well facies analysis

Carbon dust and plant stem fossils occur commonly in the coring intervals of the Sha2 member in the study area. The mudstone is mainly gray and grayish-green, indicating that the Sha2 member was primarily deposited in shallow water and is dominated by coastal–shallow lacustrine deposits.

Based on the observation of cores taken from several key wells, fan delta front deposits are developed in the Sha2 member in the northern study area, e.g., gray mudstone dominates the lithology of Well W9 and massive gray packsand is developed in the middle and lower parts. Core observations also reveal plant stem carbon dust, boulder clay, and slump deformation structures along with sedimentary structures such as low-angle oblique bedding, trough cross bedding, and parallel bedding (Fig. 7). The lithological electric property features a tooth-like bell-type and tooth-like funnel-type GR log. The core facies is interpreted as a fan delta front deposit. The western and southwestern parts are dominated by beach bar sedimentary facies, where channel scouring structure is not obvious, the sandstone exhibits fine granularity and good sorting, and boulder clay with bioclasts resulting from storm events can be seen.

4.4 Seismic data conversion

The efficacy of seismic amplitudes as lithological markers depends on the wave impedance difference between the different lithologies of layered media. Petrophysical analysis of several wells in the Qibei sub-sag indicates an insignificant wave impedance difference between different lithologies of the Sha2 member in the study area. The wave impedance of mudstone ranges from 7,200 to 14,400 (m/s g/cm^3), with two peak values at 9,000 (m/s g/cm^3) and 10,800 (m/s g/cm^3). The wave impedance of sandstone ranges from 8,400 to 13,200 (m/s g/cm^3), with a peak value at 11,400 (m/s g/cm^3). For calcareous mudstone, the wave impedance ranges from 10,800 to 12,600 (m/s g/cm^3), overlapping the peak value of sandstone. The reflection events in the 90° phase data cube fail to reach a close correspondence with different lithologies. Because the sand

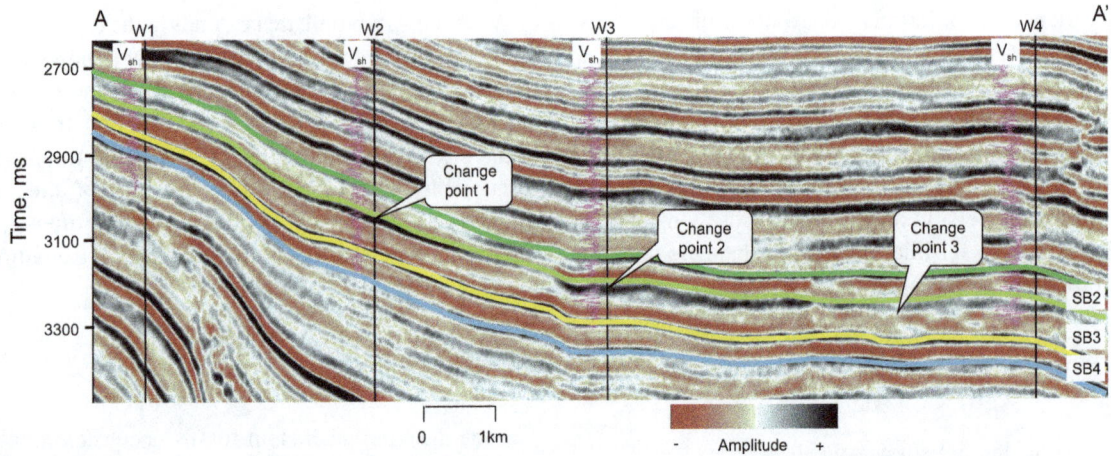

Fig. 6 Sequence division of the Sha2 member based on the AA′ well-tie seismic profile in the Qibei sub-sag

Fig. 7 Log features and core photos of Well W9

bodies are generally thin (less than 1/4 λ) and the wave impedance difference between sandstone and mudstone is insignificant, a reflection event may correspond to sandstone or mudstone on the 90° phase profile. Therefore, the 90° phase data cube is not a good lithological criterion for the complex lithology zones where the impedances of sandstone and mudstone are difficult to distinguish. The sedimentary evolutionary features are unlikely to be reflected in the stratal slices obtained from the 90° phase data cube. Figure 8 shows three stratal slices extracted from the 90° phase data cube

(see Fig. 9 for their positions). Slices 1, 2, and 3 correspond to the lowstand system tract, lake transgressive system tract, and highstand system tract of the BinIII sequence (SB3), respectively. In the lake transgressive system tract, the sand body is not developed in Wells W2 and W3, which are located within the high-amplitude and low-amplitude zones of the stratal slices, respectively. In the lowstand system tract of the BinIII sequence, a thick sand body is developed in Wells W4, W5, and W6. In addition, the sand body is not developed in the lake transgressive system tract, whereas a

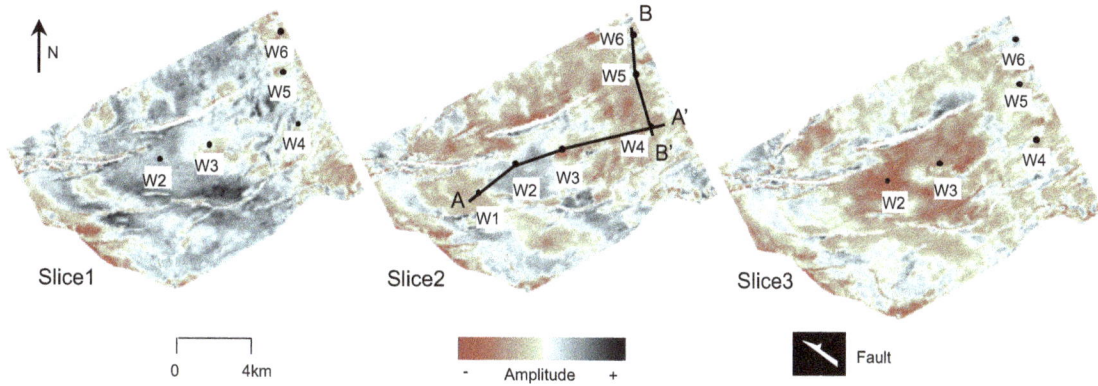

Fig. 8 Three stratal slices of the BinIII sequence extracted from the 90° phase data cube. Slices 1, 2, and 3 correspond to a lowstand system tract, lake transgressive system tract, and highstand system tract, respectively

Fig. 9 NS-trending well-tie profile

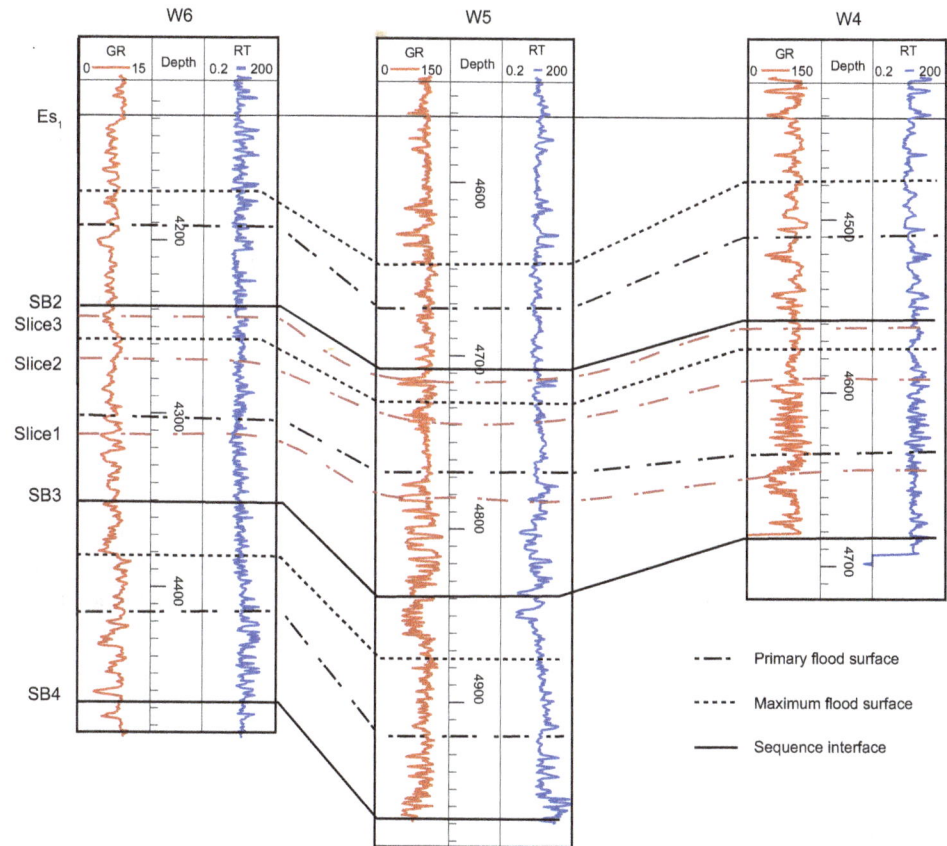

thin sand body is developed in the highstand system tract, but the sedimentary evolution features are not reflected on the stratal slices.

The co-kriging and fractal theory-based seismic data conversion method described above is adopted to obtain a lithological data cube applicable in sedimentation analysis. First, the logging data capable of reflecting the reservoir features of the study area are selected. In this case, the shale content curve (V_{sh}) is determined based on GR. The shale content not only helps to distinguish reservoirs and non-reservoirs but also can reveal the hydrodynamic condition

and thus contribute to the identification of the sedimentary environment. Second, detailed well-seismic calibration was performed, and seismic data were converted into a 90° phase data cube. Finally, under the constraint of the time-equivalent stratigraphic framework and taking the 90° phase shifting data cube as the constraint data cube, the co-kriging and fractal interpolation reconstruction methods were used to achieve the seismic trace lithological data inversion.

Figure 10 shows the inversion profile corresponding to Fig. 6. The inversion results obtained by the above methods have the following features: (1) Different lithologies

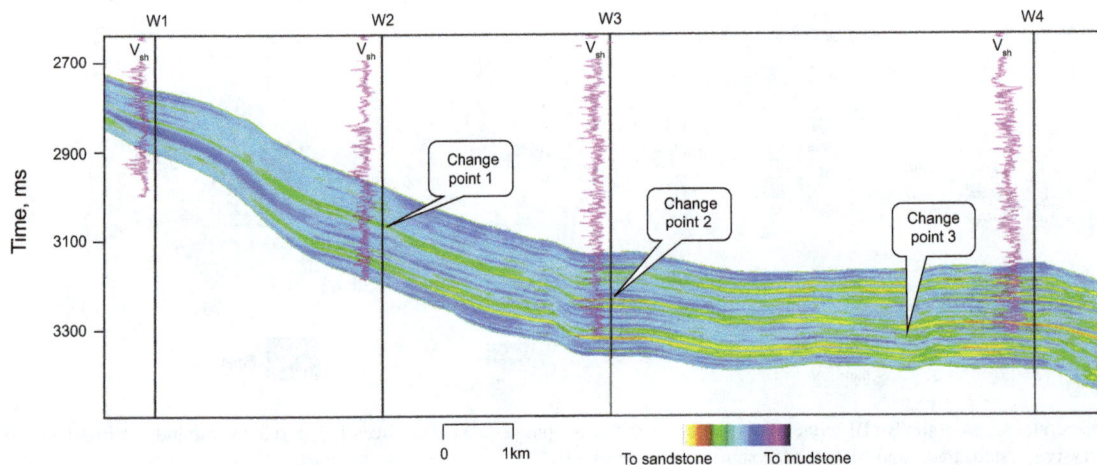

Fig. 10 AA′ inversion profile of the Qibei sub-sag

are effectively discriminated, and the details of the original seismic data's horizontal changes are kept. Because the inversion values completely depend on the *H* index, σ, and co-kriging interpolation values describing the original data, the horizontal change features of the original seismic profile are better kept in the inversion results, and the inherent modeling features of a well-constrained inversion are overcome, thereby laying a solid foundation for the lithological data cube to reflect the change in sedimentation. Figure 6 and Fig. 10 show three major change points. Change points 1 and 2 show reflection peaks on the seismic profile. As indicated by the shale content curve, the reservoir is not developed at change point 1 but is better developed at change point 2. This issue is resolved perfectly by inversion. At change point 3, the seismic reflection weakens gradually from west to east, its wave form becomes poorer in continuity, the number of sand bodies increases from west to east on the corresponding inversion profile, and the sand body is lens-like, showing a consistent change with seismic wave form. (2) The resolution is enhanced by inversion. Due to the supplementary high-frequency logging information, the resolution for various lithologies and strata is enhanced significantly on the inversion lithological profile, and the inversion results match perfectly with the logging data.

4.5 Time-equivalent seismic attribute and seismic sedimentology analysis

Inversion effectively discriminates various lithologies, enhances the resolution of the lithological data cube, greatly reduces the thickness of the minimal mapping unit, and provides conditions for using the time-equivalent slice to study the sedimentation change. To facilitate analysis and correlation, three slices of the BinIII sequence corresponding to slices 1, 2, and 3 were interpreted. These slices describe the distribution and interrelationship of the BinIII lowstand system tract, lake transgressive system tract, and highstand system tract deposits. Figure 11 shows their accurate locations on profile BB′. In this figure, red and yellow represent sandstone, green represents argillaceous sandstone, and blue represents mudstone.

Slice 1 in Fig. 12 is the stratal slice of a lowstand system tract, containing three sand bodies in a fan shape. Fan body 1 developed in the northeastern part and has the widest distribution range. This fan body is distributed from north to southeast and southwest and intersects with fan body 3 in a fault south of Well W4. Fan body 3 is mainly distributed in the southeastern part of the study area. Fan body 2 developed in the northwestern part and has three lobes extending toward the south. Drilling cores and mud logging results indicate that these fan bodies are mostly interbeds of gray packsand and siltstone, with a single-layer thickness of 4–8 m. The GR log represents a combination of tooth-like funnel-type and tooth-like bell shapes in Well W4, a tooth-like shape in W5, and a funnel-type shape in W6. Wells W4 and W6 are located in the red and yellow zones, whereas W5 is in the green zone. Depositional model analysis shows that Wells W4, W5, and W6 are a fan delta front distributary channel, river mouth bar, and interchannel deposit, respectively. A small area of argillaceous sandstone is distributed parallel to the lake shoreline in the southwestern part and is interpreted as coastal beach bar sand.

Slice 2 in Fig. 12 is the stratal slice of a lake transgressive system tract, indicating that the sand bodies retrograded toward the basin margin as the lake area increased. In the study area, sparse sand bodies are distributed only parallel to the lake shoreline in the southwestern part and are interpreted as beach bar sand. Gray shale occurs in Wells W1, W2, W3, W4, W5, and W6, which feature straight GR log curves.

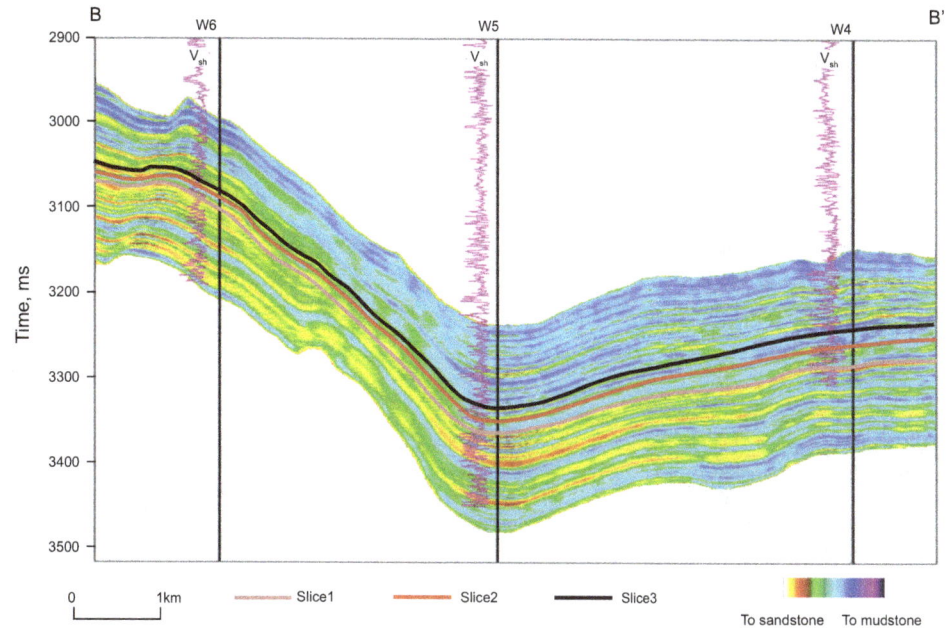

Fig. 11 NS-trending lithological profile BB′ showing the locations of three stratal slices

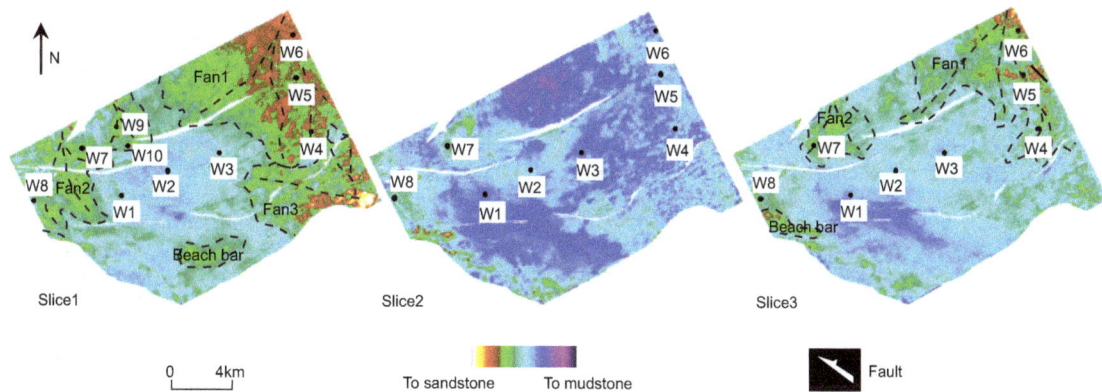

Fig. 12 Three strata slices of the BinIII sequence extracted from the lithological data cube. Slices 1, 2, and 3 correspond to a lowstand system tract, lake transgressive system tract, and highstand system tract, respectively

Slice 3 in Fig. 12 is the stratal slice of a highstand system tract, indicating that the lake level dropped and the delta prograded gradually in this period. Two fan bodies in the same location as the lowstand system tract appear in the northern part of the slice. Compared with the period of the lowstand system tract, the scale of these fan bodies decreased. The fan body in the east comprises three lobes. Log curves show a coarsening-upward funnel shape in Well W6, tooth-like funnel shape in W5, and tooth shape in W4. A beach bar sand body is developed in the southwestern part.

The depositional model observed on these three slices corresponds well with the sequence evolution stage and the drilling results, thereby demonstrating the reliability of the new seismic lithology method. Moreover, fan body 3, located in the south, was confirmed by later drilling.

5 Conclusions

Continental basins differ greatly from marine basins in the major control factors of their structure, sedimentation and sequence formation, petrophysical features, and seismic reflection features. Therefore, new methods and techniques for seismic sedimentology research on continental basins must be developed.

Seismic sedimentology research on continental basins should highlight the verification and calibration of geological data. In this paper, we present the concepts of time-equivalent seismic attributes and seismic sedimentary bodies as well as seismic data-lithology conversion processing and time-equivalent seismic attribute extraction methods and techniques. On this basis, we introduce a "four-step" approach applicable to seismic sedimentology research on continental basins.

Using the methods discussed above, seismic sedimentology research was conducted for the Qikou Sag in the Bohai Bay Basin. The research results show that the Sha2 member, where deltaic deposits are developed, is a potential site for future exploration. The above findings have been confirmed by subsequent drilling, thereby demonstrating the reliability and validity of the methods and techniques introduced in this paper.

Acknowledgments This research was supported by the Key Scientific and Technological Project "Seismic-Sedimentology Software System Investigation and Application" of PetroChina Company Limited (2012B-3709). The authors would like to take this opportunity to express our appreciation and thanks to Ni Changkuan, Cai Gang, and Zhang Zhaohui for their hard work on this study.

References

Cao MS, Ren QW, Wang HH. A method of detecting seismic singularities using combined wavelet with fractal. Chin J Geophys. 2005;48(3):672–9 (in Chinese).

Chang X, Liu YK. The generalized fractal dimension of seismic records and its application. Chin J Geophys. 2002;45(6):839–46 (in Chinese).

Dong CM, Zhang XG, Lin CY. Discussions on several issues about seismic sedimentology. Oil Geophys Prospect. 2006;41(4):405–9 (in Chinese).

Dong YL, Zhu XM, Hu TH, et al. Research on seismic sedimentology of He-3 Formation in Biyang Sag. Earth Sci Front. 2011;18(2):284–93 (in Chinese).

Fen ZD, Dai JS, Deng H, et al. Quantitative evaluation of fractures with fractal geometry in Kela-2 gas field. Oil Gas Geol. 2011;32(6):928–33 (in Chinese).

Gu JY, Guo BC, Zhang XY. Sequence stratigraphic framework and model of the continental basins in China. Pet Explor Dev. 2005;32(5):11–5 (in Chinese).

Hentz TF, Zeng HL. High-frequency Miocene sequence stratigraphy, offshore Louisiana: Cycle framework and influence on production distribution in a mature shelf province. AAPG Bull. 2003;87(2):197–230.

Huang CY, Wang H, Wu YP, et al. Analysis of the hydrocarbon enrichment regularity in the sequence stratigraphic framework of Tertiary in Qikou Sag. J Jilin Univ (Earth Sci Ed). 2010;40(5):986–95 (in Chinese).

Huang HD, Cao XH, Luo Q. An application of seismic sedimentology in predicting organic reefs and banks: A case study on the Jiannan-Longjuba region of the eastern Sichuan fold belt. Acta Pet Sin. 2011;32(4):629–36 (in Chinese).

Huang YP, Dong SH, Geng JH. Ordovician limestone aquosity prediction using nonlinear seismic attributes: case from the Xutuan coal mine. Appl Geophys. 2009;6(4):359–66.

Huang ZY, Gan LD, Dai XF, et al. Key parameter optimization and analysis of stochastic seismic inversion. Appl Geophys. 2012;9(1):49–56.

Hubbard SM, Smith DG, Nielsen H, et al. Seismic geomorphology and sedimentology of a tidally influenced river deposit, Lower Cretaceous Athabasca oil sands, Alberta, Canada. AAPG Bull. 2011;95(7):1123–45.

Jiang SH, Chen JY, Jiang Y, et al. Fractal dimension calculation of seismic traces and application to reservoir description in coastal regions. Period Ocean Univ China. 2006;36(5):841–4 (in Chinese).

Kolla V, Bourges P, Urruty JM, et al. Evolution of deep-water Tertiary sinuous channels offshore Angola (west Africa) and implications for reservoir architecture. AAPG Bull. 2001;85(8):1373–405.

Leary PC. Deep borehole log evidence for fractal distribution of fractures in crystalline rock. Geophys J Int. 1991;107(3):615–27.

Li W, Yan T, Bi XL. Mechanism of hydraulically created fracture breakdown and propagation based on fractal method. J China Univ Pet. 2008;32(5):87–91 (in Chinese).

Li XF, Li XF. Seismic data reconstruction with fractal interpolation. Chin J Geophys. 2008;51(4):1196–201 (in Chinese).

Lin CY, Zhang XG. The discussion of seismic sedimentology. Adv Earth Sci. 2006;21(11):1140–4 (in Chinese).

Lin CY, Zhang XG, Dong CM. Concepts of seismic sedimentology and its preliminary application. Acta Pet Sin. 2007;28(2):69–71 (in Chinese).

Loucks RG, Moore BT, Zeng HL. On-shelf lower Miocene Oakville sediment-dispersal patterns within a three-dimensional sequence-stratigraphic architectural framework and implications for deep-water reservoirs in the central coastal area of Texas. AAPG Bull. 2011;95(10):1795–817.

Lu JA, Li ZB. The self-similarity study of well logging curves. Well Logging Technol. 1996;20(6):422–7 (in Chinese).

Miall AD. Architecture and sequence stratigraphy of Pleistocene fluvial systems in the Malay Basin, based on seismic time-slice analysis. AAPG Bull. 2002;86(7):1201–16.

Pentland AP. Fractal-based description of natural scenes. IEEE Trans Pattern Anal Mach Intell. 1984;6(6):661–74.

Posamentier HW. Ancient shelf ridges, a potentially significant component of the transgressive systems tract: case study from offshore northwest Java. AAPG Bull. 2002;86(1):75–106.

Qian RJ. Analysis of some issues in interpretation of seismic slices. Oil Geophys Prospect. 2007;42(4):482–7 (in Chinese).

Reijenstein HM, Posamentier HW, Bhattacharya JP. Seismic geomorphology and high-resolution seismic stratigraphy of inner-shelf fluvial, estuarine, deltaic, and marine sequences, Gulf of Thailand. AAPG Bull. 2011;95(11):1959–90.

Su MJ, Wang XW, Yuan SQ. Seismic sedimentologic analysis and its application in areas with complex lithology—Case study on Qibei Sag in Huanghua Depression. SEG International Exposition and 83rd Annual Meeting. 2013, 22-27 September, Houston, Texas.

Wei PS, Li XB, Yong XS, et al. Discussion on petroleum seismogeology. Lithol Reserv. 2011;23(3):1–6 (in Chinese).

Wu YP, Yang CY, Wang H, et al. Integrated study of tectonics—sequence stratigraphy—sedimentation in the Qikou Sag and its application. Geotectonica et Metallogenia. 2010;34(4):451–60 (in Chinese).

Xu HN, Yang SX, Zheng XD, et al. Seismic identification of gas hydrate and its distribution in Shenhu Area, South China Sea. Chin J Geophys. 2010;53(7):1691–8 (in Chinese).

Xu Y, Shu P, Ji XY. A research on object-control geology modeling of volcano reservoir with seismic and logging data in Xushen gas field. Chin J Geophys. 2011;54(2):336–42 (in Chinese).

Yang K, Ai DF, Geng JH. A new geostatistical inversion and reservoir modeling technique constrained by well-log, crosshole and surface seismic data. Chin J Geophys. 2012;55(8):2695–704 (in Chinese).

Zeng HL. Geologic significance of anomalous instantaneous frequency. Geophysics. 2010;75(3):P23–30.

Zeng HL. Seismic sedimentology in China: a review. Acta Sedimentol Sin. 2011;29(3):417–26 (in Chinese).

Zeng HL, Backus MM. Interpretive advantages of 90 degrees–phase wavelets: part 1—modeling. Geophysics. 2005a;70(3):C7–15.

Zeng HL, Backus MM. Interpretive advantages of 90 degrees–phase wavelets: part 2—seismic applications. Geophysics. 2005b;70(3):C17–24.

Zeng HL, Hentz TF. High-frequency sequence stratigraphy from seismic sedimentology: applied to Miocene, Vermilion Block50, Tiger Shoal area, offshore Louisiana. AAPG Bull. 2004; 88(2):153–74.

Zeng HL, Backus MM, Barrow KT. Stratal slicing, part I: realistic 3-D seismic model. Geophysics. 1998a;63(2):502–13.

Zeng HL, Henry SC, Riola JP. Stratal slicing, part II: real seismic data. Geophysics. 1998b;63(2):514–22.

Zeng HL, Zhu XM, Zhu RK, et al. Guidelines for seismic sedimentologic study in non-marine postrift basins. Pet Explor Dev. 2012;39(3):275–84 (in Chinese).

Zeng HL, Zhu XM, Zhu RK, et al. Seismic prediction of sandstone diagenetic facies: applied to Cretaceous Qingshankou Formation in Qijia Depression, Songliao Basin. Pet Explor Dev. 2013;40(3):266–74 (in Chinese).

Zhang JH, Zhou ZX, Tan MY, et al. Discussions on several issues in seismic slice interpretation. Oil Geophys Prospect. 2007;42(3): 348–52 361 (in Chinese).

Zhang XG, Lin CY, Zhang T. Seismic sedimentology and its application in shallow sea area, gentle slope belt of Chengning Uplift. J Earth Sci. 2010;21(4):471–9.

Zhao WZ, Zou CN, Chi YL, et al. Sequence stratigraphy, seismic sedimentology, and lithostratigraphic plays: Upper Cretaceous, Sifangtuozi area, southwest Songliao Basin, China. AAPG Bull. 2011;95(2):241–65.

Zhou LH, Pu XG, Zhou JS, et al. Sand-gathering and reservoir-controlling mechanisms of Paleogene slope-break system in Qikou Sag, Huanghua Depression, Bohai Bay Basin. Pet Geol Exp. 2011;33(4):371–7 (in Chinese).

Zhu XM, Dong YL, Hu TH, et al. Seismic sedimentology study of fine sequence stratigraphic framework: a case study of the Hetaoyuan Formation in the Biyang Sag. Oil Gas Geol. 2011;32(4):615–24 (in Chinese).

Zhu XM, Li Y, Dong YL, et al. The program of seismic sedimentology and its application to Shahejie Formation in Qikou Depression of North China. Geol China. 2013;40(1):152–62 (in Chinese).

Zhu XM, Liu CL, Zhang YN, et al. Application of seismic sedimentology in prediction of non-marine lacustrine deltaic sand bodies. Acta Sedimentol Sin. 2009;27(5):915–21 (in Chinese).

12

Research into the inversion of the induced polarization relaxation time spectrum based on the uniform amplitude sampling method

Pu Zhang[1] · Sheng Wang[1] · Kai-Bo Zhou[1] · Li Kong[1] · Hua-Xiu Zeng[2]

Abstract The induced polarization relaxation time spectrum (RTS) reflects the distribution of rock pore size, which is a key factor in estimating the oil or water storage capacity of strata. However, as the data acquisition and transmission abilities of well logging instruments are much limited due to the underground environment, it is necessary to explore suitable sampling methods which can be used to obtain an accurate RST with less sampling data. This paper presents a uniform amplitude sampling method (UASM), and compares it with the conventional uniform time sampling method (UTSM) and logarithm time sampling method (LTSM) in terms of the adaptability to different strata, RTS inversion accuracy, and stratum vertical resolution. Numerical simulation results show that the UASM can obtain high inversion accuracy of RTS with different kinds of pore size distribution formation, with high dynamic ranges of pore size, and with a small number of sampling points. The UASM, being able to adapt to the attenuation speed of polarization curve automatically, thus has the highest vertical resolution. The inversion results of rock samples also show that the UASM is superior to the UTSM and LTSM.

Keywords Uniform amplitude sampling · Relaxation time spectrum · Stratum pore distribution · Induced polarization

✉ Kai-Bo Zhou
zhoukb@hust.edu.cn

[1] School of Automation, Huazhong University of Science and Technology, Wuhan 430074, Hubei, China

[2] China Petroleum Logging Co., Ltd, Xi'an 710077, Shaanxi, China

Edited by Jie Hao

1 Introduction

The size distribution of rock pores is a key factor in determining the storage capacity of oil or water in the strata. Compared with the NMR logging mode, the IP logging mode has greater probing depth, higher signal-to-noise ratio, and less cost, so it has wider and more favorable application prospects (Titov et al. 2002, 2010; Wang 2004). The IP logging instrument can obtain the polarization decay potential (Guan et al. 2010; Li et al. 2011) which contains very rich underground information. The RTS inverted from the IP decay potential can be used to analyze the pore size distribution of the strata, and then to make assessments on the rock permeability, the amount of oil or water storage, and so on (Liu et al. 2014a, b; Rezaee et al. 2006; Binley et al. 2005; Zimmermann et al. 2008; Gurin et al. 2013; Beckett and Augarde 2013; Li et al. 2010). A lot of research work has been carried out on the RTS, including inversion algorithms (Moody and Xia 2004; Xiao et al. 2012; Chen et al. 2013; Jang et al. 2014; Florsch et al. 2014), determining the appropriate number of relaxation time constants and its distribution (Tong et al. 2004; 2005; Buecker and Hoerdt 2013; Revil et al. 2014). The results of this work have a certain relevance for the inversion of IP RTS (Liu et al. 2014a). However, the above research was mostly done in the laboratory, and very little was in the in situ environment (Nie et al. 1987; Wang et al. 2011).

Restricted by underground environments, data storage space, calculation ability, and data transmission ability of the logging instruments has been quite limited (Wang et al. 2010; Wu et al. 2009; Zhang et al. 2007), so it is practically meaningful to find a new downhole data acquisition mode by which relatively complete IP relaxation process can be represented with less acquired data.

The conventional sampling method is UTSM which is based on equal time interval sampling. Although the amount of data can be reduced by increasing the sampling interval, it will lose high-frequency components which are used to accurately describe the fast change of polarizability. If the sampling interval is too short, the amount of data will be too large, and it will overload the data flow of the logging instrument. So UTSM is not practical in the applications of in situ well logging. LTSM takes sampling points based on logarithmic time interval, and can match the attenuation characteristics of polarization curve, so it greatly reduces the amount of sampling data. However, since the lithology of actual rock strata is complex, the length of IP relaxation process changes greatly. The LTSM can only be set to a fixed sampling length, which leads to the difficulty of improving the vertical resolution. This paper presents a new way of sampling—UASM, based on equal amplitude interval sampling, which is expected to describe the polarization decay curve accurately with less sampling data. We compare the three methods in terms of the adaptability of the downhole instrument to different pore sizes and distributions of strata, the relationship between number of sampling points and the inversion error, and the vertical resolution to different lithology of strata. Lastly, we give the most suitable number of sampling points by using UASM.

2 The IP RTS and uniform amplitude sampling

2.1 IP discrete equations description and the RTS

We assume that the porous medium has the same pore diameter, then the measured IP decay potential will satisfy a single exponential decay law (Tong and Tao 2007; Liu et al. 2015). The corresponding relationship between the relaxation time constant and the pore diameter is as follows:

$$T = \frac{\Phi^2}{D} \tag{1}$$

where T is the relaxation time constant, Φ is the pore diameter, and D is the diffusion coefficient of ions in the pore fluid. Equation (1) shows that the relaxation time constant and the pore diameter have one-to-one correspondence.

However, the actual strata contain a series of pores with different diameters. Therefore, the actual relaxation process represents a superposition of a series of relaxation processes with different time constants. The decay potential $V(t)$ is composed of a series of exponential functions with different relaxation time constants. In order to eliminate the influence of the exciting current on the

measurement, the ratio of decay potential $V(t)$ measured after supply current is off and the potential V_p measured when polarization process approaches stability, is regarded as the polarizability decay curve, and it can be described as follows:

$$\eta(t) = \frac{V(t)}{V_p} = \int_{T_{\min}}^{T_{\max}} f(T) \exp\left(-\frac{t}{T}\right) dT \tag{2}$$

The corresponding discrete matrix equation is

$$\eta_{m \times 1} = J_{m \times n} f_{n \times 1} \quad i = 1, 2, \ldots, m, \quad j = 1, 2, \ldots, n \tag{3}$$

where, $J_{ij} = \exp\left(-\frac{t_i}{T_j}\right)$, J is the $m \times n$ dimensional coefficient matrix; t_i is the time moment of measuring the IP decay signal; T_j is the relaxation time constant; η_i is the IP decay signal at the time moment t_i; f_j is the weight coefficient corresponding to T_j.

There is no exact solution for the discrete matrix equations, but we can use a damped least square method to obtain an approximate solution f which makes the error $\|Jf - \eta\|^2$ the smallest. In order to obtain the best least-squares solution, the regularization method is used to seek the minimum value of $\|Jf - \eta\|^2 + \alpha^2 \|f\|^2$. Assume

$$L = \|Jf - \eta\|^2 + \alpha^2 \|f\|^2 = \sum_{i=1}^{m} \left[\sum_{j=1}^{n} (J_{ij} f_j) - \eta_i \right]^2 + \alpha^2 \sum_{j=1}^{n} f_j^2 \tag{4}$$

where α^2 is a damping factor, which is used to control the weight of regularization term $\|f\|^2$ to the fitting term $\|Jf - \eta\|^2$. Then calculate the extreme value point of Eq. (4), let

$$\frac{\partial L}{\partial f_k} = 2 \sum_{i=1}^{m} \left[J_{ik} \sum_{j=1}^{n} (J_{ij} f_j) \right] - 2 \sum_{i=1}^{m} (J_{ik} \eta_i) + 2\alpha^2 f_k = 0,$$
$$k = 1, 2, \ldots, n \tag{5}$$

After reorganization, we can get

$$(J^T J + \alpha^2 I) f = J^T \eta \tag{6}$$

Then the least squares solution can be obtained as

$$f = (J^T J + \alpha^2 I)^{-1} J^T \eta \tag{7}$$

Because f_j cannot be negative, we put non-negative constraints on Eq. (6). Calculate repeatedly until the solution meets the condition $f_j \geq 0, j = 1, 2, \ldots, n$. The specific calculation steps are as follows:

Step 1 Calculate the initial solution f^k $(k = 0)$ according to Eq. (7);
Step 2 Calculate the polarizability $\eta' = Jf^k$ according to Eq. (3);

Step 3 Calculate the polarizability bias $\Delta\eta = \eta - \eta'$;

Step 4 Obtain the solution's correction Δf with $\Delta\eta$ according to Eq. (7);

Step 5 Correct the spectral components according to the equation $f^{k+1} = f^k + \Delta f$;

Step 6 If every component of the solution f^{k+1} meets the constraint $f_j \geq 0$, then f^{k+1} is the expected solution. Otherwise, set the negative component to zero, then let $\eta = \eta'$, and then repeat step 2 to step 6 until the solution f^{k+1} meets the requirement or the k reaches the set iteration number.

2.2 Uniform amplitude sampling

The IP potential curve, after spontaneous potential compensation, is shown in Fig. 1, where I_0 is the supplied current, V_p^+ is the potential when the IP potential excited by positive current achieves a stable state, $V_t^+(t)$ is the IP decay potential after the positive current is switched off. V_p^- is the potential when the IP process excited by negative current achieves stability, $V_t^-(t)$ is the IP decay potential after the negative current is switched off.

In the beginning after the current is switched off, the IP potential curve usually has a high decay speed and as time goes on the decay speed slows down. If we select the UTSM, it will be very difficult to solve the contradiction between the accuracy of the IP decay curve and the small number of sampling points. Since the UASM can track the change of decay voltage as shown in Fig. 2c, it can provide better accuracy with small sample numbers. Although LTSM matches the decay curve, it takes a longer sampling time compared with UASM as shown in Fig. 2b, c.

The polarization decay curve is drawn again in Fig. 3 with the timeline in logarithmic coordinates. If the number of sampling points is m, then the normalized equal-interval amplitude scale is obtained as $\{A_i\} = \left(1, \frac{m-1}{m}, \ldots, \frac{m-i+1}{m}, \ldots, \frac{2}{m}, \frac{1}{m}\right)$, $i = 1, 2, \ldots, m$. And then the time signal

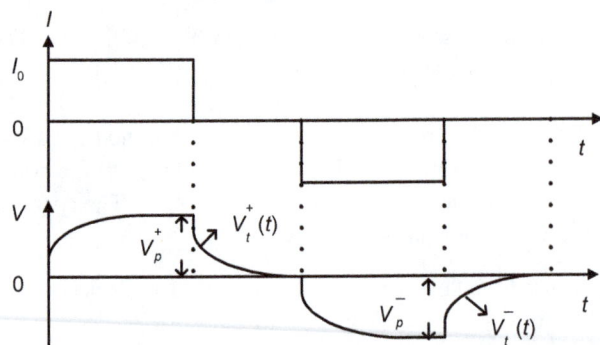

Fig. 2 Sampling methods for the polarization decay curve. **a** Uniform time sampling, **b** Uniform logarithm sampling, **c** Uniform amplitude sampling

sequence according to the corresponding amplitude scale can be determined as shown in Fig. 3.

Since in the actual measuring process, there is disturbance at the moment when supply current is switched off, the logging instrument usually begins to acquire polarization decay signals after a short period of pause. The A/D convertor of the instrument works at a fixed high frequency, but not all the sampled signal data will be saved.

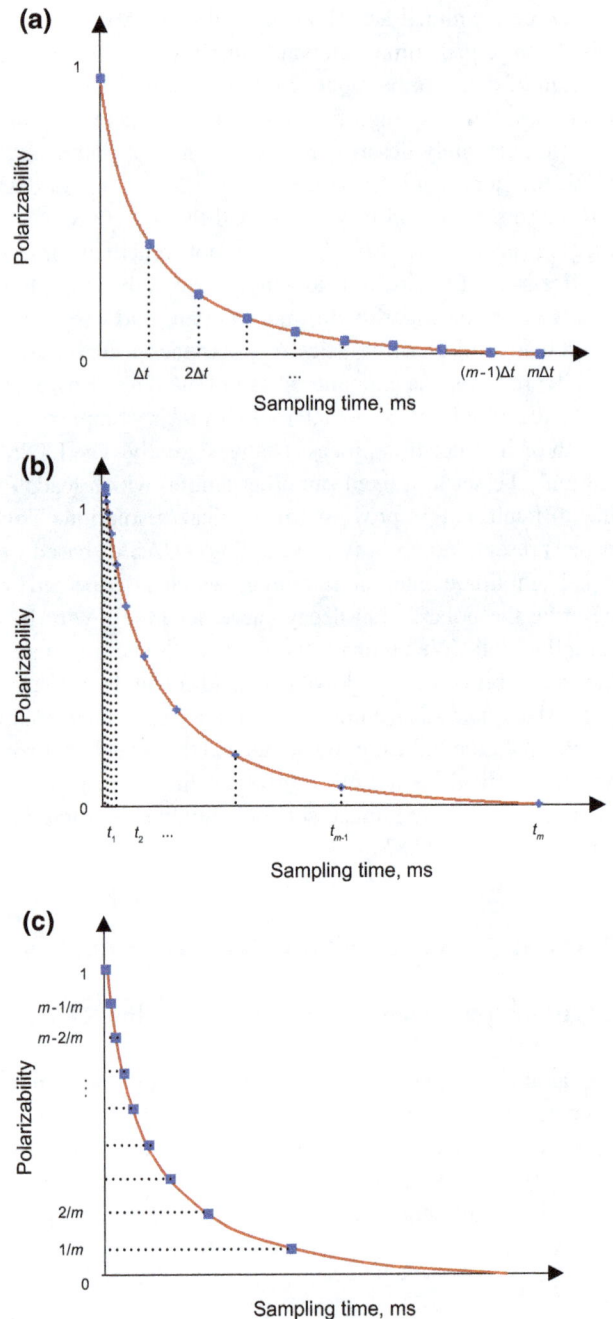

Fig. 1 Induced polarization potential curve

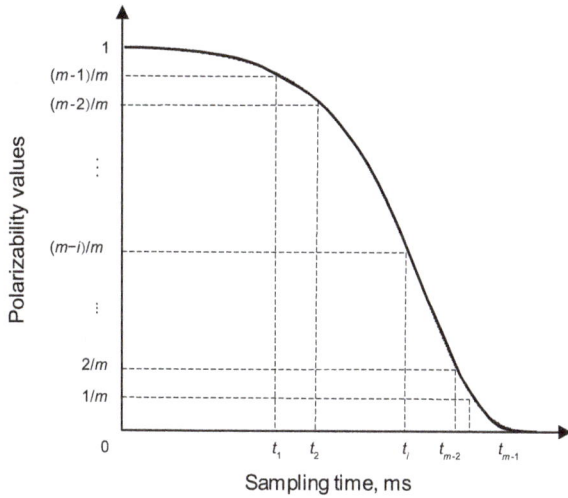

Fig. 3 Uniform amplitude sampling of the normalized polarizability curve

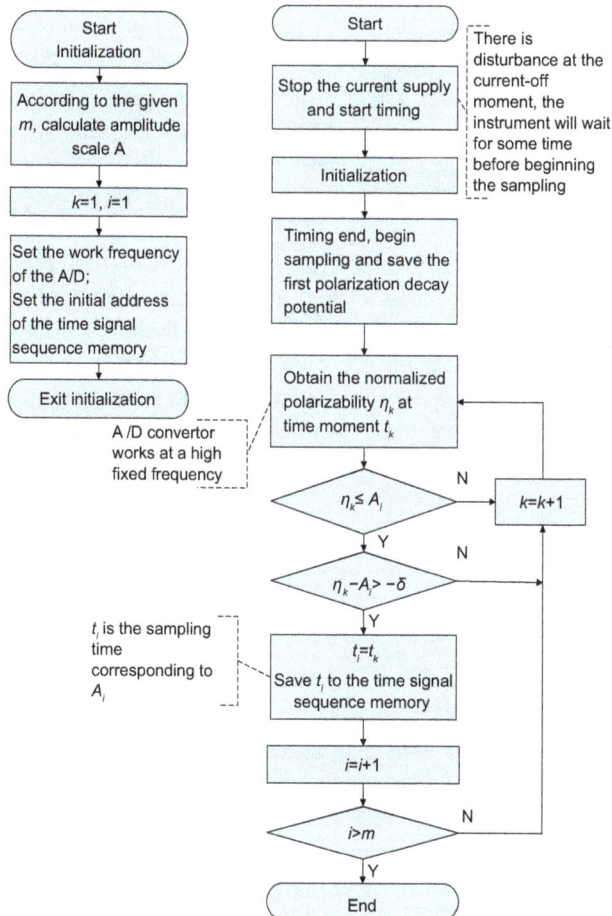

Fig. 4 Operation flow diagram of uniform amplitude sampling

Suppose the first time moment of A/D conversion of the IP decay potential be the zero of the time signal. The first A/D converted value is saved in the register, and is set as a standard for normalizing each follow-up A/D conversion value. We compare the sampled value of normalized polarization decay potential with amplitude scale value A_i. If the sampled value is bigger than A_i, the sampling time will not be saved, and the measurement process goes into the next A/D conversion cycle. If the sampled value is not bigger than A_i, and meets the condition $\eta_k - A_i > -\delta$, the sampling time is saved as the ith time signal t_i in the corresponding time signal sequence memory. If the sampled value does not meet the above condition, that value is set as interference, then this sampling time is not saved, and the measurement process goes into the next A/D conversion cycle. The process goes on until all the time signal values corresponding to the m amplitude scale values are recorded in turn. The specific uniform amplitude sampling operation flow diagram is shown in Fig. 4.

As the UASM is very sensitive to the kind of disturbance which causes the amplitude to decrease sharply, while the actual normalized polarizability signal between two consecutive A/D values is far less than $\frac{1}{m}$, we set a threshold $\delta < \frac{1}{m}$. Only when the normalized polarizability value η_k meets the condition of $-\delta < \eta_k - A_i \leq 0$, the sampling time is regarded as time signal t_i corresponding to A_i, and it is then saved. Setting of this condition can completely inhibit the disturbance which led to a decrease of the amplitude greater than δ in the sampling process.

3 The simulation

3.1 The comparison of the three sampling methods

Since the lithology of actual rock strata is complex, and the rock pore sizes of different regions are very different (Yang et al. 2014; Zhu et al. 2014) even in the same well, the rock pore sizes of different depths may also be very different (Tan et al. 2010; Shou and Zhu 1998). We hope that the IP logging instrument can adapt to a wide range of strata with different pore diameters, and obtain higher inversion accuracy of the polarization relaxation time spectrum from a smaller number of sampling points. In order to compare the relative merits of the three sampling methods, we designed several relaxation spectrum models which meet the logarithm normal distribution, and do research on sampling of the polarization curve and on inverting the polarization relaxation time spectra of these models.

3.1.1 Single-peak model

Two kinds of single-peak relaxation time spectra are set as shown in Fig. 5. The relaxation time constants of model A are small, corresponding to the small diameters of the rock

Fig. 6 Polarization decay curves of model A and model B

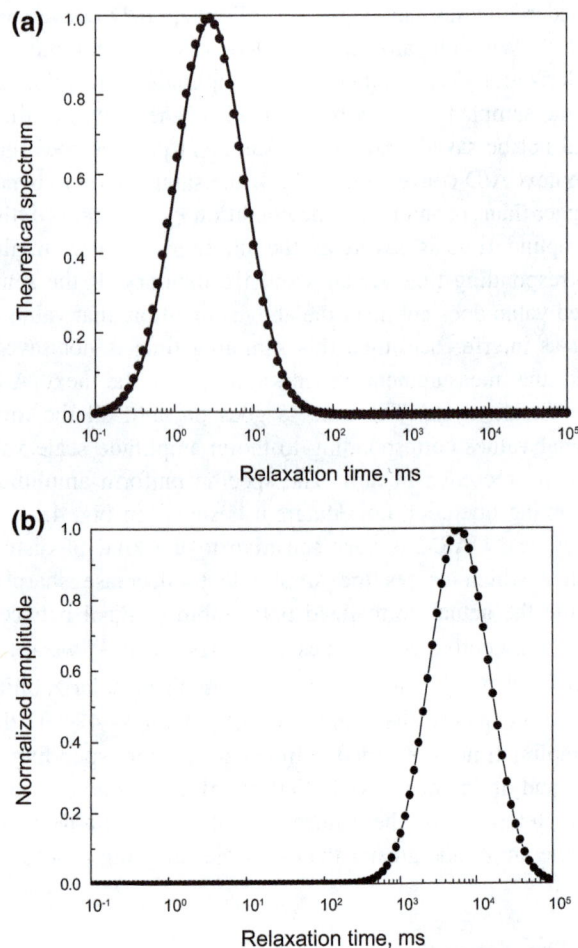

Fig. 5 The models of single-peak relaxation spectra. **a** Model A, **b** Model B

pores. The relaxation time constants of model B are large, corresponding to the large diameters of the rock pores. Polarization decay curves of two models are shown in Fig. 6, indicating that the curve of model A decays fast and the curve of model B decays slowly.

Assume that the sampling lengths of the two corresponding decay curves are both 100,000 ms, and the numbers of sampling points are set as 30, 60, 100, 200, 300, respectively; the three sampling methods will be used to sample the two corresponding decay curves. Suppose the relaxation time constants following a uniform logarithmic distribution between 0.1 ms and 100,000 ms be 100 for the above five situations, the relaxation time spectra can be inverted as shown in Figs. 7 and 8 with the nonnegative damping least squares method.

The root mean square error (RMSE) of inversion varies with the number of sampling points as shown in Tables 1 and 2.

The logging instrument usually rises from the well bottom to the surface at a constant speed when measuring

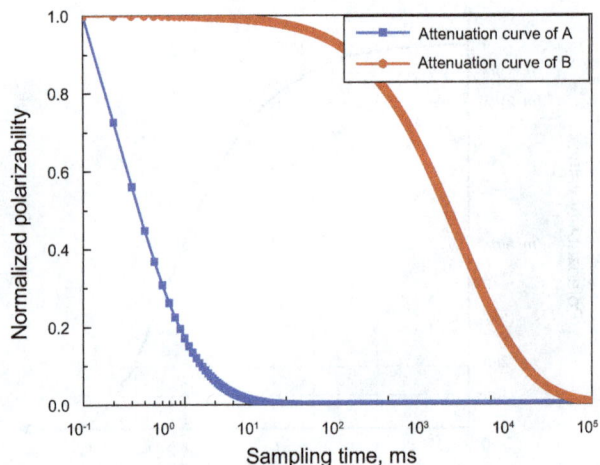

strata, thus the shorter the sampling time is, the thinner the corresponding stratum is, and the thinner the measured stratum is, the higher the vertical resolution will be. When using UTSM and LTSM, the sampling time is set as 100,000 ms. However when using UASM, the data acquisition process ends once the last time point is recorded. The actual sampling time of UASM is shown in Table 3.

Figures 7, 8 and Tables 1, 2 show that, with the UTSM, the inversion accuracy of RTS is high only when the pore diameter of the rock formation is big. With the LTSM and UASM, the inversion accuracy of RTS is high with both big and small pore diameters of rock formation, and meanwhile the required accuracy can be approached with less sampling points. Table 3 shows that the UASM can track the amplitude variety of the polarizability curve very well, and thus it has the best vertical resolution among the three methods.

3.1.2 Double-peak model

In order to verify whether the above result is universal, we simulated the double-peak spectrum models C, D, and E over a wide range of pore diameter distribution. The models are shown in Fig. 9, where model C has more small pores and model D has more large pores, the high peak amplitude is about twice as high as the low peak amplitude. Model E has the same height of peak amplitude, but its pore size distribution is not continuous and the mean size values of the peaks are very different.

The simulation conditions are the same as single-peak models, and the simulation results are shown in Figs. 10, 11, and 12. The inversion errors are shown in Tables 4, 5,

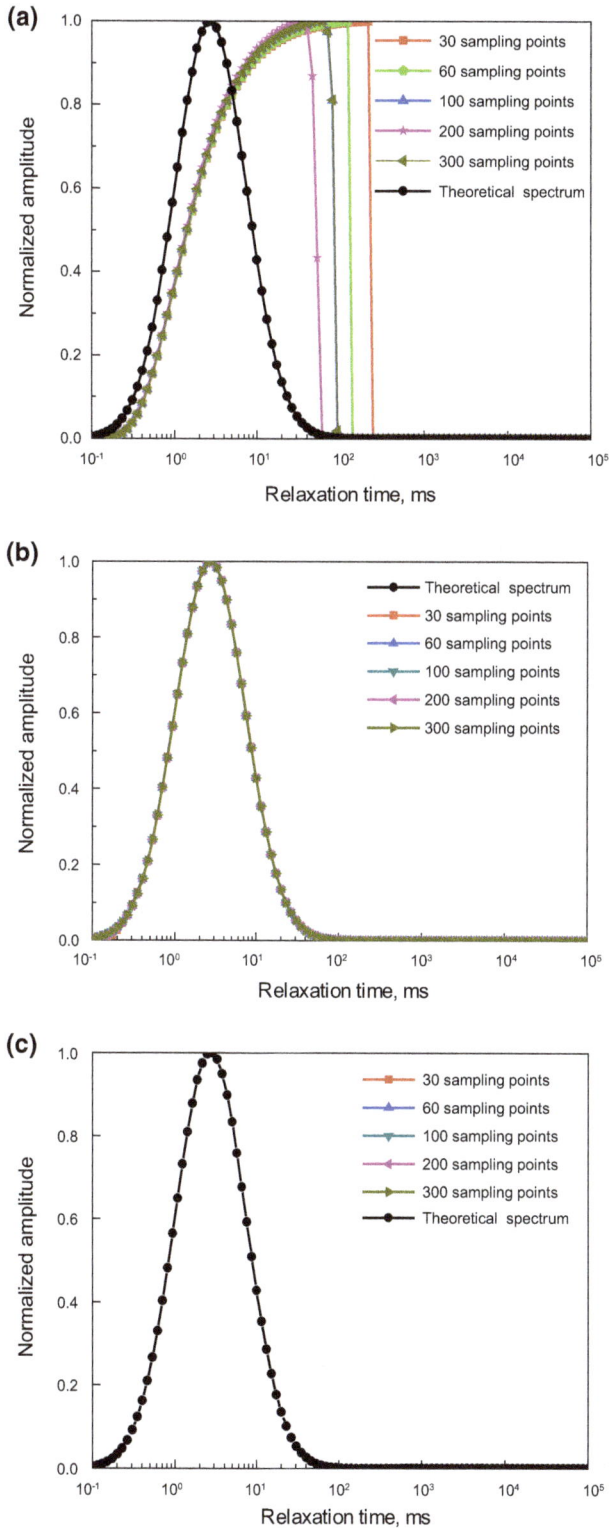

Fig. 7 The inverted relaxation time spectra of model A. **a** The relaxation time spectrum with the UTSM, **b** The relaxation time spectrum with the LTSM, **c** The relaxation time spectrum with the UASM

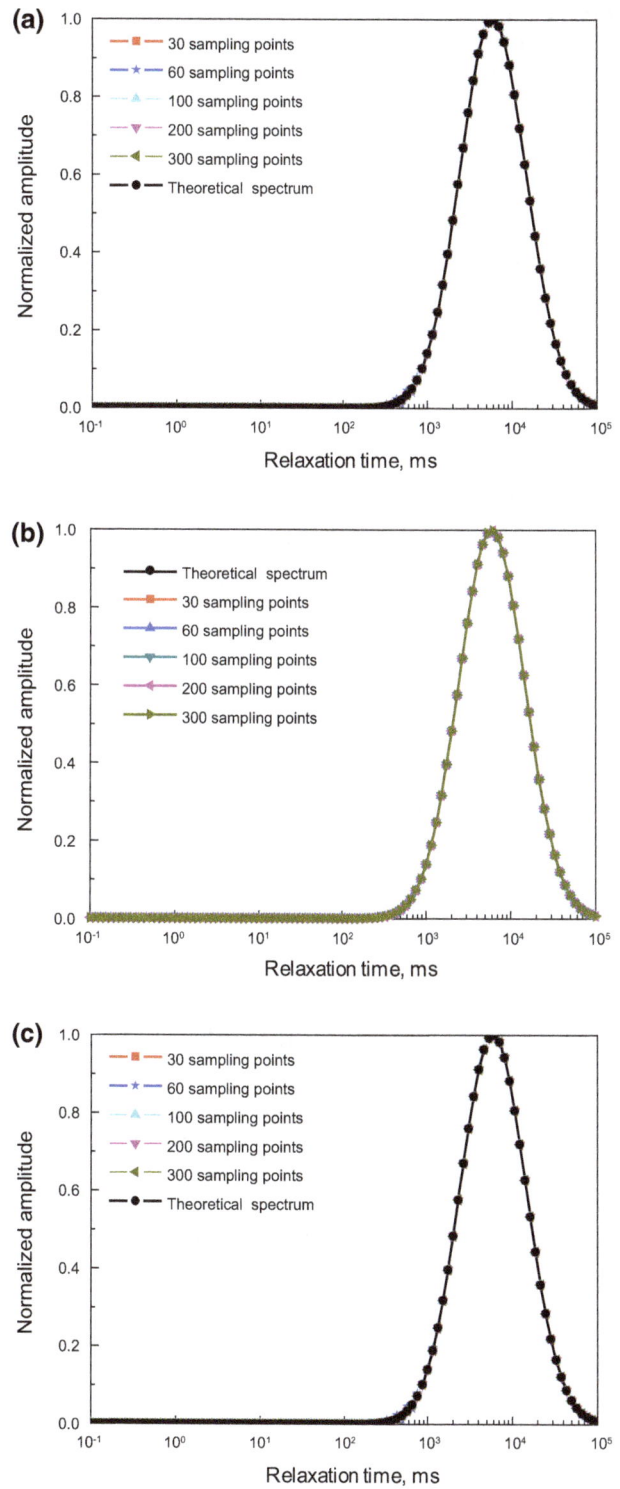

Fig. 8 The inverted relaxation time spectra of model B. **a** The relaxation time spectrum with the UTSM, **b** The relaxation time spectrum with the LTSM, **c** The relaxation time spectrum with the UASM

Table 1 The RMSE of model A's inversion with three sampling methods

Sampling points	30	60	100	200	300
UTSM	0.292	0.28	0.266	0.245	0.23
LTSM	2.17×10^{-3}	1.76×10^{-4}	3.19×10^{-4}	2.19×10^{-4}	3.18×10^{-5}
UASM	3.1×10^{-4}	1.1×10^{-4}	4.5×10^{-5}	2.5×10^{-4}	7×10^{-5}

Table 2 The RMSE of model B's inversion with three sampling methods

Sampling points	30	60	100	200	300
UTSM	2.1×10^{-3}	7.8×10^{-4}	2.6×10^{-4}	1.3×10^{-4}	1.5×10^{-4}
LTSM	3.33×10^{-5}	5.24×10^{-6}	1.6×10^{-4}	2.18×10^{-4}	1.13×10^{-4}
UASM	1.4×10^{-4}	1.6×10^{-4}	1.25×10^{-4}	1.8×10^{-4}	1.6×10^{-4}

Table 3 Sampling time of model A and model B with UASM

Sampling points	30	60	100	200	300
Model A, ms	38	52	63	70	77
Model B, ms	54,298	74,339	87,897	98,373	84,532

and 6. The actual sampling time of model C, D, and E with UASM is shown in Table 7.

Figures 10, 11, 12, and Tables 4, 5, and 6 show that there exists a large deviation between the inverted spectrum and the original theoretical spectrum using the UTSM. The deviation is especially larger for models C and E which have more small pores. However, by using the LTSM and the UASM, the inversion results are in good agreement with the theoretical spectrum for all three models. Table 7 shows that the actual sampling time decreases a lot when using UASM, which means that the logging instrument can measure more samples and improve the vertical resolution.

The simulation results of the double-peak spectrum are consistent with those of the single-peak spectrum, which can prove that the results in Sect. 3.1.1 have a certain universality. All the above simulation results of the rock formation models show that the UASM is better than the UTSM and the LTSM.

3.2 Determine the best number of sampling points when using UASM

We hope to get a qualified RTS with few sampling points. It can be found in the simulation in Sect. 3.1 that with the UASM, when the sampling point number is more than 100,

the inversion accuracy will not increase significantly. Therefore, the sampling point numbers are set as 15, 30, 40, 50, 60, and 80, respectively, in the following simulations. Since the actual pore size of strata mostly follows a multi-peak distribution, then a double-peak model F, a three-peak model G, and a four-peak model H are set as the simulation objects. The inverted spectra are shown in Fig. 13.

It is difficult to determine which number of sampling points is the best by observing the inverted spectra, so we calculate the RMSEs of the inverted spectra with different numbers of sampling points, and the results are shown in Fig. 14.

Figure 14 shows that the inversion error decreases slowly when the number of sampling points is more than 30, and the inversion error may also fluctuate with the increase of the number of sampling points. Meanwhile, overmuch sampling will increase the data transfer burden and inversion computation time. So in summary, with the UASM, the appropriate range of sampling points to get the IP decay curve is 30–60.

4 Experiment

Rock samples were obtained from a Chinese oilfield, and the sorting characteristics of clastic particles inside the rock samples are good or medium. Rock pores are mainly composed of intergranular dissolved pores, accounting for about 90 %, followed by primary intergranular pores and intragranular dissolved pores. The pore distributions are mainly double-peak or multimodal. For a double-peak sample A and a three-peak sample B, their porosity values are 30.2 % and 24.5 %, and the

(a)

(b)

(c)

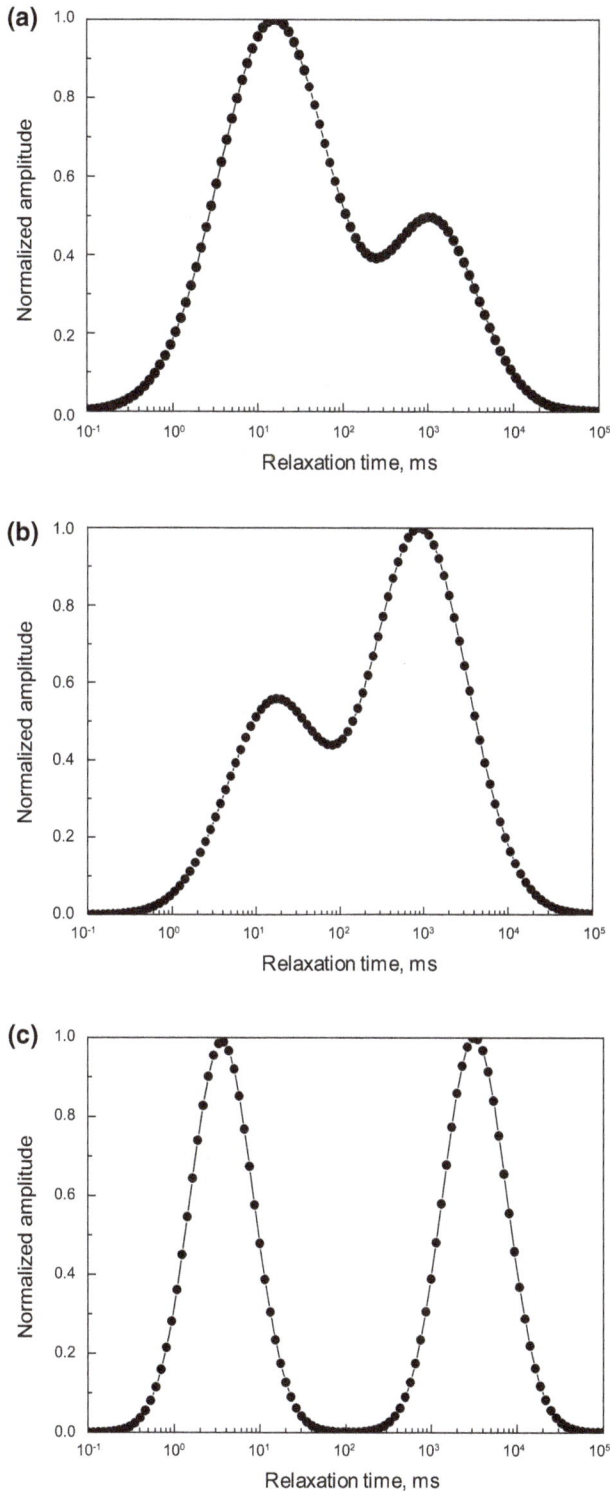

Fig. 9 Theoretical double-peak relaxation time spectrum models. **a** Model C, **b** Model D, **c** Model E

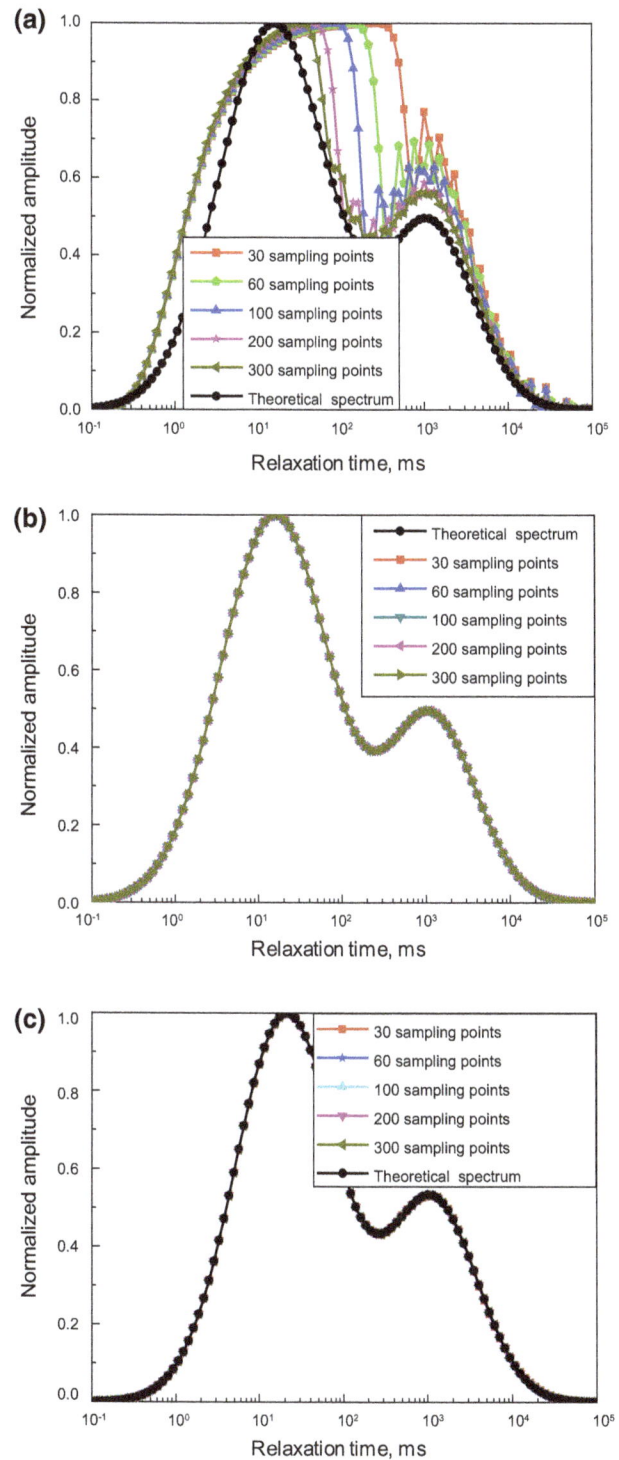

Fig. 10 The inverted relaxation time spectrum of Model C. **a** The relaxation time spectrum with the UTSM, **b** The relaxation time spectrum with the LTSM, **c** The relaxation time spectrum with the UASM

permeability values are 430×10^{-3} μm^2 and 69.4×10^{-3} μm^2. The pore distributions of casting thin sections which were made from the rock samples were recorded and are shown in Fig. 15, where the ordinate indicates the volume ratio of pores with a certain diameter to the total pores.

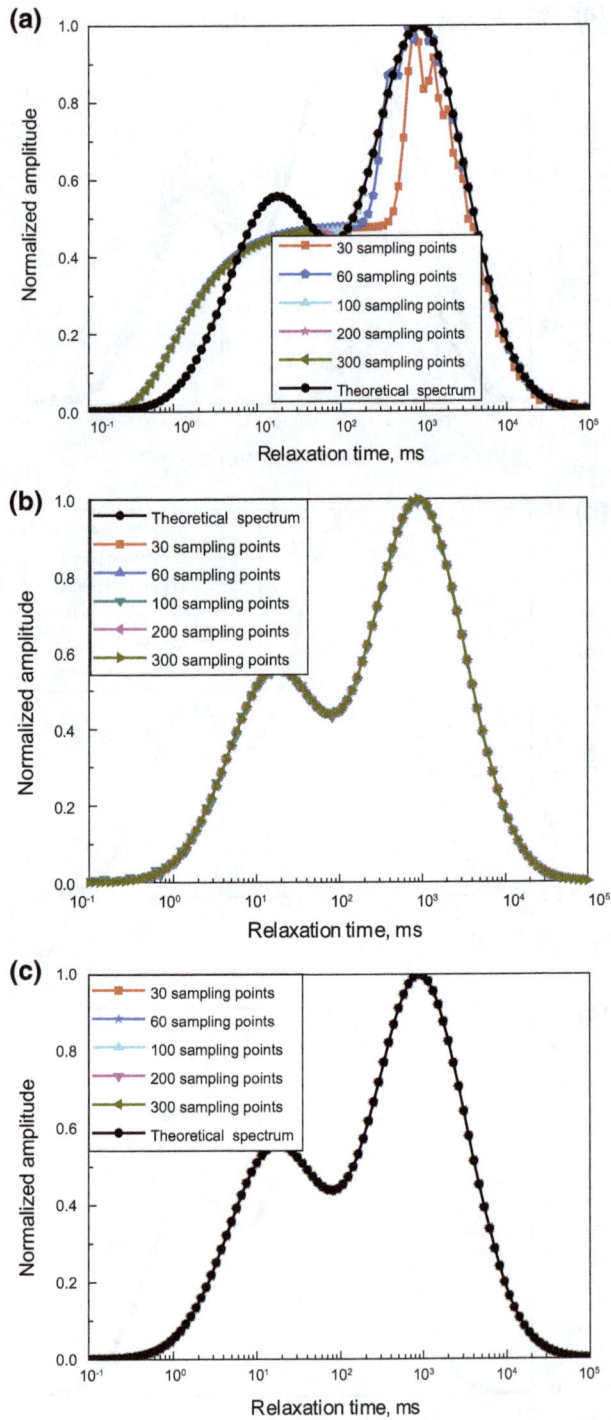

Fig. 11 The inverted relaxation time spectrum of Model D. **a** The relaxation time spectrum with the UTSM, **b** The relaxation time spectrum with the LTSM, **c** The relaxation time spectrum with the UASM

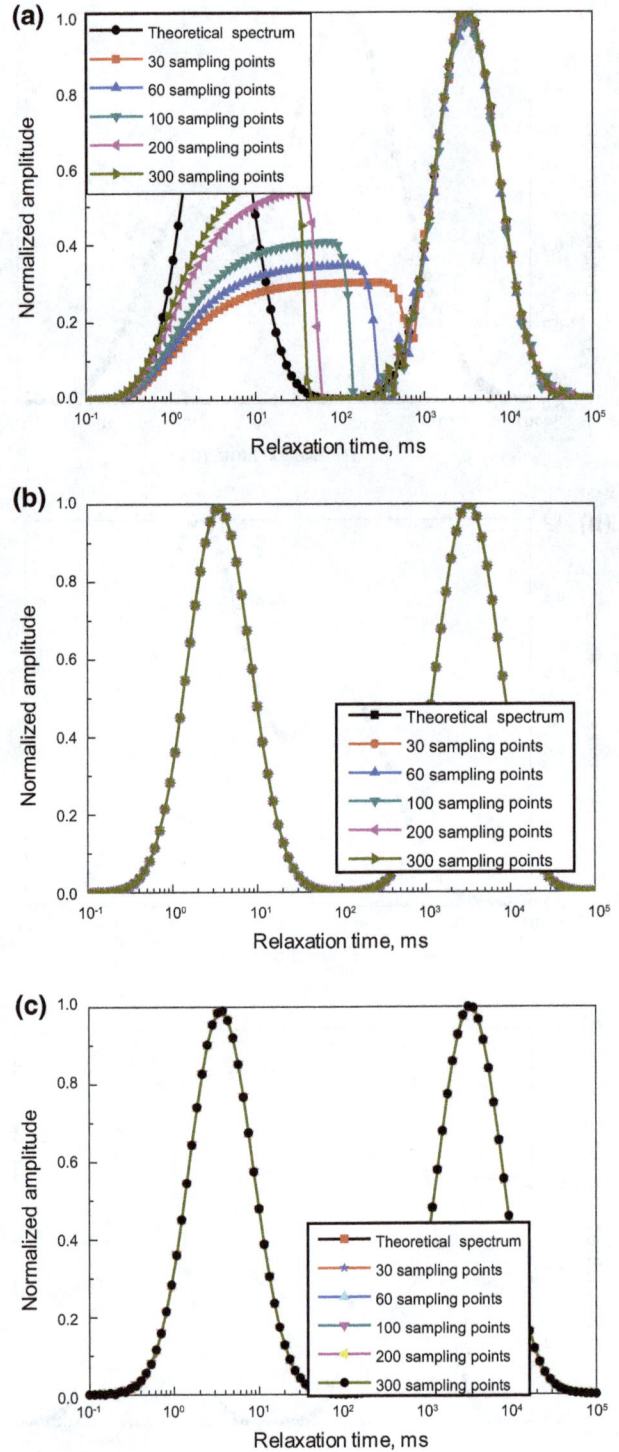

Fig. 12 The inverted relaxation time spectrum of Model E. **a** The relaxation time spectrum with the UTSM, **b** The relaxation time spectrum with the LTSM, **c** The relaxation time spectrum with the UASM

The pore diameter distribution can be transformed into RTS by Eq. (1), and based on it, the actual effect of the three sampling methods can be judged. We select 100

relaxation time constants following an even logarithmic distribution between 0.1 ms and 100,000 ms, and then set the length of sampling time as 100,000 ms and the number

Table 4 The RMSE of Model C's inversion with three methods

Sampling points	30	60	100	200	300
UTSM	0.15	0.129	0.1	0.053	0.038
LTSM	1.13×10^{-4}	1.58×10^{-5}	4.26×10^{-4}	4.53×10^{-4}	7.55×10^{-4}
UASM	1.9×10^{-3}	6.2×10^{-4}	3.8×10^{-4}	4.4×10^{-4}	5.8×10^{-4}

Table 5 The RMSE of model D's inversion with three methods

Sampling points	30	60	100	200	300
UTSM	0.091	0.078	0.055	0.039	0.031
LTSM	1.13×10^{-3}	1.65×10^{-4}	1.34×10^{-3}	4.3×10^{-4}	2.1×10^{-4}
UASM	8.5×10^{-4}	5.9×10^{-4}	7.5×10^{-4}	2.3×10^{-4}	1.8×10^{-4}

Table 6 The RMSE of model E's inversion with three methods

Sampling points	30	60	100	200	300
UTSM	0.271	0.262	0.248	0.215	0.195
LTSM	9.56×10^{-5}	5.61×10^{-6}	1.35×10^{-3}	1.25×10^{-4}	3.7×10^{-4}
UASM	1.31×10^{-3}	5.11×10^{-4}	1.39×10^{-3}	7.8×10^{-4}	5.1×10^{-4}

Table 7 Sampling time of model C, D, and E with UASM

Sampling points	30	60	100	200	300
Model C, ms	12,259	19,121	24,250	28,462	32,087
Model D, ms	8263	13,359	17,184	20,332	23,044
Model E, ms	21,193	29,715	35,548	40,097	43,872

of sampling points as 50. The two rock samples were saturated with 3000 mg/L NaCl solution. We cut the power off after sufficient power supply for the rock samples, and then the polarizability decay signal was obtained respectively with the three sampling methods. The inverted spectra are shown in Fig. 16. The sampling time of the three methods is shown in Table 8.

Figure 16 shows clearly, with UTSM the inverted spectra have a large deviation with actual relaxation time spectra of the two samples. With LTSM and UASM, the inverted spectra are essentially coincident with actual spectra. Table 8 shows that sampling by using UASM takes only about half of the time lengths of the other two methods, so the vertical resolution can be improved effectively.

5 Conclusions

UASM can adapt well to measuring the strata with different kinds of pore size distribution. It has a good dynamic range for measuring both small pore size strata with fast decay IP voltage and large pore size strata with slow decay IP voltage with high RTS accuracy. A small number of sampling points from 30 to 60 are sufficient enough to get high RTS inversion accuracy thus the data traffic of the logging instrument is reduced effectively. UASM can

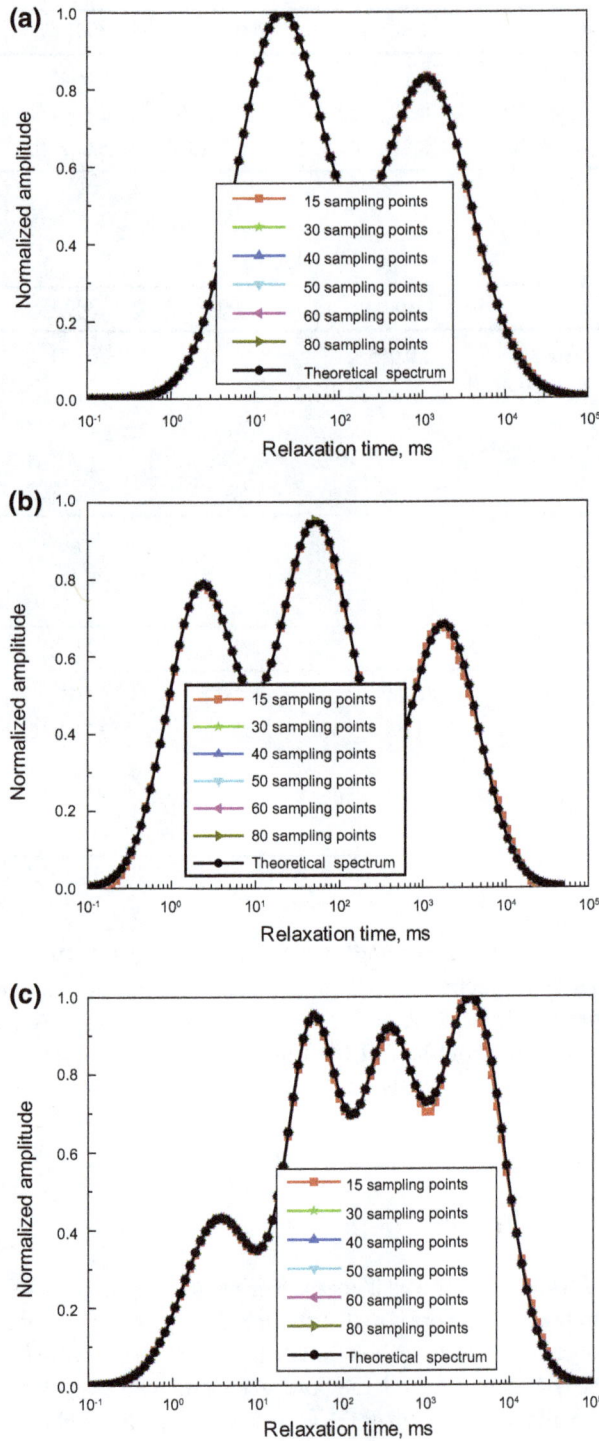

Fig. 13 The inverted spectra of the three models. **a** Model F, **b** Model G, **c** Model H

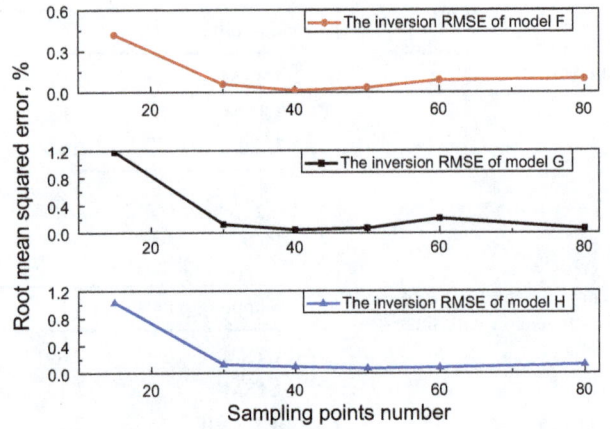

Fig. 14 The relationship between inversion RMSEs and sampling point numbers

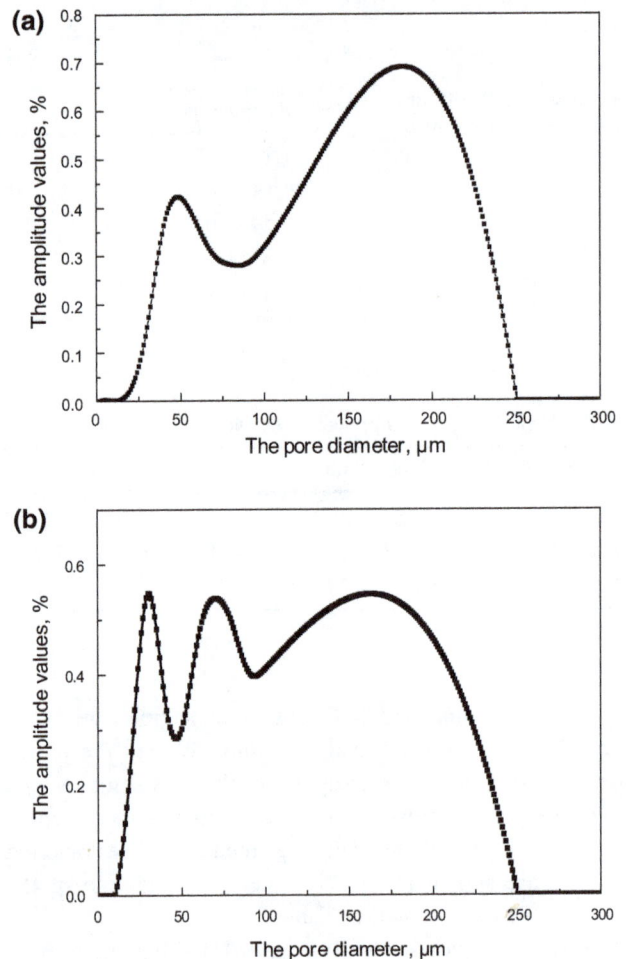

Fig. 15 The pore diameter distribution curve of the samples. **a** Sample A, **b** Sample B

automatically adapt to the attenuation speed of the polarization curve, and the actual sampling time changes along with the rock pore size, which is helpful to improve the vertical resolution of the strata measurement. The simulation and experimental results show that the UASM is superior to the UTSM and LTSM.

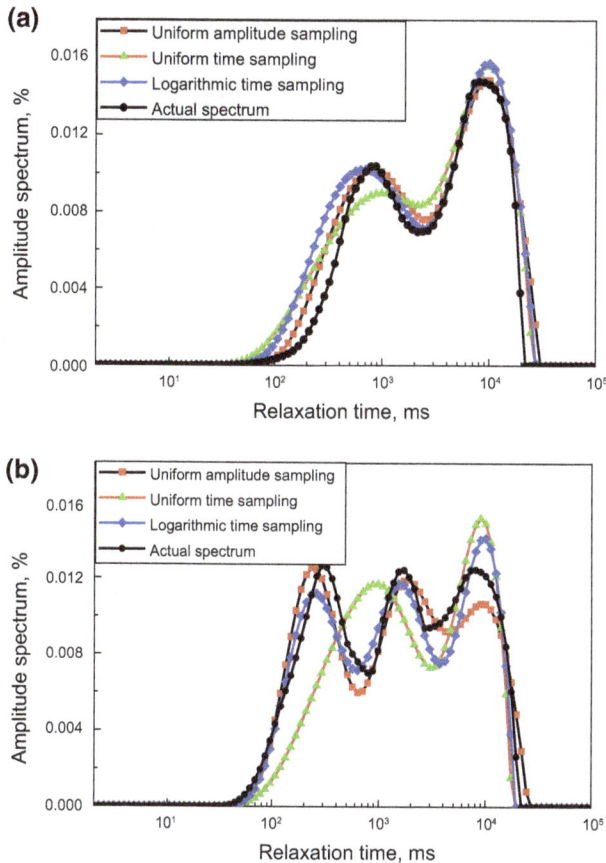

Fig. 16 The inverted relaxation time spectra of rock samples. **a** Rock sample A, **b** Rock sample B

Table 8 Sampling time of rock samples with the three sampling methods

Sampling time	UTSM	LTSM	UASM
Sample A, ms	100,000	100,000	58,802
Sample B, ms	100,000	100,000	56,720

Acknowledgments This work was partially supported by a project from the National Natural Science Foundation of China (No. 61401168).

References

Beckett CTS, Augarde CE. Prediction of soil water retention properties using pore-size distribution and porosity. Can Geotech J. 2013;50(4):435–50.

Binley A, Slater LD, Fukes M, et al. Relationship between spectral induced polarization and hydraulic properties of saturated and unsaturated sandstone. Water Resour Res. 2005;41(12):W12417.

Buecker M, Hoerdt A. Long and short narrow pore models for membrane polarization. Geophysics. 2013;78(6):299–314.

Chen SS, Wang HZ, Zhang XL. Study on multi-exponential inversion method for NMR relaxation signals with Tikhonov regularization. Engineering. 2013;5(1):32–7.

Florsch N, Revil A, Camerlynck C. Inversion of generalized relaxation time distributions with optimized damping parameter. J Appl Geophys. 2014;109:119–32.

Guan JT, Wang Q, Fan YH, et al. Study on the mechanisms of electrochemical logging response in shaly sandstone based on capillary model. Chin J Geophys. 2010;53(1):214–23 (in Chinese).

Gurin G, Tarasov A, Ilyin Y, et al. Time domain spectral induced polarization of disseminated electronic conductors: Laboratory data analysis through the Debye decomposition approach. J Appl Geophys. 2013;98:44–53.

Jang H, Park S, Kim HJ. A simple inversion of induced-polarization data collected in the Haenam area of Korea. J Geophys Eng. 2014;11(1):015011.

Li CL, Zhou CC, Li X, et al. A novel model for assessing the pore structure of tight sands and its application. Appl Geophys. 2010;7(3):283–91.

Li Y, Lin PR, Xiao Y, et al. Induced polarization effect on frequency domain electromagnetic sounding with electric dipole source. Chin J Geophys. 2011;54(7):1935–44 (in Chinese).

Liu HQ, Yan J, Deng YM. Study of the properties of non-gas dielectric capacitors in porous media. Pet Sci. 2015;12(1):104–13.

Liu XN, Kong L, Zhang P, et al. Permeability estimation using relaxation time spectra derived from differential evolution inversion. J Geophys Eng. 2014a;11(1):1–8.

Liu XN, Kong L, Zhou KB, et al. A time domain induced polarization relaxation time spectrum inversion method based on a damping factor and residual correction. Pet Sci. 2014b;11(4):519–25.

Moody JB, Xia Y. Analysis of multi-exponential relaxation data with very short components using linear regularization. J Magn Reson. 2004;167(1):36–41.

Nie XW, Zhou AC, Yao GD, et al. Prospecting for oil and gas with induced polarization method—a discussion on the effect and anomaly model. Chin J Geophys. 1987;30(4):412–22.

Revil A, Florsch N, Camerlynck C. Spectral induced polarization porosimetry. Geophys J Int. 2014;198(2):1016–33.

Rezaee MR, Jafari A, Kazemzadeh E. Relationships between permeability porosity and pore throat size in carbonate rocks using regression analysis and neural networks. J Geophys Eng. 2006;3(4):370–6.

Shou JF, Zhu GH. Study on quantitative prediction of porosity preservation in sandstone reservoirs. Chin J Geol (Sci Geol Sin). 1998;32(2):244–50 (in Chinese).

Tan FQ, Li HQ, Xu CF, et al. Quantitative evaluation methods for waterflooded layers of conglomerate reservoir based on well logging data. Pet Sci. 2010;7(4):485–93.

Titov K, Komarov V, Tarasova V, et al. Theoretical and experimental study of time domain-induced polarization in water-saturated sands. J Appl Geophys. 2002;50(4):417–33.

Titov K, Tarasov A, Ilyin Y, et al. Relationships between induced polarization relaxation time and hydraulic properties of sandstone. Geophys J Int. 2010;180(3):1095–106.

Tong MS, Li L, Wang W, et al. Estimation of pore size distribution and permeability of shaly sands from induced polarization time spectra. Chin J Geophys. 2005;48(3):785–91 (in Chinese).

Tong MS, Tao HG. Experimental study of induced polarization relaxation time spectra of shaly sands. J Pet Sci Eng. 2007;59(3–4):239–49.

Tong MS, Wang WN, Jiang YZ, et al. Estimation of permeability of shaly sand reservoir from induced polarization relaxation time spectra. J Pet Sci Eng. 2004;45(1–2):31–40.

Wang CJ. Induced polarization potential logging technique. Spec Oil Gas Reservoirs. 2004; 11(4): 57–9, 66 (**in Chinese**).

Wang WN, Yu QF, Jian YZ, et al. Induced polarization potential attenuation spectrum well logging continuous recording method. OGP. 2010;45(Supplement 1):214–6 (**in Chinese**).

Wang WN, Yu QF, Tong MS, et al. Electrode configuration design of induced polarization potential decay spectrum logging. Progr Geophys. 2011;26(1):371–5 (**in Chinese**).

Wu RQ, Chen W, Chen TQ, et al. A time-driven transmission method for well logging networks. Pet Sci. 2009;6(3):239–45.

Xiao LZ, Zhang HR, Liao GZ, et al. Inversion of NMR relaxation in porous media based on Backus-Gilbert theory. Chin J Geophys. 2012;55(11):3821–8 (**in Chinese**).

Yang RF, Wang Y, Cao J. Cretaceous source rocks and associated oil and gas resources in the world and China: a review. Pet Sci. 2014;11(3):331–45.

Zhang W, Shi YB, Tang JY. Application of an improved ADSL technique in the high-speed wireline logging telemetry system. ICEMI. 2007;2:108–12.

Zhu XM, Zhu SF, Xian BZ, et al. Reservoir differences and formation mechanisms in the Ke-Bai overthrust belt, northwestern margin of the Junggar Basin, China. Pet Sci. 2014;7(1):40–8.

Zimmermann E, Kemna A, Berwix J, et al. A high-accuracy impedance spectrometer for measuring sediments with low polarizability. Meas Sci Technol. 2008;19(10):105603.

13

Numerical analysis of rock fracturing by gas pressure using the extended finite element method

Majid Goodarzi[1] · Soheil Mohammadi[2] · Ahmad Jafari[3]

Abstract High energy gas fracturing is a simple approach of applying high pressure gas to stimulate wells by generating several radial cracks without creating any other damages to the wells. In this paper, a numerical algorithm is proposed to quantitatively simulate propagation of these fractures around a pressurized hole as a quasi-static phenomenon. The gas flow through the cracks is assumed as a one-dimensional transient flow, governed by equations of conservation of mass and momentum. The fractured medium is modeled with the extended finite element method, and the stress intensity factor is calculated by the simple, though sufficiently accurate, displacement extrapolation method. To evaluate the proposed algorithm, two field tests are simulated and the unknown parameters are determined through calibration. Sensitivity analyses are performed on the main effective parameters. Considering that the level of uncertainty is very high in these types of engineering problems, the results show a good agreement with the experimental data. They are also consistent with the theory that the final crack length is mainly determined by the gas pressure rather than the initial crack length produced by the stress waves.

Keywords Gas fracturing · Numerical modeling · Extended finite element · Fracture mechanics

1 Introduction

High energy gas fracturing (HEGF) is a technique to stimulate wellbores by producing several radial cracks around the holes. The cracks are generated by high pressure gas produced from burning a propellant. This approach creates multiple fractures and avoids the inherent limitations of other common well stimulating techniques such as hydraulic fracturing (HF) and explosive fracturing (EF). Hydraulic fractures are generated using a fluid which needs pumping equipment on the top of the well, and the result is usually in the form of two fractures perpendicular to the minimum principle stress orientation. Explosive fracturing can also generate several fractures, but releasing a very high amount of energy in a few milliseconds may cause considerable crushing of rock and leaving a residual compressive stress zone around the wellbore. HEGF produces a higher pressure in a shorter time than HF but a significantly lower pressure in a longer time than EF, so multiple cracks can be generated without causing substantial damage to the rock structure.

Since a higher recovery is obtained by HF due to the possibility of having very long fractures, HEGF has not been accepted as the first choice for increasing the recovery. Despite the disadvantageous of short crack lengths, HEGF has its own applications and advantages: no need for special pumping equipment, low overall costs, simple and fast procedure, and the possibility of having multiple fractures without causing an extensive damage. Krilove et al. (2008) investigated the capability of this technique by applying it on petrophysical laboratory samples and inside

✉ Soheil Mohammadi
smoham@ut.ac.ir

[1] Department of Civil Engineering and Geoscience, University of Newcastle, Newcastle upon Tyne NE1 7RU, UK

[2] School of Civil Engineering, University of Tehran, Tehran 11365-4563, Iran

[3] School of Mining Engineering, University of Tehran, Tehran 515-14395, Iran

Edited by Yan-Hua Sun

production wells. They concluded that HEGF is an effective and efficient method which can increase the oil production rate by a factor of 2 to 3. It has also been experimentally observed that HEGF is rather suitable for exploratory wells or wells with natural fissures around them (Yang et al. 1992; Wu et al. 2012). In addition, this method has been successfully implemented in other applications such as enhancing the injectivity of gas injection wells (Salazar et al. 2002), prefracturing before hydraulic fracturing to reduce the friction pressure losses near the wellbore (Jaimes et al. 2012), stimulating geothermal wells (Chu et al. 1987), extracting gas from coal seams (Chao et al. 2013), etc.

The procedure of crack initiation and propagation has been comprehensively studied for blasting applications, and the role of different effects has been determined through numerous experimental and numerical investigations which will be briefly discussed in this section. According to these studies, one can conclude that a conventional blasting process has two major stages which contribute to crack propagation and rock fragmentation: (a) stress wave and (b) gas pressure. The role of the stress wave is to create initial cracks, while the gas pressure leads to crack propagation. In fact, the stress wave can only initiate limited cracking and crushing of the rock near the borehole which would not exceed more than several hole diameters (Kutter and Fairhurst 1971). Based on some field and laboratory experiments, McHugh (1983) concluded that the effect of gas pressure could be more noticeable than the effect of stress wave. The same result was confirmed by Daehnke et al. (1997). The peak pressure of propellant in HEGF is not as high as an explosive charge and this pressure is released over a longer period of time, as a result, the HEGF procedure can be assumed to be very similar to the second stage of blasting (Nilson et al. 1985).

Possibility of unexpected results during such complicated and fast engineering actions, which may cause major safety and economic problems, motivates implementation of numerical and analytical simulations to predict a wide range of problems. Several attempts have been devoted to simulating the complex process of blasting, but only those related to this research are briefly reviewed. Nilson et al. (1985) developed equations of conservation of mass and momentum for penetration of a gas through a crack. These equations were solved numerically, while analytical solutions were implemented for analyzing the solid media. Munjiza et al. (2000) suggested a simple model for evaluation of gas pressure through cracks. Gas pressure was only considered in a specific area around the source, and the combined finite-discrete element method was used for the analysis of the cracked solid. The Nilson equations were implemented by Cho et al. (2004b) to investigate the dynamic fracture process of rock. A dynamic FEM code

equipped with a re-meshing algorithm was used to consider crack growth, and the gas pressure was estimated as a one-dimensional flow through cracks. In a different approach, Mohammadi and Bebamzadeh (2005) developed an approach to model gas–solid interaction. This model used two separate but coupled meshes for the computation of solid and gas phases based on the mechanics of porous media. Then Mohammadi and Pooladi (2007) improved the method proposed by Munjiza et al. (2000) to non-uniform gas flow through fractures to account for the effects of cracking and deformation induced by blasting on the pressure and density of the gas. The same idea was used and further developed by Mohammadi and Pooladi (2012) to efficiently simulate the process of gas flow through a complex system of fractures. Different benchmark examples were simulated to assess the performance of their proposed approach. Other numerical techniques such as discrete element method (DEM) for particulate media have also been implemented to simulate rock fragmentation by high energy gas (Ruest et al. 2006). This method can handle highly complex fracture networks but the computational cost is extremely high.

Similar to blasting, the gas fracturing procedure can be classified into two stages; rapid rising of gas pressure which causes some cracking around the hole and the gas penetration which leads to crack extension. The crack initiation step can be simulated using sophisticated rate-dependent constitutive models. Several models have been proposed for random generation of cracks in rocks under dynamic loading, including Cho et al. (2004a, 2008), Cho and Kaneko (2004), Zhu et al. (2004, 2007), and Ma and An (2008). The second stage of gas fracturing, gas penetration into existing cracks, is of great importance because it predominantly determines the final crack extension. It has been considered as a quasi-static phenomenon due to a lower rate of loading (Paine and Please 1995; Nilson et al. 1985).

HEGF can be regarded as an engineering problem in highly complicated conditions with several uncertainties. The final results depend on many factors such as rock strength (tensile strength, toughness), in situ stresses, type of propellant and quality of sealing. The first stage of HEGF procedure is not in the scope of this study and the main focus is to simulate the process of gas penetration and crack extension to obtain quantitatively acceptable results. For simulating the solid medium, the powerful extended finite element method (XFEM) is implemented. This method simulates the existing and propagating cracks independent of the generated mesh, so avoiding the difficult re-meshing and stress transfer algorithms. This method has been used to study hydraulic fractures in concrete dams by (Ren et al. 2009), in which the fluid pressure was applied as a uniform constant pressure through the entire crack surfaces. Different coupled hydro-mechanical formulations of XFEM were also proposed in several studies to simulate hydraulic fracturing in porous media, while the injected fluid can permeate into the surrounding rocks (Mohammadnejad and Khoei 2013; Gordeliy and Peirce 2013; Gholami et al. 2013). Here, to consider the gas flow through the fractures, a one-dimensional transient flow model governed by conservation of mass and momentum (Nilson and Griffiths 1983; Nilson et al. 1985) is adopted. These equations are solved using an explicit finite difference method (FDM). In each time step, the geometrical parameters of fractures are given to the FDM code, and the resultant solution for the gas pressure along the crack is applied as the boundary conditions on the solid medium. These equations were previously used by Cho et al. (2004b) and Goodarzi et al. (2011, 2013) to simulate a laboratory scale experiment conducted by Cho et al. (2002) to study the gas flow inside a crack. Applicability of these equations was confirmed by the good agreement obtained between numerical and experimental values of the average gas velocity inside the crack.

In this paper, after introducing the gas flow and XFEM equations, the provided XFEM code is validated against an analytical solution, and the effects of different numerical parameters are assessed in order to achieve a reasonable accuracy for the numerical results. The proposed algorithm is then evaluated by simulating two field experiments of gas fracturing, with comprehensive sensitivity analyses to investigate the effect of each parameter.

2 Numerical modeling of gas flow

After a blast, a small zone with many cracks would appear around the blast-hole, and just a few of them can surpass the others and extend. Experimental investigations have also shown that the number of major cracks around a blast-hole is between 3 and 8 (Garnsworthy 1990). Accordingly, in this research, the gas flow is only considered in those surpassing fractures. The gas penetration through the cracks is assumed to be a one-dimensional transient flow. Moreover, because of the insignificant loss of mass and heat into the surrounding rock, it is reasonable to presume that the gas expansion is an adiabatic process, and the rock is impermeable (Nilson et al. 1985).

The one-dimensional equations of gas flow, governed by the laws of conservation of mass and momentum, can be written as follows:

$$\frac{\partial(\rho h)}{\partial t} + \frac{\partial(\rho v h)}{\partial x} = 0, \tag{1}$$

$$\rho h \left(\frac{1}{\rho} \frac{\partial P}{\partial x} + \psi \right) = 0, \tag{2}$$

where ρ is the density; v is the velocity; P is the gas pressure; and Ψ is the viscous shear stress, which can be approximated by Eq. (3a) and (3b) for laminar and turbulent flow, respectively (Paine and Please 1995),

$$\psi = \frac{12\mu v}{\rho h^2}, \tag{3a}$$

$$\psi = a \left(\frac{\varepsilon}{h} \right)^b \frac{v^2}{h}, \tag{3b}$$

where μ is the viscosity of fluid; h is the fracture opening; ε is the fracture roughness; and a and b are experimental constants: $a = 0.1$ and $b = 0.5$ (Nilson et al. 1985). Cho et al. (2004b) showed that the turbulent model for gas flow through the fracture is much more reasonable, so Eq. (3b) is chosen for the rest of this study.

Replacing the viscous shear stress in Eq. (2) by Eq. (3b) and after simple manipulations, the velocity can be determined from

$$v = \sqrt{\frac{h}{f_t \rho} \left(-\frac{\partial P}{\partial x} \right)}; \quad f_t = a(\varepsilon/h)^b. \tag{4}$$

Substituting Eq. (4) into Eq. (1), the discretized form of Eq. (1) on the mesh shown in Fig. 1, can be written as follows:

$$
\begin{aligned}
\rho_N^{t+\Delta t} - \rho_N^t = & -\frac{4\Delta t}{(\Delta x_R + \Delta x_L)(h_w + h_x)} \\
& \times \left(h_x \sqrt{-\rho_x^t \frac{h_x}{f} \frac{(P_L^t - P_N^t)}{\Delta x_L}} \right. \\
& \left. - h_w \sqrt{-\rho_w^t \frac{h_w}{f} \frac{(P_N^t - P_M^t)}{\Delta x_R}} \right),
\end{aligned}
\tag{5}
$$

where h is a constant input associated with an element, and the density of elements is calculated as the average of the densities of their nodes. Despite the fact that an advanced equation of state such as JWL can better predict the explosive pressure, the JWL parameters for the propellant used in our verification examples are not available in the literature. As a result, to estimate the detonation gas pressure along the fractures, an ideal gas equation of state is implemented (Paine and Please 1995; Mortazavi and Katsabanis 2001).

$$P = P_0 \left(\frac{\rho}{\rho_0} \right)^\gamma, \tag{6}$$

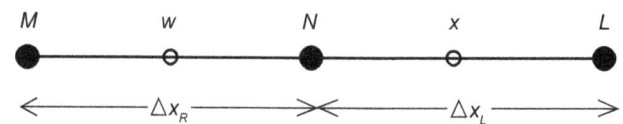

Fig. 1 The finite difference mesh for one-dimensional gas flow, w and x are in the *middle* of the elements

where P_0 and ρ_0 are the initial pressure and density of the gas; P and ρ are the current values and γ is the coefficient of the ideal gas.

3 Extended finite element method

3.1 Formulation

The finite element method (FEM) is one of the most powerful methods in engineering analyses, frequently used to model various problems in solid media. One of the main approaches of FEM in modeling crack propagation problems is to use the technically difficult and time-consuming adaptive re-meshing approach. The extended finite element method, on the other hand, simulates the cracks by enriching the shape functions of the elements which are involved with cracks. In this way, after each step of crack propagation, there is absolutely no need to change the initial mesh and only the new involved elements should be detected for proper enrichments.

When an element takes part in a crack simulation, its XFEM displacement approximation can be defined as follows (Mohammadi 2008):

$$\mathbf{u}(\mathbf{x}) = \sum_{j=1}^{n} N_j(\mathbf{x})\mathbf{u}_j + \sum_{h=1}^{m} N_h(\mathbf{x})H(\xi(\mathbf{x}))\mathbf{a}_h$$
$$+ \sum_{k=1}^{mt} N_k(\mathbf{x}) \left(\sum_{l=1}^{mf} F_l(\mathbf{x})\mathbf{b}_k^l \right). \qquad (7)$$

Here $\mathbf{u}$ is the conventional FEM nodal displacements; n is the number of nodes of the element; m is the number of nodes which are involved with the crack length; mt is the number of nodes being related to the crack tip; mf is the number of functions that are used for enriching the crack tip element; and $\mathbf{a}_h$ and $\mathbf{b}_k^l$ are the additional degrees of freedom associated with crack discontinuity and crack tip singularity enrichments, respectively; N is the conventional shape functions of FEM; and H is the Heaviside function for simulation of displacement discontinuity across a crack,

$$H(\xi(x)) = \begin{cases} 1, & \xi(x) \geq 0 \\ 0, & \xi(x) < 0 \end{cases}. \qquad (8)$$

In Eq. (7), F is a set of functions which are obtained from analytical solution of displacement around a crack tip. The crack tip enrichment function F for an isotropic elastic material can be defined as follows:

$$F_\alpha(r,\theta) = \left\{ \sqrt{r}\sin\frac{\theta}{2}, \sqrt{r}\cos\frac{\theta}{2}, \sqrt{r}\sin\theta\sin\frac{\theta}{2}, \sqrt{r}\sin\theta\cos\frac{\theta}{2} \right\}. \qquad (9)$$

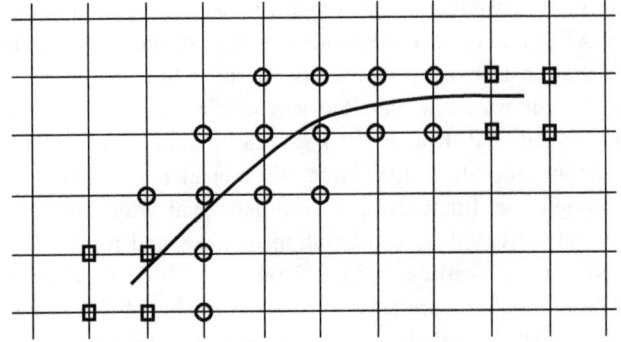

Fig. 2 Selection of the nodes for enrichments, *squares* show the crack tip enrichment, and *circles* are related to the Heaviside enrichment

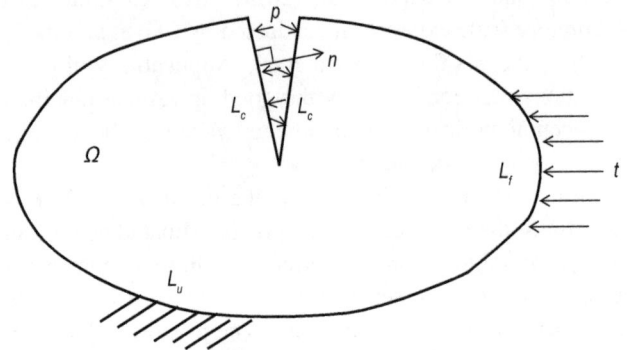

Fig. 3 A cracked solid domain subjected to an internal pressure on crack surfaces

Selection of the enriched nodes is performed according to the crack position, as shown in Fig. 2.

The conventional FEM formulation should be updated to account for the additional degrees of freedom. If a cracked body subjected to body force $\mathbf{b}$ and internal pressure $\mathbf{p}$ on the crack surfaces is assumed, as depicted in Fig. 3, the global governing equation for determining the unknown vectors $\mathbf{u}$ can be defined as follows:

$$\mathbf{Ku} = \mathbf{f}, \qquad (10)$$

where the unknowns vector $\mathbf{u}$, the stiffness matrix $\mathbf{K}$, and the external force vector $\mathbf{f}$, for each element, can be determined from the equations as follows:

$$\mathbf{u}^e = \left\{ \mathbf{u}_j, \mathbf{a}_h, \mathbf{b}_k^l \right\}^T, \qquad (11)$$

$$\mathbf{k}_{ij}^e = \begin{bmatrix} k_{ij}^{uu} & k_{ij}^{ua} & k_{ij}^{ub} \\ k_{ij}^{au} & k_{ij}^{aa} & k_{ij}^{ab} \\ k_{ij}^{bu} & k_{ij}^{ba} & k_{ij}^{bb} \end{bmatrix}, \quad (r,s = u,a,b), \qquad (12)$$

$$\mathbf{f}_i^e = \left\{ \mathbf{f}_i^u, \mathbf{f}_i^a, \mathbf{f}_i^{bl} \right\}. \tag{13}$$

Considering **B** and **D** as the matrix of the shape function derivatives and the constitutive matrix, respectively, different terms in Eq. (12) and (13) can be determined as following:

$$k_{ij}^{rs} = \int_{\Omega^e} \left(\mathbf{B}_i^r \right)^T \mathbf{D} \mathbf{B}_j^s \, d\Omega, \tag{14}$$

$$\mathbf{f}_i^u = \int_\Omega N_i \mathbf{b} d\Omega + \int_{L_f} N_i \mathbf{t} d\Gamma, \tag{15}$$

$$\mathbf{f}_i^a = \int_\Omega N_i H \mathbf{b} d\Omega + \int_{L_f} N_i H \mathbf{t} d\Gamma + 2 \int_{L_c} n. N_i \mathbf{p} d\Gamma, \tag{16}$$

$$\mathbf{f}_i^{bl} = \int_\Omega N_i F_l \mathbf{b} d\Omega + \int_{L_f} N_i F_l \mathbf{t} d\Gamma + 2 \int_{L_c} n \sqrt{r}. N_i \mathbf{p} d\Gamma, \tag{17}$$
$$(l = 1 - 4).$$

3.2 Numerical integration

Despite the simple idea of XFEM, specific details are required for its implementation. One of them, which is critical to achieve proper accuracy, is the integration on the elements that are involved with a crack. The Gauss quadrature method is usually adopted for this purpose in conventional FEM simulation. However, it may not be accurate enough for singular or discontinuous functions usually encountered in XFEM simulations. One way to improve the results is to subdivide the both sides of the enriched element into subtriangles in such a way that their edges conform to the geometry of the crack and the element (Mohammadi 2008). Figure 4 shows a simple typical procedure for subdividing a crack element and a crack tip element; a larger number of triangles may be required to achieve sufficient accuracy. It should be noted that integration in each triangle is performed by a standard Gauss quadrature rule.

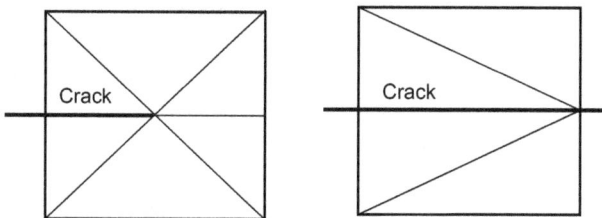

Fig. 4 Subtriangles for integration on a crack tip element (*left*) and a cracked element (*right*)

More details about the formulation, implementation, and applications of XFEM can be found in Mohammadi (2008, 2012).

4 Coupling process and crack propagation

The two numerical approaches for solving the gas flow and the cracked solid medium have to be coupled. At first, initial lengths of cracks are assumed as a result of the first phase of blasting (the shock wave propagation), which is not directly simulated in this paper. The initial FDM mesh is generated on the existing cracks and the gas flow algorithm is performed for a small time span (time step). Then the calculated gas pressure is applied as the boundary conditions into the XFEM code for simulating the cracked domain. The new crack lengths and the crack opening displacements (COD) are computed and exported to the gas flow algorithm for the next step of calculation.

A criterion is also required for crack propagation. The stress intensity factor (SIF) is calculated and compared with the critical value in each step. There are several methods for numerical evaluation of SIF, but due to the assumption of linear elasticity in this study, the computationally inexpensive displacement extrapolation method is adopted. As the problem is solved in a quasi-static condition, cracks propagate and extend to a specific value when the criterion is satisfied. In other word, a pseudo-velocity is assumed for crack propagation, and the specific value of propagation extent for each step is obtained from this velocity multiplied by the time step. The proposed algorithm is described in Fig. 5.

Assuming a linear elastic analysis, the SIF can be calculated using the analytical solution of displacements around the crack tip (Eq. 18). Rewriting these expressions in terms of SIF and substituting the numerically obtained displacements for several points on a radial line emanating from the crack tip, a set of data for SIF in mode I (K_I) or mode II (K_{II}) with respect to the distance r from the crack tip is generated. The SIF at the crack tip is the extrapolated value for $r = 0$. Figure 6 shows the procedure of this approach,

$$4G\sqrt{\frac{2\pi}{r}}\left\{\begin{matrix} u \\ v \end{matrix}\right\} = K_I \left\{\begin{matrix} (2k-1)\cos\frac{\theta}{2} - \cos\frac{3\theta}{2} \\ (2k+1)\sin\frac{\theta}{2} + \sin\frac{3\theta}{2} \end{matrix}\right\},$$
$$4G\sqrt{\frac{2\pi}{r}}\left\{\begin{matrix} u \\ v \end{matrix}\right\} = K_{II} \left\{\begin{matrix} -(2k+3)\sin\frac{\theta}{2} - \sin\frac{3\theta}{2} \\ (2k-3)\cos\frac{\theta}{2} + \cos\frac{3\theta}{2} \end{matrix}\right\}. \tag{18}$$

Fig. 5 The flowchart for the proposed algorithm

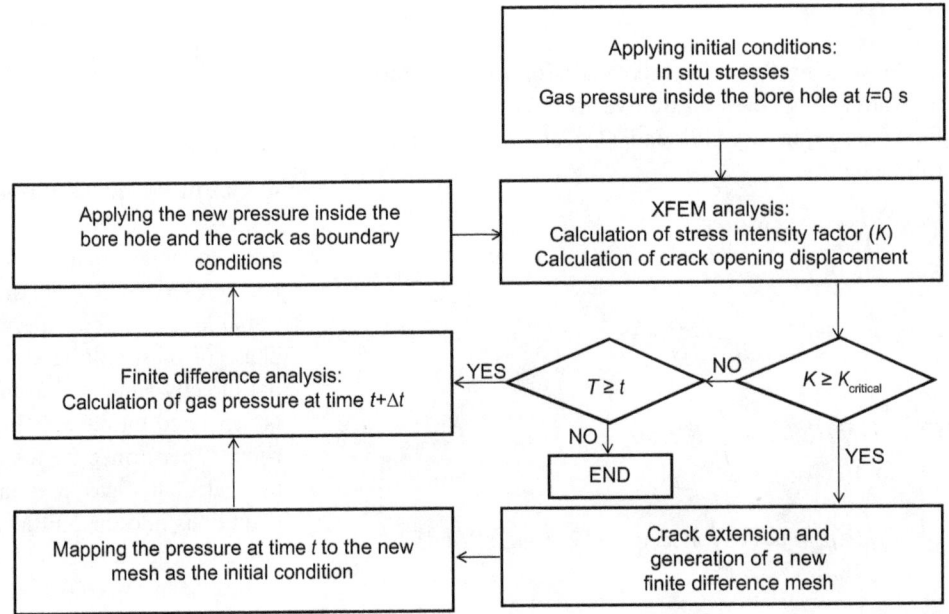

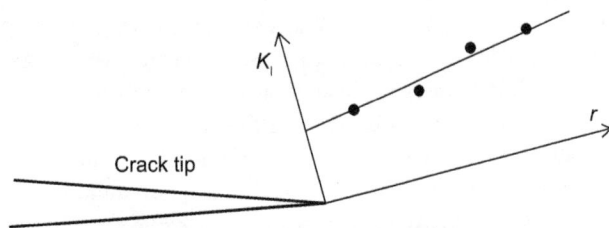

Fig. 6 Displacement extrapolation method; the SIF at the crack tip is estimated from the best *fitted line* on the sampling points

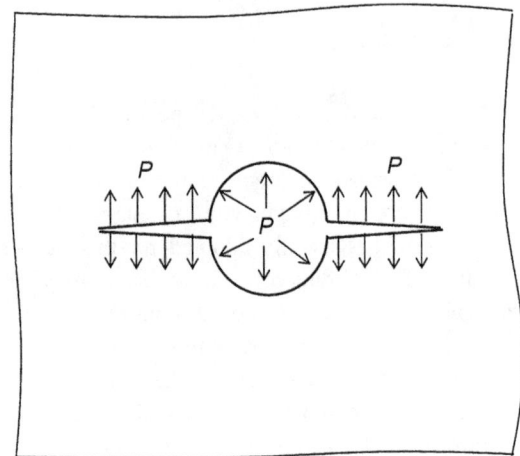

Fig. 7 Geometry of the hole with two radiating cracks

5 Numerical results

5.1 Validation of XFEM code

To verify the accuracy of the presented XFEM code, a classic problem with available analytical solution is simulated. A pressurized hole with two radiating cracks in an infinite plate (Fig. 7) has the following closed-form solution for the stress intensity factor,

$$K = \beta P \sqrt{\pi a}, \tag{19}$$

where β is a coefficient related to the ratio of crack tip distance from the center of the hole to the hole radius. A hole with 5 cm radius and two 15 cm radiating cracks is assumed and a uniform internal pressure of 1 MPa is applied inside the hole and the cracks. β for this problem is 0.9976 (Saouma 2000), so the analytical stress intensity factor (Eq. 19) is computed, 7.91 MPa m$^{0.5}$.

Due to the axial symmetry of the problem, one half of the geometry is simulated with the developed XFEM code. Figure 8 shows the generated mesh of 2200 nodes, the enriched nodes and the distribution of Gauss points around

the cracks. It is noted that only 54 extra degrees of freedom are required to simulate the crack. Increasing the number of Gauss points around the crack tip can reduce the error but increases the computational time, so an optimum distribution should be obtained for each type of problem. The numerically predicted stress intensity factor for this model is 7.7 MPa m$^{0.5}$ with an acceptable error of about 2.7 %.

5.2 Gas fracturing simulation

In order to investigate the capability of the proposed approach to simulate gas fracturing problems, the experimental studies conducted by the Sandia National laboratory (Nilson et al. 1985) in deep tunnels excavated in a homogenous Tuff with 10 MPa hydrostatic stress are

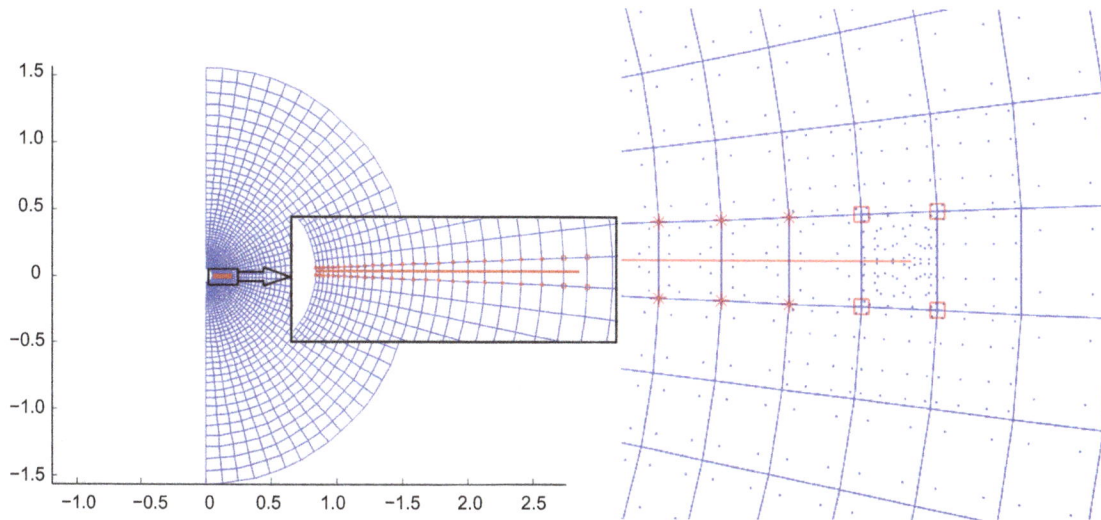

Fig. 8 Generated mesh and extra degrees of freedom for enrichments (*left*) and the distribution of Gauss points (*right*)

Table 1 The details and results of the experiments (after Nilson et al. 1985)

Experiment ID	D1	GF2
Pressure		
Peak, MPa	90	40
Rise time, ms	0.5	3
Decay time, ms	16	18
Wellbore		
Diameter, m	0.2	0.048
Propellant		
Diameter, m	0.2	0.04
Density, g/cm^3	0.5	0.5
Type	M5B	M5B
Cracks		
Number	7	2
Length range, m	0.9–2.5	0.4–0.9
Length mean, m	1.7	0.7
Cracking pattern		

Table 2 Properties of the host rock (Nilson et al. 1985)

Parameters	Values
Toughness, MPa m$^{0.5}$	0.5
Shear modulus, GPa	3
Poisson's ratio	0.3
Crack roughness, mm	0.4
Shear wave velocity, m/s	1200

fracturing with 6 major radiating cracks. The details of the experiments and the rock properties are presented in Tables 1 and 2.

The gas produced from the propellant burning is considered as an ideal gas, while its expansion is assumed as an adiabatic expansion. These are reasonable assumptions for high temperature gases produced by blasting (Mortazavi and Katsabanis 2001). Initial tiny cracks around the borehole are assumed to initiate the crack propagation. In addition, the rapid phase of pressure rise is ignored. In fact, the simulation starts immediately after the peak pressure is reached. The fluid pressure acts normal to the crack surface and in these particular examples, the stress state is hydrostatic; therefore, the crack propagation occurs in pure Mode I which means no change in the direction of the crack during its propagation. It should be noted that it will not be the case when the in situ stress state becomes anisotropic. In addition, there are two unknown parameters in this simulation which are determined based on calibration of the experiments: the constant of equation of state (γ) and the crack propagation velocity.

Figure 9 shows the generated model for the first experiment (D1) which contains 3000 nodes. The initial crack

modeled. Two of these examples are chosen for this study. One of them is a low-power fracturing, which produced only two fractures and the other one is a high energy

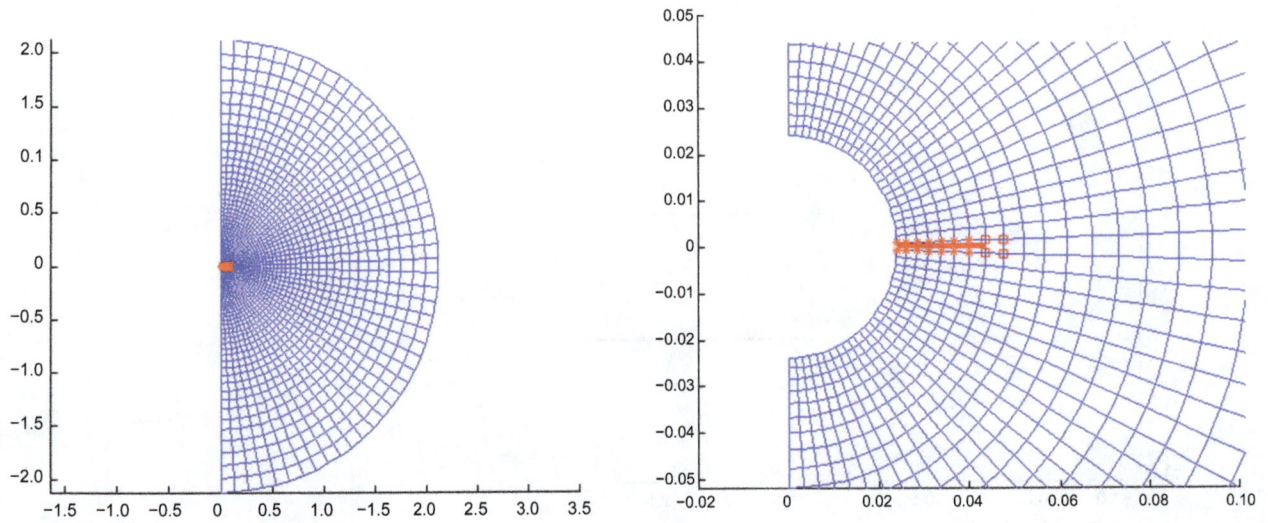

Fig. 9 Generated mesh (*left*) and enriched nodes (*right*) for the D1 experiment

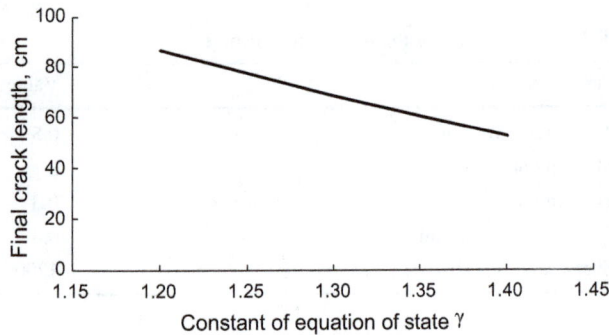

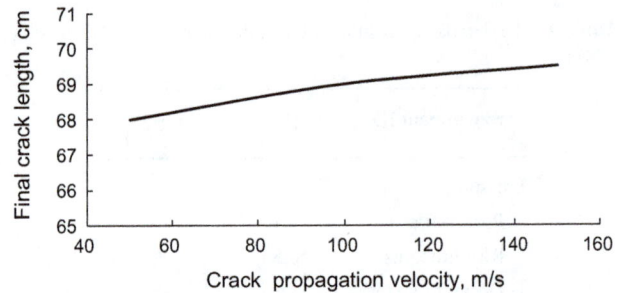

Fig. 10 Calibration of the constant of equation of state

Fig. 11 Sensitivity of the final crack length with respect to the initial crack length, crack propagation velocity, and the time step

length is 2 cm, the time step for analyzing the solid medium is 100 μs, and the crack velocity is assumed to be 100 m/s. According to the observed cracking pattern for this experiment, the average final crack length is equal to 0.7 m, so the constant of equation of state can be calibrated. Performing a back analysis on the results, the value of 1.29 is obtained which is in the expected range of 1.2–3 for blast-induced high temperature and high density gases, as proposed by Mortazavi and Katsabanis (2001) (Fig. 10). The calibrated value for the constant of equation of state might not be exactly equal to the real value, due to so many unavoidable uncertainties in these complex problems and the simplifications and assumptions that are essential to make the simulation possible. The obtained value may cover some of them but it can generate the same overall result.

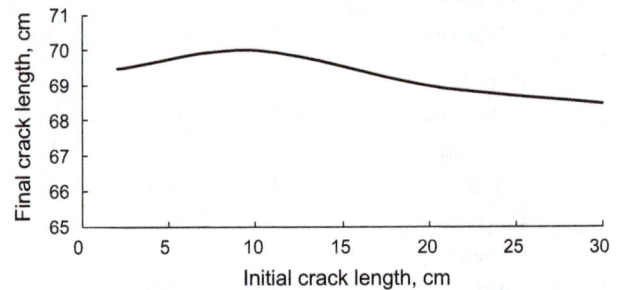

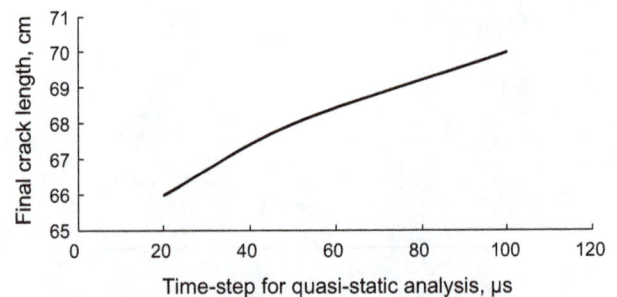

To better investigate the effects of other parameters, a sensitivity analysis is carried out on the crack propagation

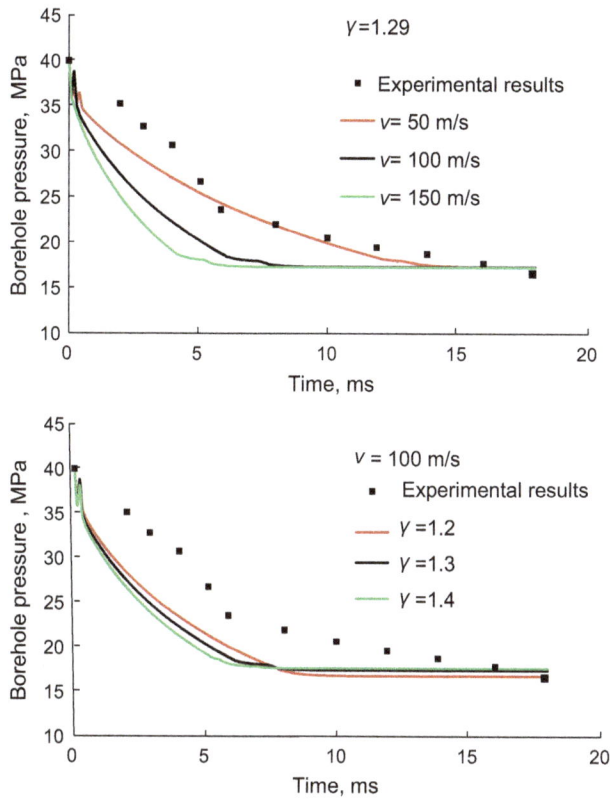

Fig. 12 The effect of crack propagation velocity and the constant of EOS (γ) on the decay of borehole pressure

Fig. 13 The effect of in situ stress on the final crack length

Table 3 Sensitivity analysis of the final crack length with respect to the mesh size

Number of nodes	3600	3000	2400	1144
Final crack length, m	0.695	0.70	0.76	0.55

velocity, the initial crack length and the time step. Values for these parameters are changed in reasonable ranges and their effects on the final crack length are studied. Figure 11 clearly shows that the results for the final crack length remain practically insensitive to these numerical assumptions.

Despite the fact that the final solution is not sensitive to the assumption of the crack propagation velocity, its value should be set in a logical range. Nilson et al. (1985) argued that in a dynamic state, the maximum velocity for crack propagation mostly depends on the mechanical properties of the solid medium and it can be roughly estimated around 50 % of the Rayleigh wave speed in the medium. In contrast, in hydraulic or gas fracturing, the fluid-dynamic considerations control the crack propagation velocity and it depends on how fast the driving pressure can push fluid into the fracture, so the crack speed becomes slower than the dynamic mode. As the Rayleigh wave speed is slightly less than the shear wave speed which is 100 m/s (Table 2) for this rock. The assumed pseudo-crack propagation velocity should be less than 600 m/s. Figure 12 shows the effect of crack velocity and the constant of equation of state on the borehole pressure decay of the D1 experiment. It can be concluded that it is the crack propagation velocity that mainly determines the rate of pressure drop in the

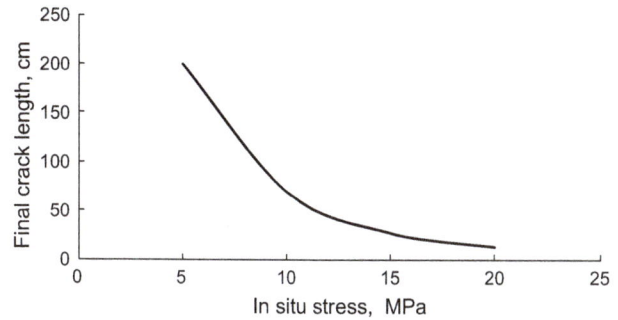

borehole. The crack extension velocity of 50 m/s can well be matched with the field data, which is also in agreement with the description provided by Nilson et al. (1985) that fluid-driven fracturing is slower than dynamic fracturing.

Another important issue that should be clarified is the effect of in situ stress, as HEGF might be applied in different depths. The final results of this test with different in situ stresses (Fig. 13) reveal that this parameter has a significant non-linear effect on the final results and it should be considered in the design procedure of a successful HEGF operation.

Additionally, to investigate the effect of mesh size on the results, the same problem is simulated by different number of nodes. The results are summarized in Table 3 which indicates that for around 3000 nodes or more, for this particular simulation, the final result will converge and become mesh insensitive. It should be noted that the mesh size can also slightly change the loading evaluated inside the crack and consequently affects to some extent the accuracy of the predicted stress intensity factor.

The calibrated parameters obtained from the first experiment (D1) are now used to simulate the second experiment (GF2) because the host rock and the propellant are the same for both tests. Figure 14 shows the adopted mesh for the GF2 experiment which has 2750 nodes. The crack propagation velocity and the initial crack length are assumed to be 200 m/s and 5 cm, respectively, which may not be the real values, but the results are expected to be insensitive to them, as it was investigated in the previous simulation.

After 16 ms, the final crack length becomes 2.3 m. According to the reference description and the borehole pressure sensor results, after this time, the test sealing had broken, and the gas pressure was lost so, this time is considered as the end of the simulation. The stress states at this time are shown in Fig. 15.

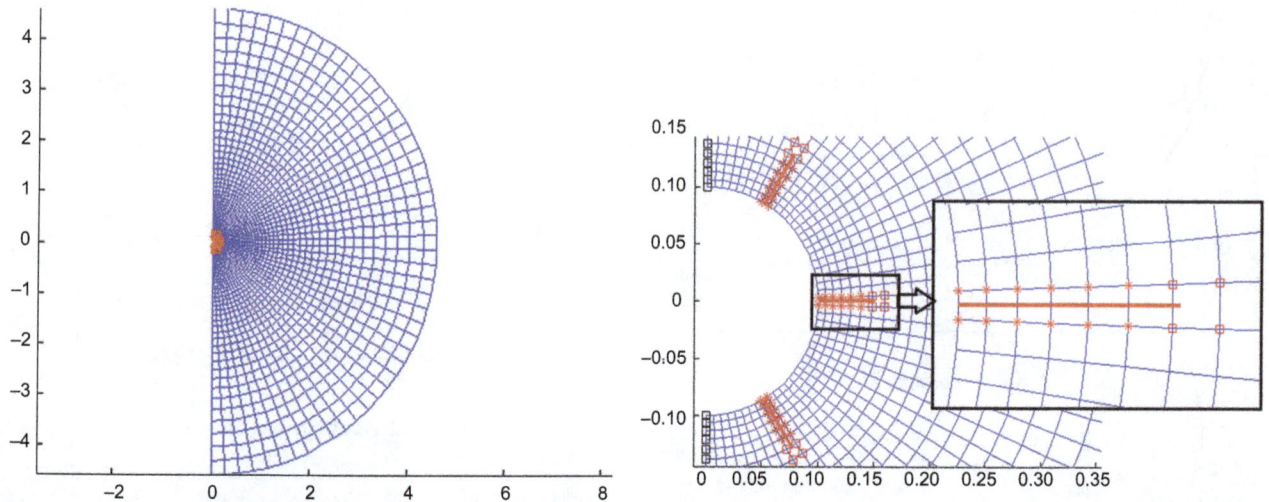

Fig. 14 Generated mesh for the GF2 experiment (*left*), crack positions, and the enriched nodes (*right*)

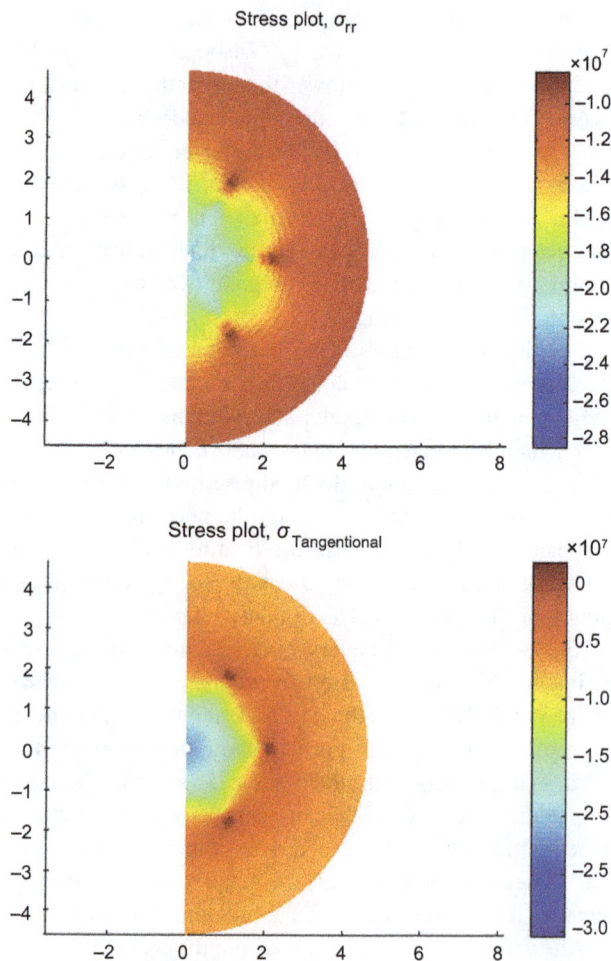

Fig. 15 Radial and tangential stresses at GF2 model after 16 ms (stress dimension is Pa)

The observed difference between the average of final crack lengths in the numerical model (2.3 m) and the experimental test (1.7 m) can be discussed on the basis of some unavoidable sources of error. Firstly, the evaluated material properties, especially the toughness, involve some level of uncertainties. Secondly, due to the short rise time (0.5 ms), small cracks are generated around the hole which absorb a portion of the gas energy through its penetration into these small spaces. As a result, the final crack length is shortened. Generally, because of high level of uncertainties, some authors believe that a precise quantitatively prediction of such a complex problem with a conventional computational model seems very difficult and unlikely, and they may even accept a model that can predict with twofold difference for practical purposes (Nilson et al. 1985).

6 Conclusions

In this paper, a simple model was proposed and evaluated to predict the final length of gas fractures. The available equations of gas flow through a crack were implemented and solved on a 1D finite difference mesh. The novel and computationally efficient XFEM approach was used for simulation of the cracked solid. A simple fracture mechanics problem with an analytical solution was also utilized for validation of the XFEM code. Moreover, the details of XFEM integration and evaluation of the stress intensity factor were determined in such a way that with a reasonable computational effort, a sufficient accuracy could be obtained. Two experimental studies were chosen to calibrate and evaluate the model. Although the predicted results of numerical simulation could not be perfectly matched with the experimental study, the error remained in an acceptable level for practical purposes. In addition, comprehensive sensitivity analyses were performed and the influential parameters were clarified. The in situ stress was found to be a very critical factor on the final result which should be considered for HEGF designs. The final lengths of the cracks were found to be independent of the initial crack lengths. As a result, in a blasting process, the role of gas pressure on the extension of cracks remained more important than the role of the stress wave. This model can be easily improved for complicated geometries, stress states, and material properties as a general computational tool for checking the initial or finalized designs of HEGF operations. Moreover, similar engineering problems such as control blasting in mining industry can be simulated with this algorithm.

Acknowledgments The authors would like to acknowledge the technical support of the High Performance Computing Lab, School of Civil Engineering, University of Tehran. The support of Iran National Science Foundation is also gratefully appreciated.

References

Chao Z, Baiquan L, Yan Z, et al. Study of fracturing-sealing integration technology based on high-energy gas fracturing in single seam with high gas and low air permeability. Int J Min Sci Technol. 2013;23:841–6.

Cho SH, Risei K, Kato M, et al. Development of numerical simulation method for dynamic fracture propagation due to gas pressurization and stress wave. In: Proceedings of 2002 ISRM regional symposium (3rd Korea-Japan joint symposium) on rock engineering problem and approaches in underground construction, Jul 22–24, Seoul; 2002. p. 755–62.

Cho SH, Ogata Y, Kaneko K. Strain-rate dependency of the dynamic tensile strength of rock. Int J Min Sci Technol. 2004a;40:763–77.

Cho SH, Nakamura Y, Kaneko K. Dynamic fracture process analysis of rock subjected to stress wave and gas pressurization. Int J Min Sci Technol. 2004b;41:433–40.

Cho SH, Kaneko K. Influence of the applied pressure waveform on the dynamic fracture processes in rock. Int J Min Sci Technol. 2004;41:771–84.

Cho SH, Nakamura Y, Mohanty B, et al. Numerical study of fracture plane control in laboratory-scale blasting. Eng Fract Mech. 2008;75:3966–84.

Chu TY, Jacobson RD, Warpiniski N. Geothermal well stimulation using high energy gas fracturing. In: Proceedings of 12th workshop on geothermal reservoir engineering, Jan 20–22, Stanford, CA; 1987.

Daehnke A, Rossmanith HP, Napier AL. Gas pressurization of blast-induced conical cracks. Int J Rock Mech Min Sci. 1997;34(3–4):263.e1–17.

Garnsworthy RK. The mathematical modeling of rock fragmentation by high pressure arc discharges. In: 3rd international symposium on rock fragmentation blasting, Aug 26–31, Brisbane; 1990. p. 143–7.

Gholami A, Rahman SS, Natarajan S. Simulation of hydraulic fracture propagation using XFEM. In: EAGE symposium, sustainable earth sciences, Sept 30–Oct 4, Pau; 2013.

Goodarzi M, Mohammadi S, Jafari A. Analysis of gas-driven crack propagation around a blasthole with the extended finite element method. In: Proceedings of the 2nd international symposium on computational geomechanics (COMGEO II), April 27–29, Croatia; 2011. p. 425–33.

Goodarzi M, Salmi EF, Mohammadi S, et al. Numerical modeling of gas fracturing with the extended finite element method. In: Proceedings of the 3rd international symposium on computational geomechanics (COMGEO III). August, 21–23, Poland; 2013. p. 706–16.

Gordeliy E, Peirce A. Coupling schemes for modeling hydraulic fracture propagation using the XFEM. Comput Methods Appl Mech Eng. 2013;253:305–22.

Jaimes MG, Castillo RD, Mendoza SA. High energy gas fracturing: a technique of hydraulic prefracturing to reduce the pressure losses by friction in the near wellbore—a Colombian field application. In: The SPE Latin American and Caribbean petroleum engineering conference, April 16–18, Mexico City (SPE 152886); 2012.

Krilove Z, Kavedzija B, Bukovac T. Advanced well stimulation method applying a propellant technology. Wiertnictwo Nafta Gaz. 2008;25(2):405–16.

Kutter HK, Fairhurst C. On the fracture process in blasting. Int J Rock Mech Min Sci. 1971;8:181–202.

Ma GW, An XM. Numerical simulation of blasting-induced rock fractures. Int J Rock Mech Min Sci. 2008;75:966–75.

McHugh S. Crack extension caused by internal gas pressure compared with extension caused by tensile stress. Int J Fract. 1983;21:163–76.

Mohammadi S, Bebamzadeh A. A coupled gas-solid interaction model for FE/DE simulation of explosion. Finite Elem Anal Des. 2005;41:1289–308.

Mohammadi S, Pooladi A. Non-uniform isentropic gas flow analysis of explosion in fractured solid media. Finite Elem Anal Des. 2007;43:478–93.

Mohammadi S. Extended finite element method for fracture analysis of structure. London: Blackwell Publishing; 2008.

Mohammadi S. XFEM fracture analysis of composites. London: Wiley; 2012.

Mohammadi S, Pooladi A. A two-mesh coupled gas flow–solid interaction model for 2D blast analysis in fractured media. Finite Elem Anal Des. 2012;50:48–69.

Mohammadnejad T, Khoei AR. An extended finite element method for hydraulic fracture propagation in deformable porous media with the cohesive crack model. Finite Elem Anal Des. 2013;73:77–95.

Mortazavi A, Katsabanis PD. Modeling burden size and strata dip effects on the surface blasting process. Int J Rock Mech Min Sci. 2001;38:481–98.

Munjiza A, Latham JP, Andrews KRF. Detonation gas model for combined finite-discrete element simulation of fracture and fragmentation. Int J Numer Method Eng. 2000;49:1495–520.

Nilson RH, Griffiths SK. Numerical analysis of hydraulically-driven fractures. Comput Method Appl Mech Eng. 1983;36:359–70.

Nilson RH, Proffer WJ, Duff RE. Modeling of gas-driven fracture induced by propellant combustion within a borehole. Int J Rock Mech Min Sci Geomech Abstr. 1985;22(1):3–19.

Paine AS, Please CP. An improved model of fracture propagation by gas during rock blasting—some analytical results. Int J Rock Mech Min Sci Geomech Abstr. 1995;31(6):699–706.

Ren QW, Dong YW, Yu TT. Numerical modeling of concrete hydraulic fracturing with extended finite element method. Sci China Ser E. 2009;52(3):559–65.

Ruest M, Cundall P, Guest A, et al. Developments using the particle flow code to simulate rock fragmentation by condensed phase explosives. In: FRAGBLAST—8 (8th international symposium on rock fragmentation by blasting), May 8–11, Santiago; 2006. p. 140–51.

Salazar A, Almanza E, Folse K. Application of propellant high-energy gas fracturing in gas-injector wells at El Furrial field in Northern Monagas State–Venezuela. In: The SPE international symposium and exhibition on formation damage control, Feb 20–21, Lafayette (SPE 73756); 2002.

Saouma VE. Lecture notes in fracture mechanics (Chapter 7). Department of Civil Environmental and Architectural Engineering, University of Colorado; 2000.

Wu J, Liu L, Zhao G, et al. Research and exploration of high energy gas fracturing stimulation integrated technology in Chinese shale gas reservoir. Adv Mater Res. 2012;524–527:1532–6.

Yang W, Zhou C, Qin F, et al. High-energy gas fracturing (HEGF) technology: research and application. In: European petroleum conference, Nov 16–18, Cannes (SPE 24990); 1992.

Zhu Z, Mohanty B, Xie H. Numerical investigation of blasting-induced crack initiation and propagation in rocks. Int J Rock Mech Min Sci. 2007;44:412–24.

Zhu WC, Tang CA, Huang ZP, et al. A numerical study of the effect of loading conditions on the dynamic failure of rock. Int J Rock Mech Min Sci. 2004;41(3):424.

14

A review of closed-loop reservoir management

Jian Hou · Kang Zhou · Xian-Song Zhang ·
Xiao-Dong Kang · Hai Xie

Abstract The closed-loop reservoir management technique enables a dynamic and real-time optimal production schedule under the existing reservoir conditions to be achieved by adjusting the injection and production strategies. This is one of the most effective ways to exploit limited oil reserves more economically and efficiently. There are two steps in closed-loop reservoir management: automatic history matching and reservoir production optimization. Both of the steps are large-scale complicated optimization problems. This paper gives a general review of the two basic techniques in closed-loop reservoir management; summarizes the applications of gradient-based algorithms, gradient-free algorithms, and artificial intelligence algorithms; analyzes the characteristics and application conditions of these optimization methods; and finally discusses the emphases and directions of future research on both automatic history matching and reservoir production optimization.

J. Hou
State Key Laboratory of Heavy Oil Processing, China University of Petroleum, Qingdao 266580, Shandong, China

J. Hou (✉) · K. Zhou (✉)
School of Petroleum Engineering, China University of Petroleum, Qingdao 266580, Shandong, China
e-mail: houjian@upc.edu.cn

K. Zhou
e-mail: zhoukang_upc@163.com

X.-S. Zhang · X.-D. Kang
CNOOC Research Institute, Beijing 100027, China

H. Xie
ZTE Corporation, Shenzhen 518055, Guangdong, China

Edited by Yan-Hua Sun

Keywords Closed-loop reservoir management ·
Automatic history matching · Reservoir production optimization · Gradient-based algorithm · Gradient-free algorithm · Artificial intelligence algorithm

1 Introduction

With the rapid development of the world economy, the depletion of oil resources increases year by year. It is now difficult to find additional large oil fields. Therefore, the demand for exploiting the limited oil reserves efficiently and economically becomes increasingly significant and has attracted more global attention in recent years. To achieve this goal, an important technique proposed is closed-loop reservoir management. It consists of two steps: automatic history matching and reservoir production optimization.

Automatic history matching is a sequential model updating method, where the estimate of uncertain reservoir properties is updated continuously according to the production measurements available at the time. Reservoir production optimization is a complete or partial automation process for maximizing the development effect within the lifecycle of a reservoir by optimizing operational parameters. The main idea is to exploit the oil reserves as near to the desired optimum as possible. Both automatic history matching and reservoir production optimization are optimization problems as mentioned by researchers such as Brouwer and Jansen (2004), Sarma et al. (2005) and Wang et al. (2009). These problems can be solved by optimization theories.

Automatic history matching has been studied since the 1960s (Jahns 1966; Wasserman and Emanuel 1976; Yang and Watson 1988), but it is still a very difficult problem at present. The existing history matching methods can be broken into two categories. One category is based on gradient, including finite difference approximation of derivative, adjoint gradient-based methods, and gradient simulator methods. The other category is based on gradient-free optimization, such as simultaneous perturbation stochastic approximation, genetic algorithm, particle swarm optimization, and pattern search methods (PSMs). Oliver and Chen (2011) have summarized the recent progress on automatic history matching.

The origin of solving reservoir production problems using optimization theories can be traced back to Lee and Aronofsky (1958). They used a linear programming method to maximize the net present value of production for a homogeneous reservoir. Later, some other papers appeared in journals such as Operation Research, Management Science, and Journal of Petroleum Technology. However, most papers published before the 1980s did not pay enough attention to optimization algorithms, and successful applications were very rare (Aronofsky and Williams 1962; Wattenbarger 1970; McFarland et al. 1984). With the advances in optimization algorithms and com-

2 Problem descriptions

2.1 Automatic history matching

The estimate of unknown geological properties using production measurements is recognized as history matching. It is an ill-posed inverse problem with many unknown reservoir parameters that could be adjusted to obtain a match against a relatively smaller number of measurements. Traditionally, the unknown parameters are adjusted manually by trial and error. This method is time-consuming and often yields a reservoir numerical model which may be unrealistic or not consistent with geological properties. To address these problems, automatic history matching has been studied for several decades. As shown in Fig. 1, automatic history matching is an iterative procedure where the unknown reservoir parameters are adjusted automatically with an optimizer to match the observed production or pressure data. In fact, automatic history matching is an optimization problem and the most commonly used objective can be written as follows:

$$O(m) = \sum_{k=1}^{K} \left[(g_k - d_k)^T (C_D)_k^{-1} (g_k - d_k) + (m_k - \hat{m}_k)^T (C_M)_k^{-1} (m_k - \hat{m}_k) \right], \tag{1}$$

puting power, research has increased greatly since the 1980s (Sequeira et al. 2002; Chacón et al. 2004; Barragán et al. 2005; Gunnerud and Foss 2010; Knudsen and Foss 2013; Tavallali et al. 2013).

In order to find out under which operational parameters at current reservoir conditions the oil production might be most efficient and profitable, automatic history matching and reservoir production optimization should be combined together. This combination forms a concept of closed-loop reservoir management, where the geological model will be updated once the production measurements are available and the operational parameters will be optimized based on the newly updated reservoir model. Representative papers on this concept include Brouwer and Jansen (2004), Jansen et al. (2005), Nævdal et al. (2006), and Bieker et al. (2007).

This paper gives a general review of research on automatic history matching and reservoir production optimization; analyzes the characteristics and application conditions of gradient-based algorithms, gradient-free algorithms, and artificial intelligence algorithms; and finally discusses the emphases and directions of future research on both automatic history matching and reservoir production optimization.

where the subscript k is the discrete time step; K is the total number of time steps; m is the vector of reservoir parameters to be estimated; d is the vector of observed historical data; g is the vector of corresponding data to be matched;

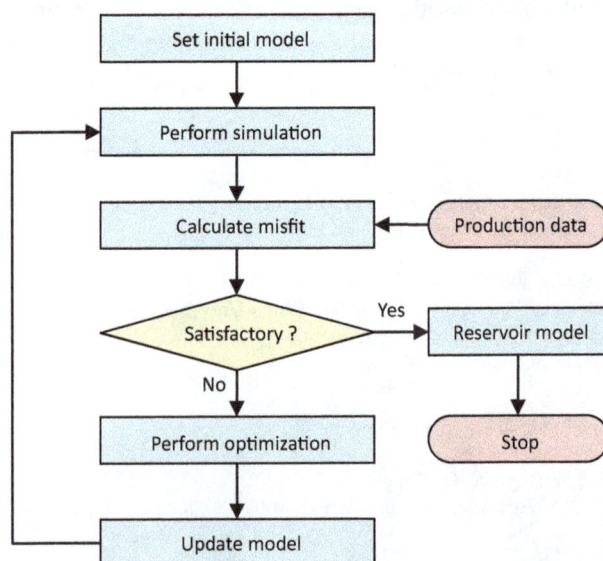

Fig. 1 Flow chart for automatic history matching

$\hat{m}$ is the prior model information; C_D is the covariance matrix of measurement errors; C_M is the covariance matrix of prior probability density function. These matrices determine the weight of individual terms in the objective function.

Although the least-square error has been used successfully to match the observed production data, it does not work well when the seismic data are history matched at the same time. To address this problem, Tillier et al. (2013) presented an appropriate objective function for history matching of seismic attributes based on image segmentation and a modified Hausdorff metric. The objective function of history matching commonly has a complex shape and multiple local minima (Oliver and Chen 2011). This is mainly because unknown parameters are always much more numerous than available production measurements.

In order to carry out history matching in a lower space and reduce the necessity of an explicit regularization term in the objective function, various parameterization methods have been presented, including the zonation method, pilot point method, subspace method, spectral decomposition, discrete cosine transform, truncated singular value decomposition, and the multiscale method. Jacquard (1965) and Jahns (1966) applied the zonation method to reduce variables in their automatic history matching study. The gradzone method in Bissell et al. (1994) and the adaptive multiscale method in Grimstad et al. (2003, 2004) were variations of the original zonation method. Marsily et al. (1984) presented the pilot point method, in which variables were estimated only at the pilot points and others were calculated by the Kriging method. Abacioglu et al. (2001) adopted a subspace method based on segmentation of the objective function. Oliver (1996a, b) applied the parameterization method based on spectral decomposition of the prior covariance matrix to history match a two-dimensional (2D) permeability field. Jafarpour and McLaughlin (2008), Jafarpour et al. (2010) used the discrete cosine transform as a parameterization method in their history matching study. Tavakoli and Reynolds (2010, 2011) provided a theoretical basis for parameterization based on truncated singular value decomposition of the dimensionless sensitivity matrix. He et al. (2013) applied the orthogonal decomposition method to transform the high-dimensional states into a low-dimensional subspace. They also described geological models in reduced terms by the Karhunen–Loève expansion of the log-transmissibility field.

The parameterization methods should be performed carefully, or results may sometimes mislead us. For example, Oliver et al. (2008) showed a small number of variables may underestimate the uncertainty in automatic history matching problem. They adjusted a spatially varying porosity with a uniform permeability or adjusted a spatially varying permeability with a uniform porosity to history match a same set of well-test data. Both of the two history matching results are good enough, but the assumptions on uncertainty are definitely different.

More and more attention has been paid to the generation and history matching of geologically realistic non-Gaussian reservoir models. Researchers have presented several methods including Gaussian mixture models, truncated pluri-Gaussian method, level set method, discrete cosine transform, and other principal component analysis methods. Dovera and Rossa (2011) proposed expressions of conditional means, covariances, and weights for Gaussian mixture models, so that the ensemble Kalman filter algorithm (EnKF) became usable in this case. Liu and Oliver (2005a, b) combined EnKF with a truncated pluri-Gaussian method for history matching of reservoir facies. Agbalaka and Oliver (2008) extended this method to a three-dimensional (3D) reservoir case. Chang et al. (2010) proposed a methodology to combine a level set method with EnKF for history matching of facies distribution in a 2D reservoir model. Hu et al. (2013) introduced a new method to update complex facies models generated by multipoint simulation while preserving their geological and statistical consistency. Jafarpour and McLaughlin (2008) tested a discrete cosine transform method on two 2D, two-phase reservoir models. Other principal component analysis methods can also be used to deal with non-Gaussian reservoir models. However, their computational cost may be high.

During the development of automatic history matching, various optimization algorithms have been introduced and modified (Bissell et al. 1994; Lee and Seinfeld 1987; Gomez et al. 2001). Streamline-based techniques were also used to improve computational efficiency of history matching by researchers such as Agarwal and Blunt (2003), Cheng et al. (2004, 2005), and Gupta and King (2007). With streamline-based techniques, a reservoir simulation model was automatically decoupled into a series of one-dimensional models along streamlines, which could minimize numerical dispersion and the effects of grid generation while maintaining a sharp displacement front. Caers (2003) and Negrete et al. (2008) combined streamline simulators with a deformation method and EnKF, respectively, to carry out automatic history matching. These streamline-based history matching techniques inherit the shortcomings of streamline methods such as inability to model very complex physics at the same time. Therefore, they are not suitable to all history matching cases.

In order to characterize the uncertainty of unknown geological properties, researchers introduced a Bayesian framework, with which one can formally construct a posterior density function. The books of Tarantola (2005) and Oliver et al. (2008) provide a detailed description about Bayesian framework. Generally, Bayesian estimation depends on a prior Gaussian model. To address this limitation, Sarma et al. (2008b) transformed this problem into feature space using kernel principal component analysis. Other important research works involved Bayesian framework include Chu et al. (1995), He et al. (1997), Zhang et al. (2002, 2005), Liu and Oliver (2003), Oliver et al. (2008), and Emerick and Reynolds (2012).

Although numerous papers on automatic history matching have been published, most of them have concentrated on a limited type of estimated parameters such as permeability and porosity. They are impractical at present for industrial field application in which we have to calibrate a myriad of other discrete and continuous parameters, including fluid contacts, rock compressibility, and relative permeability. However, automatic history matching is a meaningful research direction and much more effort will be necessary in the future.

2.2 Reservoir production optimization

Production optimization aims at achieving the best development performance for a given reservoir by optimizing well controls. Figure 2 shows the schematic for this optimization process. In order to evaluate the performance of different development programs, various objectives have been proposed during the long research into production optimization. For example, Rosenwald and Green (1974) minimized the difference between the production-demand curve and the flow curve actually attained. Babayev (1975) provided minimum total cost per unit output. Lasdon et al. (1986) maximized the deliverability of a gas reservoir at a specified time, minimized the total gas withdrawal shortfall between the demand schedule and the amount of gas that can actually be delivered in each month, and they also optimized the weighted combinations of the above two objectives. When optimizing production strategies, one often encounters multiple local maxima. This phenomenon may be a good thing sometimes because it means that there are extra degrees of freedom in the optimization problem, which can be used to accomplish other optimization objectives. For instance, Van Essen et al. (2011) incorporated short-term goals into the life-cycle optimization problem and proposed a hierarchical production optimization structure with multiple objectives. Chen et al. (2012) also optimized both long-term and short-term net present value. As more and more oilfields enter the high water cut period, the production costs increase gradually. Therefore, the net present value is commonly selected as the objective function for production optimization. In terms of water flooding projects, it is defined as

$$J(\boldsymbol{u}) = \sum_{n=1}^{L} \left[\sum_{j=1}^{N_P} \left(r_o q_{o,j}^n - r_w q_{w,j}^n \right) - \sum_{i=1}^{N_I} r_{wi} q_{wi,i}^n \right] \frac{\Delta t^n}{(1+i_c)^{t^n}},$$

$$(2)$$

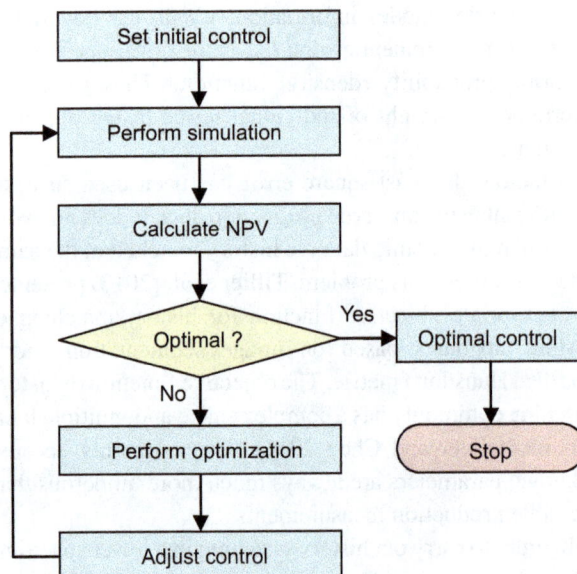

Fig. 2 Flow chart for reservoir production optimization

where J is the net present value and is a function of the control vector $\boldsymbol{u}$; L is the total number of simulation time steps; N_P is the total number of producers; N_I is the total number of injectors; r_o is the oil price; r_w is the produced water treatment cost; r_{wi} is the water injection cost; $q_{o,j}^n$ and $q_{w,j}^n$ are the average oil and water production rates of the jth producer during the nth simulation time step, respectively; $q_{wi,i}^n$ is the average water injection rate of the ith water injection well during the nth simulation time step; i_c is the annual discount rate; Δt^n is the length of the nth simulation time step; t^n is the cumulative time up to the nth simulation time step in years.

According to the general form of optimal control problems, a mathematical model for reservoir production optimization can be written as

$$\max \ J(\boldsymbol{u}) = \sum_{n=1}^{L} \left[\sum_{j=1}^{N_P} \left(r_o q_{o,j}^n - r_w q_{w,j}^n \right) - \sum_{i=1}^{N_I} r_{wi} q_{wi,i}^n \right] \frac{\Delta t^n}{(1+i_c)^{t^n}}$$

$$(3)$$

$$\boldsymbol{A}\boldsymbol{u} \leq \boldsymbol{b} \tag{4}$$

$$\boldsymbol{u}^{\text{low}} \leq \boldsymbol{u} \leq \boldsymbol{u}^{\text{up}}, \tag{5}$$

where Eq. (4) represents the linear or nonlinear constraints; Eq. (5) gives the boundary constraints.

Asheim (1988) maximized the net present value for waterflooding with multiple vertical injectors and a vertical producer by optimizing rate allocation based on the product of permeability and thickness. Brouwer and Jansen (2004) studied static and dynamic waterflooding optimization. For the static one, they kept inflow control valves constant during the displacement process until water breakthrough.

For the dynamic one, they applied gradients calculated with an adjoint method to dynamically optimize the production performance and considered a simple constraint where the total injection was equal to the total production. In order to increase the displacement efficiency, Sudaryanto and Yortsos (2000) optimized the front shape of injected fluid by controlling injection rates. Results showed that the waterflooding optimization was a "bang–bang" control problem, where each control variable took either its minimum or maximum allowed values. Zandvliet et al. (2007) further investigated why and under what conditions waterflooding problems had optimal solutions under bang–bang control. They concluded that waterflooding optimization with simple boundary constraints sometimes had bang–bang optimal solutions, while problems with other general inequality or equality constraints would have a smooth optimal solution. Gao and Reynolds (2006) proposed a log-transformation method to deal with boundary constraints. Alhuthali et al. (2007) also achieved optimal waterflooding management using rate control.

Many pilot tests and commercial projects using enhanced oil recovery (EOR) methods have been performed during the past few decades to improve the development effect of waterflooding. The challenges of huge investment, high cost, and high risk promote research into production optimization for EOR methods. As early as in 1972, Gottfried (1972) proposed a nonlinear programming model for a cyclic steam injection process. He maximized the net present value by optimizing steam injection volume and cycle length. Ramirez et al. (1984) and Fathi and Ramirez (1984) tried to maximize oil production at the minimum injection costs based on the calculus of variations and Pontryagin's weak minimum principle. They optimized development strategies for waterflooding, carbon dioxide (CO_2) flooding, and surfactant flooding. Amit (1986) formulated a two-phase dynamic optimization model which incorporated the relationships between extraction rates, investment decisions, and cumulative oil recovery. Wackowski et al. (1992) applied rigorous decision analysis methodology to find the optimal development strategy for a 20 years CO_2 flooding project. The control variables included CO_2 recycle capacity, CO_2 purchase contract, processing rate, water-to-gas ratio, and slug size. Wu (1996) found that chemical flooding performance was sensitive to operational parameters such as chemical slug size, concentration and adsorption, price of oil and chemicals, annual discount rate, and reservoir permeability. Results showed that the optimal design was a large slug injection of low concentration surfactant and polymer, followed by a small slug of subsequent polymer injection. One of the best examples about gas lift optimization has been achieved by McKie et al. (2001), in which gas lift injection rates, compressor settings, and field fuel consumption for more than 500 wells were optimized to maximize liquid production. Codas et al. (2012) integrate simplified well deliverability models, vertical lift performance relations, and flowing pressure behavior of the surface gathering system to develop a framework of integrated production optimization for complex oil fields.

The above review of production optimization showed that the exploitation method involved has extended from waterflooding to EOR methods, the control variables optimized have extended from simple to complex operational parameters, and that the optimization algorithms applied have extended from gradient-based to gradient-free methods. In addition, production optimization has improved from an open adjustment process to a closed-loop management workflow (Mochizuki et al. 2006), which will be discussed in detail in the next section.

2.3 Closed-loop reservoir management

As more and more production and pressure measurements become available, the reservoir model can be updated to achieve a better estimate of the unknown geological properties. Then the operational parameters should be optimized again based on the newly updated reservoir model. This cycle of model update and production optimization is repeated during the whole process of the reservoir development. This forms the concept of closed-loop reservoir management. The schematic of this closed-loop process is shown in Fig. 3.

Nævdal et al. (2006) applied the EnKF method to update the reservoir simulation model and then optimized operational parameters with an adjoint formulation in order to maximize the economic profits. Sarma et al. (2005) presented a closed-loop management approach for efficient real-time production optimization. Their approach consisted of three key elements: adjoint models for gradient calculations, polynomial chaos expansions for uncertainty propagation, Karhunen–Loeve expansions and Bayesian inversion theory for history matching. Results showed that their approach increased the net present value by 25 %. For a similar problem, Saputelli et al. (2005) proposed a model predictive control method. It was a class of computer control algorithms that explicitly used a plant model for online prediction of future behavior, and computation of appropriate control action subjected to various constraints through online optimization of a cost objective. The method addressed the overwhelming complexity of overall optimization problems by suggesting an oilfield operations hierarchy which entailed different time scales. Due to the rapid development of ensemble-based optimization algorithms, Chen et al. (2009a, b) provided a new closed-loop reservoir management method which integrated an ensemble-based optimization method with the EnKF

Fig. 3 Schematic of closed-loop reservoir management

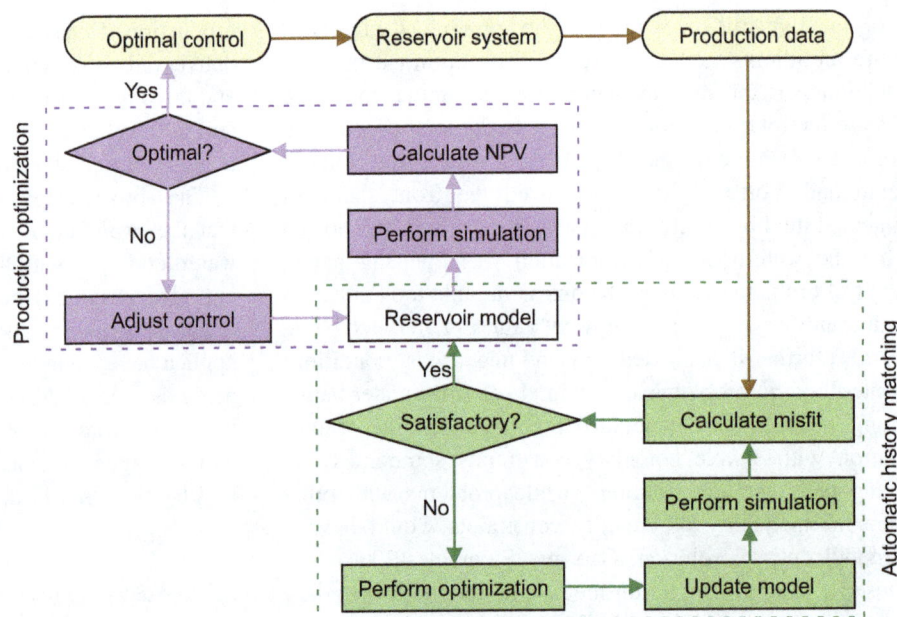

algorithm. Jansen et al. (2008) and Jansen (2011) discussed an emerging technique to increase oil recovery. Their technique is an operational use of model-based optimization which requires a combination of long-term and short-term objectives through multi-level optimization strategies. Moridis et al. (2013) established a self-teaching expert system to increase oil production by improving flooding efficiency and reducing geological uncertainty.

When reviewing closed-loop reservoir management, it is necessary to present the Brugge test case which was prepared for SPE (Society of Petroleum Engineers) Applied Technology Workshop held in Brugge in June 2008. In the test case, well-log data, reservoir structure, 10 years' production data, inverted time-lapse seismic data, and other information necessary are given by TNO (Netherlands Organisation for Applied Science Research) to estimate unknown parameters such as permeability, porosity, and net-to-gross thickness. After history-matched reservoir models were created, water flooding strategies for 20 producers and 10 injectors were optimized. Peters et al. (2010) have summarized in detail the results of the Brugge test case obtained by nine research groups. Briefly, Table 1 compares the reservoir simulators, optimization methods, and the net present values optimized in Year 10. In the table, the net present value was obtained using the optimal strategy of each participant in the Brugge test model. As can be seen, three participants who applied ensemble-based methods in history matching step achieved a similar highest net present value, although their optimization algorithms used in the production optimization step were different. This is mainly because ensemble-based optimization methods had two distinct advantages. First, the search direction was approximated through the correlations provided by ensemble members. Second, the objective function was the expectation of each ensemble member. Therefore, the ensemble-based optimization method was fairly robust with respect to the uncertainty of the estimated geological models.

With the rapid increase in research interests and great improvements of optimization techniques, the original Brugge test case shows its weakness in the low frequency of the feedback loop. In order to keep the Brugge test case as a challenging problem for testing and comparing different techniques in closed-loop reservoir management, Peters et al. (2013) provided a lot of additional data including well constraints and production history from individual well completions for another 20 years, as well as the updated data of oil saturation and reservoir pressure.

3 Optimization methods

According to the techniques for determining search direction and step size, the optimization methods can be classified into three categories: gradient-based algorithms, gradient-free algorithms, and artificial intelligence algorithms. In fact, artificial intelligence algorithms are also independent of gradient information. This paper considers them as an independent category because almost all of them are inspired by intelligent behaviors in nature such as inference, designing, thinking, and learning.

3.1 Gradient-based algorithms

The calculation of derivatives or the Hessian matrix is the key to solving optimization problems using gradient-based

Table 1 Research results of the Brugge test case

Participant	Simulator	History matching method	Production optimization method	Net present value, 10^9 \$
Halliburton	Nexus	Landmark's DMS™	Scatter/tabu search method	3.53[a]
International Research Institute of Stavanger	ECLIPSE 100	Ensemble Kalman filter	Ensemble Kalman filter	4.41
University of Oklahoma/ Chevron	ECLIPSE 100	Randomized maximum likelihood method	Ensemble-based gradient method	4.42
Roxar/Energy Scitech	Tempest MORE	EnABLE™	Sequential experimental design method	4.03
Shell International Exploration and Production BV	MoReS	Ensemble Kalman filter	Adjoint-based gradient method	4.12
Schlumberger	ECLIPSE 100	Selection of realizations based on their fit to the data	Artificial neural networks	4.10
Stanford University/ Chevron	GPRS; ECLIPSE 100	Sequential quadratic programming; Hooke-Jeeves direct search method	Adjoint-based gradient method	4.26
Texas A&M University	FrontSim	Streamline-based generalized travel time inversion	Sequential quadratic programming	4.22
University of Tulsa	ECLIPSE 100/300	Ensemble Kalman filter	Adjoint-based gradient method	4.47

[a] The result was achieved using one control interval per well, while others were obtained using three control intervals per well

Table 2 Comparison of calculation methods for gradients and Hessian matrix

Methods	Calculation principles	Characteristics
Numerical perturbation	Small perturbations of the model parameters and calculation of the production responses	Easy to implement; expensive computational cost; unsuitable for large-scale optimization problems
Sensitivity equation	Differentiation of the flow and transport equations	Difficult to obtain analytical expressions for nonlinear optimization problems
Adjoint method	Optimal control theories and calculus of variations	Easy to implement; dependent on reservoir simulators; hard to transplant elsewhere

algorithms. In terms of production optimization and history matching, researchers have applied various calculation methods including numerical perturbation, sensitivity equation, and adjoint method, as summarized in Table 2.

The roots of the adjoint method can be found in Jacquard (1965). Later, Carter et al. (1974) formulated their work in a better way using Frechet derivatives. He et al. (1997) further extended Carter's work to three dimensions approximately. For single-phase flow problems, Chen et al. (1974) and Chavent et al. (1975) proposed a method which was regarded as what we call the adjoint method now. Li et al. (2003) presented the first formulation of the adjoint method for three-phase flow problems and pointed out that the coefficient matrix of the adjoint equations is simply the transpose of the Newton–Raphson Jacobian matrix used in a fully implicit reservoir simulator. Therefore, the derivation of the individual adjoint equation can be avoided by extracting and saving the Jacobian matrices. Rodrigues (2006) derived the adjoint equations in a much neater way, and provided a method for multiplication of a vector by the

sensitivity matrix or its transpose. Now, the adjoint method has become one of the most efficient methods existing today to compute gradients for gradient-based algorithms.

Generally, gradient-based algorithms can be classified into two categories. One is the first-order methods, which only require the derivative information. For example, the steepest ascent algorithm and the conjugate gradient method have been widely used by many researchers such as Brouwer and Jansen (2004), Sarma et al. (2008a), and Wang et al. (2009). The other category is the second-order methods, which not only require the derivative information but also require the Hessian matrix. Representative methods include Gauss–Newton, Levenberg–Marquardt, sequential quadratic programming (Barnes et al. 2007), and the limited memory Broyden Fletcher Goldfarb Shanno method (LBFGS).

When using the Gauss–Newton method for automatic history matching problems, Wu et al. (1999) introduced an artificially high variance of measurement errors at early iterations to damp the changes in model parameters and thus avoid undershooting or overshooting. Tan and Kalo-gerakis (1991) pointed out that the standard Gauss–Newton and Levenberg–Marquardt methods require the computation of all sensitivity coefficients in order to formulate the Hessian matrix, which seems impossible in reality due to the large number of unknown parameters relative to limited available measurements. In order to eliminate this problem, the quasi-Newton method was introduced by researchers. This method only requires the gradient of the objective function which can be computed from a single adjoint solution as done in Zhang et al. (2002). In order to further improve the computational efficiency and robustness of the LBFGS method, Gao and Reynolds (2006) proposed a new line search strategy, rescaled the model parameters, and applied damping factors to the production data. They also noticed that the new line search strategy had to satisfy the strong Wolfe conditions at each iteration, or the convergence rate would decrease significantly.

The Karhunen–Loeve expansion can create a differentiable parameterization of the numerical model in terms of a small set of independent random variables and deterministic eigenfunctions. With this expansion, the gradient-based algorithms can be applied while honoring the two-point statistics of the geological models (Gavalas et al. 1976). In order to further extend the existing gradient-based history matching techniques to deal with complex geological models characterized by multiple-point geostatistics, Sarma et al. (2007) applied a kernel principal component analysis method to model permeability fields. This method can preserve arbitrary high-order statistics of random fields, and it is able to reproduce complex geology while retaining reasonable computational requirements.

Gradient-based algorithms are widely used in research on production optimization and history matching because of their high computational efficiency and fast convergence behavior. However, these algorithms require detailed knowledge of the numerical simulators. They can hardly be used without adjoint code and they are difficult to transform from one simulator to another.

3.2 Gradient-free algorithms

In order to make full use of the advantages of commercial reservoir simulators when conducting production optimization and automatic history matching, researchers have introduced many gradient-free algorithms including simultaneous perturbation stochastic approximation algorithm (SPSA), EnKF, PSM, new unconstrained optimization algorithm (NEWUOA), and quadratic interpolation model-based algorithm guided by approximated gradient (QIM-AG).

Spall (1998) proposed the SPSA method based on the Kiefer–Wolfowitz algorithm. This new method perturbs all unknown parameters stochastically and simultaneously to generate a search direction at each iteration. The expectation of stochastic SPSA gradients is true gradient and it is always a downhill direction. Therefore, the SPSA method is a stochastic version of the steepest descent algorithm. Based on the simultaneous perturbation idea, Spall further provided a second-order SPSA method, which estimates the Hessian matrix at each iteration. Later Bangerth et al. (2006) described an integer SPSA method and used this modified SPSA method to solve well placement optimization problems. To the best of our knowledge, this is the first time that SPSA has been used in optimal control problems. In terms of closed-loop reservoir management, Wang et al. (2009) discussed the application of SPSA in the step of production optimization. Gao et al. (2007) applied a modified SPSA method for automatic history matching. In their study, the approximate Hessian matrix was calculated using the inverse of the covariance matrix of the prior model. Results showed that the modified SPSA performed almost as well as the steepest descent method. Of course, gradient-based algorithms like LBFGS would be preferred when the gradient can be calculated. Otherwise, the SPSA method may be a good choice. Based on SPSA, Li and reynolds (2011) proposed a stochastic Gaussian search discretion algorithm for history matching problems, and this modified method was successfully used in the well-known PUNQ-S3 (Production forecasting with uncertainty quantification) test case. Zhou et al. (2013) integrate the finite difference method and the SPSA method to optimize polymer flooding in a heterogeneous reservoir. But currently the SPSA method has not been widely used for reservoir optimization problems.

The EnKF algorithm was first proposed by Evensen in 1994 as a Monte Carlo approximation of the Kalman filter in the ocean dynamics literature. This method obtains gradient information through correlations of ensemble members. Nævdal et al. (2002) applied EnKF to estimate near-well permeabilities. Gu and Oliver (2005) examined EnKF for combined parameter and state estimation in a standardized reservoir test case. Gao et al. (2006) found that EnKF and the randomized maximum likelihood method (RML) gave similar computational results. Reynolds et al. (2006) further presented their mathematical connections. They also showed that EnKF may be viewed as updating each ensemble member with a single Gauss–Newton iteration. Liu and Oliver (2004, 2005a, b) investigated a highly nonlinear problem of facies estimation using the EnKF method. When using the EnKF method, Lorentzen et al. (2005) discussed the choice of initial ensemble members, while Wen and Chen (2006) focused on the effect of ensemble size. Emerick and Reynolds (2012, 2013) incorporated automatic history matching in an integrated geo-modeling workflow using the ensemble smoother method.

Although it is not long since the introduction of EnKF into petroleum engineering, this method is developing rapidly and shows huge potential for solving history matching problems because it can deal with the uncertainties.

However, the toy problem in Zafari and Reynolds (2007) showed that the standard EnKF cannot handle multimodal nonlinear problems. To address this limitation, Gu and Oliver (2007) applied an iterative EnKF to assimilate multiphase flow measurements. Li and Reynolds (2009) provided two iterative EnKF procedures, with which they obtained better history matching results than that with the standard EnKF method in two examples including the toy problem. Lorentzen and Nævdal (2011) and Wang and Li (2011) also introduced an iterative extension of EnKF in order to improve estimate results in cases where the relationship between the model and the observations is nonlinear. Agbalaka et al. (2013) proposed a two-stage ensemble-based technique to improve the performance of EnKF for history matching with multiple modes. Chen et al. (2009a, b) addressed non-Gaussian effects through a change in parameterization. They demonstrated the effectiveness of the combination of their methods with the traditional EnKF by history matching of multiphase flow in a heterogeneous reservoir. Supposing that both of the multimodal priors and posteriors can be approximated by Gaussian mixture models, Dovera and Rossa (2011) presented a modified EnKF method especially for multimodal systems. Heidari et al. (2013) combined EnKF with pilot points and gradual deformation to preserve second-order statistical properties, so the departure of constrained petrophysical properties from prior information could be greatly reduced.

Custsódio and Vicente (2007) provided a modified PSM guided by simplex derivatives (SID-PSM). This new method has two modifications over the original PSM algorithm. First, the predefined search directions are ranked using the simplex gradient during the poll step and the search directions closest to the simplex gradient are tried first. Second, when there are enough objective evaluations, a quadratic interpolation model is built in the search step and then it is minimized by a trust-region method. The SID-PSM algorithm has been introduced into reservoir engineering for optimizing the settlement and adsorption parameters of asphaltene in porous media. This method can converge to a global optimum using only the objective evaluations. However, its global convergence is very sensitive to the choice of initial values. Therefore, close attention should be paid when using SID-PSM for practical problems.

For unconstrained optimization problems without derivatives or a Hessian matrix available, Powell (2008) proposed a NEWUOA method. It is a quadratic model-based gradient-free trust-region algorithm based on quadratic interpolation. At least $N_u + 2$ (N_u is the number of

control variables) objective evaluations are needed to build the initial quadratic model before the first optimization procedure can be achieved. So it can hardly be used when the number of control variables is very large. In order to improve the computational efficiency of NEWUOA, Zhao et al. (2011) developed a QIM-AG method. This method requires a minimum of only one interpolation point to build a quadratic model at each iteration. It is similar to the quasi-Newton method and converges very fast.

Taking the net present value as objective function, Zhao et al. (2011) compared the performance of various gradient-free algorithms. As shown in Fig. 4, the quadratic interpolation model-based algorithm guided by ensemble-based gradient (QIM-EnOpt) obtained the highest net present value. The quadratic interpolation model-based algorithm guided by SPSA gradient (QIM-SPSA) obtained a similar net present value but converges much faster than QIM-EnOpt and EnOpt. The SPSA and NEWUOA methods achieved similar values, but the computational efficiency of NEWUOA was much lower at the beginning of the optimization process. In this test case, the particle swarm optimization method and SID-PSM algorithm performed much worse than the other algorithms.

3.3 Artificial intelligence algorithms

Artificial intelligence algorithms have been used to solve production optimization and history matching problems for a long time. The representative methods include the simulated annealing algorithm, genetic algorithm, artificial neural networks, particle swarm optimization method, and the tabu search method.

The simulated annealing algorithm was first introduced into reservoir engineering. Farmer (1992) applied this algorithm to generate rock models with two-point geostatistics properties. Qian (1993) introduced Markov random field theory into the basic simulated annealing algorithm

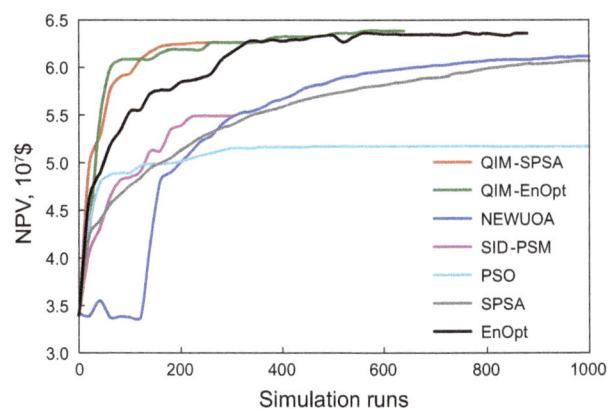

Fig. 4 Comparison of various gradient-free algorithms by Zhao et al. (2011)

and thus transformed it into a probabilistic uphill algorithm. When solving automatic history matching problems, Ouenes et al. (1993) applied a simulated annealing algorithm directly while Carter and Romero (2002) combined it with other techniques such as geostatistics, a pilot point method, and a genetic algorithm. The convergence of the simulated annealing algorithm is sensitive to the choice of initial temperature and reduction factor. If the reduction factor is too large, many extreme points will be missed. However, if the reduction factor is too small, the simulated annealing algorithm will converge very slowly.

The genetic algorithm was proposed by Holland (1975) based on the idea of population evolution. This method has been used as a global optimizer for automatic history matching problems by Tokuda et al. (2004). Compared with a simulated annealing algorithm, a genetic algorithm starts with a population of realizations rather than a single realization. However, there is no mathematical guarantee for a genetic algorithm to reach a global optimum. In spite of these drawbacks, genetic algorithms still have much appeal because they can be easily parallelized and connected to any existing reservoir simulators.

Artificial neutral networks are a rough approximation and simplified simulation of a biological neuron network system. Generally, the neutral networks are used to establish proxy models. For this purpose, a small number of simulations are run at first to build a nonlinear relationship between the objective and unknown variables. Then the relationship can be used for reservoir production optimization in order to reduce the computational cost (Elkamel 1998; Saputelli et al. 2002). Artificial neutral networks were also used directly as an optimizer by Ramgulam et al. (2007) in their studies of history matching. Although artificial neutral networks have been widely used, the procedures for network construction are still not established.

The particle swarm optimization method was proposed by Kennedy and Eberhart based on social behavior observed in nature. It has been successfully used in well placement optimization (Onwunalu and Durlofsky 2009), automatic history matching, and water flooding optimization. This method can find the global optimum with a high probability and can be connected to any numerical simulators.

In fact, artificial intelligence algorithms are stochastic gradient-free methods. These algorithms need much more numerical simulations than the deterministic methods to converge. In addition, the input parameters for implementing gradient-free algorithms are commonly determined by the trial and error method. Therefore, the gradient-free algorithms should be modified and integrated with other optimizers in practical applications.

Based on the discussions in above three sections, the main characteristics for different kinds of optimization algorithms are summarized in Table 3.

4 Future developments

In order to solve the growing demand–supply gap, many studies have been carried out in terms of closed-loop reservoir management. However, most of the studies are focused on academic research and they are far from large-scale field application. As for field application, oilfield engineers are mainly concerned about three issues: processing time, decision veracity, and operating convenience. In order to reduce the processing time, efficient optimization methods and parallel distributed systems should be paid more attention. It is important for increasing decision veracity to carry out more uncertainty analyses. Since most of the oilfield engineers are not familiar with the theoretical bases of closed-loop reservoir management, integrated software is necessary to improve their operating convenience.

4.1 Hybrid solvers

Various optimization methods have been introduced into reservoir development engineering during the past decades. Each algorithm has its own advantages as well as drawbacks. For example, the gradient-based algorithms have high computational efficiency and fast convergence rate, but the calculation of gradients requires detailed knowledge of the numerical simulators. They can hardly be used without adjoint code and they are difficult to transform from one simulator to another. The gradient-free algorithms can converge to global optimum with a high probability and they can make full use of the technical advantages of commercial reservoir simulators. However, the computational efficiency and convergence rate of gradient-free algorithms are not satisfactory. Therefore, we should combine different algorithms together to develop new hybrid solvers, which can incorporate the advantages of different optimization methods.

4.2 Parallel algorithms and distributed systems

As well known, numerical simulations are necessary in closed-loop reservoir management. However, for large-scale reservoir optimization problems, even a single numerical simulation requires several days and the total computational cost is unbearably expensive. To address this problem, it is necessary to develop parallel algorithms and distributed systems. The parallel algorithms should necessarily conserve the mathematical convergence behavior of the original algorithms. Traditionally, the distributed systems can be classified into two categories. The first category is homogeneous distributed system which is a cluster of many similar computers or work stations. The

Table 3 Characteristics of different optimization algorithms

Methods	Representive algorithms	Characteristics
Gradient-based algorithms	Steepest ascent algorithm; conjugate gradient method; LBFGS method; Levenberg–Marquardt algorithm; Gauss–Newton method	High computational efficiency; dependent on adjoint gradient which generally needs detailed knowledge of simulator numerics to implement; difficult to transform from one simulator to another
Gradient-free algorithms	SPSA; EnOpt; EnKF; SID-PSM; NEWUOA; QIM-AG	Independent of detailed knowledge of simulator numerics; capable of connecting to any reservoir simulators; implement easily
Artificial intelligence algorithms	Simulated annealing algorithm; genetic algorithm; artificial neural networks; particle swarm optimization algorithm	Low computational efficiency; independent of gradient or Hessian matrix information; expensive computation cost; no guarantee of convergence

other category is heterogeneous distributed system which is constructed by a set of work stations or computers with different architectures. Currently, with the rapid development of multicore computers and graphic processing unit (GPU) technology, many parallel programming methods have been proposed on the basis of Open Multiprocessing (OpenMp) and compute unified device architecture (CUDA). For example, Wu et al. (2014) discussed a multilevel preconditioner in a new-generation reservoir simulator and its implementation on multicore computers using OpenMP. In order to conduct large-scale closed-loop reservoir management, parallel algorithms and distributed systems should be emphasized in the future studies.

4.3 Uncertainty analyses

It is well known that automatic history matching is an underdetermined inverse problem because the number of unknown parameters is much larger than the number of available production measurements. In other words, many different parameter estimates can provide satisfactory data fits. However, some of the estimates may grossly lead to erroneous predictions of future production behavior. Therefore, developing methods that can quantify the uncertainty of parameter estimates are necessary to the success of automatic history matching. For example, Barker et al. (2001) applied the Markov chain Monte Carlo method to quantify the uncertainty. When predicting permeability from well logs, Olatunji et al. (2011) adopted a type-2 fuzzy logic system which is good at handling uncertainties in measurements and data used to calibrate the parameters. Arnold et al. (2013) carried out a hierarchical benchmark case study for history matching, uncertainty quantification, and reservoir characterization. Since automatic history matching has become increasingly important in the field of reservoir description, more efforts should be paid to uncertainty analyses to reduce the decision risks in the future.

4.4 Integrated software

In order to exploit the limited oil reserves more efficiently and economically, the field application of closed-loop reservoir management becomes increasingly significant. However, most of the oilfield engineers do not know much about its theoretical basis. Therefore, it is necessary to develop integrated software which can be used conveniently as a black box. As far as we know, some commercial simulators have done much work in this field. For example, the Computer Modelling Group Ltd has developed a CMOST studio, which integrated history matching, optimization, sensitivity analysis, and uncertainty assessment tool. Some optimization methods such as the Latin hypercube design, brute force search, random search, and PSO have been available for history matching and optimization tasks in CMOST studio. The Schlumberger developed MEPO software to deal with production optimization, uncertainty assessment, and semi-automatic history matching. Although CMOST studio and MEPO software did much work, automatic closed-loop reservoir management has not yet been achieved, especially for large-scale field applications. Therefore, much more effort should be paid on the development of integrated software, which can help the engineers to manage oilfields easily and scientifically.

5 Conclusions

Closed-loop reservoir management consisting of automatic history matching and reservoir production optimization is an effective technique to improve the technical and economic effect of reservoir development. Both of the steps are complex optimization problems which can be solved by gradient-based algorithms, gradient-free algorithms, and artificial intelligence algorithms. The computational efficiency of gradient-based algorithms is the highest but they are difficult to transform from one simulator to another. The gradient-free and artificial intelligence algorithms are independent of numerical simulators but they need much more simulations to find the optimum. More research should be conducted on hybrid optimization methods, parallel algorithms and distributed systems, uncertainty analyses and integrated software in the future.

Acknowledgments The authors greatly appreciate the financial supports from the Important National Science & Technology Specific Projects of China (Grant No. 2011ZX05024-004), the Natural Science Foundation for Distinguished Young Scholars of Shandong Province, China (Grant No. JQ201115), the Program for New Century Excellent Talents in University (Grant No. NCET-11-0734), the Fundamental Research Funds for the Central Universities (Grant No. 13CX05007A, 13CX05016A), and the Program for Changjiang Scholars and Innovative Research Team in University (IRT1294).

References

Abacioglu Y, Oliver DS, Reynolds AC. Efficient reservoir history matching using subspace vectors. Comput Geosci. 2001;5(2):151–72.

Agarwal A, Blunt MJ. Streamline-based method with full-physics forward simulation for history-matching performance data of a North Sea Field. SPE J. 2003;8(2):171–80 (paper SPE 84952).

Agbalaka CC, Oliver DS. Application of the EnKF and localization to automatic history matching of facies distribution and production data. Math Geosci. 2008;40(4):353–74.

Agbalaka CC, Stern D, Oliver DS. Two-stage ensemble-based history matching with multiple modes in the objective function. Comput Geosci. 2013;55:28–43.

Alhuthali A, Oyerinde D, Datta-Gupta A. Optimal waterflood management using rate control. SPE Reserv Eval Eng. 2007;10(5):539–51 (paper SPE 102478).

Amit R. Petroleum reservoir exploitation: switching from primary to secondary recovery. Oper Res. 1986;34(4):534–49.

Arnold D, Demyanov V, Tatum D, et al. Hierarchical benchmark case study for history matching, uncertainty quantification and reservoir characterisation. Comput Geosci. 2013;50:4–15.

Aronofsky JS, Williams AC. The use of linear programming and mathematical models in underground oil production. Manage Sci. 1962;8(4):394–407.

Asheim H. Maximization of water sweep efficiency by controlling production and injection rates. Paper SPE 18365 presented at European petroleum conference, London; 16–19 October 1988.

Babayev DA. Mathematical models for optimal timing of drilling on multilayer oil and gas fields. Manage Sci. 1975;21(12):1361–9.

Bangerth W, Klie H, Wheeler M, et al. On optimization algorithm for the reservoir oil well placement problem. Comput Geosci. 2006;10(3):303–19.

Barker JW, Cuypers M, Holden L. Quantifying uncertainty in production forecasts: another look at the PUNQ-S3 problem. SPE J. 2001;6(4):433–41 (paper SPE 74707).

Barnes RJ, Kokossis A, Shang ZG. An integrated mathematical programming approach for the design and optimisation of offshore fields. Comput Chem Eng. 2007;31(5–6):612–29.

Barragán HV, Vázquez RR, Rosales ML, et al. A strategy for simulation and optimization of gas and oil production. Comput Chem Eng. 2005;30(2):215–27.

Bieker HP, Slupphaug O, Johansen TA. Real-time production optimization of oil and gas production systems: a technology survey. SPE Prod Oper. 2007;22(4):382–91 (paper SPE 99446).

Bissell RC, Sharma Y, Killough JE. History matching using the method of gradients: two case studies. Paper SPE 28590 presented at SPE annual technical conference and exhibition, New Orleans, Louisiana; 25–28 September 1994.

Brouwer DR, Jansen JD. Dynamic optimization of water flooding with smart wells using optimal control theory. SPE J. 2004;9(4):391–402 (paper SPE 78278).

Caers J. Efficient gradual deformation using a streamline-based proxy method. J Petrol Sci Eng. 2003;39(1–2):57–83.

Carter J, Romero C. Using genetic algorithm to invert numerical simulations. Presented at the 8th European conference on the mathematics of oil recovery, Freiberg, Germany; 3–6 September 2002.

Carter RD, Kemp LF, Pierce AC, et al. Performance matching with constraints. SPE J. 1974;14(2):187–96 (paper SPE 4260).

Chacón E, Besembel I, Hennet JC. Coordination and optimization in oil and gas production complexes. Comput Ind. 2004;53(1):17–37.

Chang H, Zhang D, Lu Z. History matching of facies distributions with the EnKF and level set parameterization. J Comput Phys. 2010;229(20):8011–30.

Chavent GM, Dupuy M, Lemonnier P. History matching by use of optimal theory. SPE J. 1975;15(1):74–86 (paper SPE 4627).

Chen C, Li G, Reynolds AC. Robust constrained optimization of short- and long-term net present value for closed-loop reservoir management. SPE J. 2012;17(3):849–64 (paper SPE 141314).

Chen WH, Gavalas GR, Seinfeld JH, et al. A new algorithm for automatic history matching. SPE J. 1974;14(6):593–608 (paper SPE 4545).

Chen Y, Oliver DS, Zhang D. Data assimilation for nonlinear problems by ensemble Kalman filter with reparameterization. J Petrol Sci Eng. 2009a;66(1–2):1–14.

Chen Y, Oliver DS, Zhang D. Efficient ensemble-based closed-loop production optimization. SPE J. 2009b;14(4):634–45 (paper SPE 112873).

Cheng H, Kharghoria A, He Z, et al. Fast history matching of finite-difference models using streamline derived sensitivities. SPE Reserv Eval Eng. 2005;8(5):426–36 (paper SPE 89447).

Cheng H, Wen X, Milliken WJ et al. Field experiences with assisted and automatic history matching using streamline models. Paper SPE 89857 presented at SPE annual technical conference and exhibition, Houston, Texas; 26–29 September 2004.

Chu L, Reynolds AC, Oliver DS. Computation of sensitivity coefficients for conditioning the permeability field to well-test data. In Situ. 1995;19(2):179–223.

Codas A, Campos S, Camponogara E, et al. Integrated production optimization of oil fields with pressure and routing constraints: the Urucu field. Comput Chem Eng. 2012;46:178–9.

Custódio AL, Vicente LN. Using sampling and simplex derivatives in pattern search methods. SIAM J Optim. 2007;18(2):537–55.

Dovera L, Rossa ED. Multimodal ensemble Kalman filtering using Gaussian mixture models. Comput Geosci. 2011;15(2):307–23.

Elkamel A. An artificial neural network for predicting and optimizing immiscible flood performance in heterogeneous reservoirs. Comput Chem Eng. 1998;22(11):1699–709.

Emerick AA, Reynolds AC. Combining the ensemble Kalman filter with markov-chain monte carlo for improved history matching and uncertainty characterization. SPE J. 2012;17(2):418–40 (paper SPE 141336).

Emerick AA, Reynolds AC. Ensemble smoother with multiple data assimilation. Comput Geosci. 2013;55:3–15.

Farmer CL. Numerical rocks, the mathematics of oil recovery. Oxford: Clarendon Press; 1992.

Fathi Z, Ramirez WF. Optimal injection policies for enhanced oil recovery: part 2-surfactant flooding. SPE J. 1984;24(3):333–41 (paper SPE 12814).

Gao G, Reynolds AC. An improved implementation of the LBFGS algorithm for automatic history matching. SPE J. 2006;11(1):5–17 (paper SPE 90058).

Gao G, Zafari M, Reynolds AC. Quantifying the uncertainty for the PUNQ-S3 problem in a Bayesian setting with RML and EnKF. SPE J. 2006;11(4):506–15 (paper SPE 93324).

Gao G, Li G, Reynolds AC. A stochastic algorithm for automatic history matching. SPE J. 2007;12(2):196–208 (paper SPE 90065).

Gavalas GR, Shah PC, Seinfeld JH. Reservoir history matching by Bayesian estimation. SPE J. 1976;16(6):337–50 (paper SPE 5740).

Gomez S, Gosselin O, Barker JB. Gradient-based history matching with a global optimization method. SPE J. 2001;6(2):200–8 (paper SPE 71307).

Gottfried BS. Optimization of a cyclic steam injection process using penalty functions. SPE J. 1972;12(1):13–20 (paper SPE 3329).

Grimstad AA, Mannseth T, Aanonsen S, et al. Identification of unknown permeability trends from history matching of production data. SPE J. 2004;9(4):419–28 (paper SPE 77485).

Grimstad AA, Mannseth T, Nævdal G, et al. Adaptive multiscale permeability estimation. Comput Geosci. 2003;7(1):1–25.

Gu Y, Oliver DS. History matching of the PUNQ-S3 reservoir model using the ensemble Kalman filter. SPE J. 2005;10(2):217–24. (paper SPE 89942).

Gu Y, Oliver DS. An iterative ensemble Kalman filter for multiphase fluid flow data assimilation. SPE J. 2007;12(4):438–46 (paper SPE 108438).

Gunnerud V, Foss B. Oil production optimization-A piecewise linear model, solved with two decomposition strategies. Comput Chem Eng. 2010;34(11):1803–12.

Gupta AD, King MJ. Streamline simulation: theory and practice. Textbook series, SPE textbook series Vol. 11. Richardson, Texas; 2007.

He J, Sarma P, Durlofsky LJ. Reduced-order flow modeling and geological parameterization for ensemble-based data assimilation. Comput Geosci. 2013;55:54–69.

He N, Reynolds AC, Oliver DS. Three-dimensional reservoir description from multiwell pressure data and prior information. SPE J. 1997;2(3):312–27.

Heidari L, Gervais V, Ravalec ML, et al. History matching of petroleum reservoir models by the Ensemble Kalman Filter and parameterization methods. Comput Geosci. 2013;55:84–95.

Holland JH. Adaptation in natural and artificial systems: an introductory analysis with applications to biology, control, and artificial intelligence. Michigan: University of Michigan Press; 1975.

Hu LY, Zhao Y, Liu Y, et al. Updating multipoint simulations using the ensemble Kalman filter. Comput Geosci. 2013;51:7–15.

Jacquard P. Permeability distribution from field pressure data. SPE J. 1965;5(4):281–94 (paper SPE 1307).

Jafarpour B, McLaughlin DB. History matching with an ensemble Kalman filter and discrete cosine parameterization. Comput Geosci. 2008;12(2):227–44.

Jafarpour B, Goyal VK, McLaughlin DB, et al. Compressed history matching: exploiting transform-domain sparsity for regularization of nonlinear dynamic data integration problems. Math Geosci. 2010;42(1):1–27.

Jahns HO. A rapid method for obtaining a two-dimensional reservoir description from well pressure response data. SPE J. 1966;6(4):315–27 (paper SPE 1473).

Jansen JD. Adjoint-based optimization of multi-phase flow through porous media—a review. Comput Fluids. 2011;46(1):40–51.

Jansen JD, Bosgra OH, Van den Hof PMJ. Model-based control of multiphase flow in subsurface oil reservoirs. J Process Control. 2008;18(9):846–55.

Jansen JD, Brouwer DR, Naevdal G, et al. Closed-loop reservoir management. First Break. 2005;23(1):43–8.

Knudsen BR, Foss B. Shut-in based production optimization of shale-gas systems. Comput Chem Eng. 2013;58:54–67.

Lasdon L, Coffman PE, Macdonald R, et al. Optimal hydrocarbon reservoir production policies. Oper Res. 1986;34(1):40–54.

Lee AS, Aronofsky JS. A linear programming model for scheduling crude oil production. J Petrol Technol. 1958;10(7):51–4.

Lee TY, Seinfeld JH. Estimation of two-phase petroleum reservoir properties by regularization. J Comput Phys. 1987;69(2):397–419.

Li G, Reynolds AC. Iterative ensemble Kalman filters for data assimilation. SPE J. 2009;14(3):496–505 (paper SPE 109808).

Li G, Reynolds AC. Uncertainty quantification of reservoir performance prediction using a stochastic optimization algorithm. Comput Geosci. 2011;15(3):451–62.

Li R, Reynolds AC, Oliver DS. History matching of three-phase flow production data. SPE J. 2003;8(4):328–40 (paper SPE 87336).

Liu N, Oliver DS. Evaluation of Monte Carlo methods for assessing uncertainty. SPE J. 2003;8(2):188–95 (paper SPE 84936).

Liu N, Oliver DS. Automatic history matching of geologic facies. SPE J. 2004;9(4):429–36 (paper SPE 84594).

Liu N, Oliver DS. Critical evaluation of the ensemble kalman filter on history matching of geologic facies. SPE Reserv Eval Eng. 2005a;8(6):470–7 (paper SPE 92867).

Liu N, Oliver DS. Ensemble Kalman filter for automatic history matching of geologic facies. J Petrol Sci Eng. 2005b;47(3–4):147–61.

Lorentzen RJ, Nævdal G. An iterative ensemble Kalman filter. IEEE Trans Autom Control. 2011;56(8):1990–5.

Lorentzen RJ, Nævdal G, Vallès B et al. Analysis of the ensemble Kalman filter for estimation of permeability and porosity in reservoir models. Paper SPE 96375 presented at SPE annual technical conference and exhibition, Texas, Dallas; 9–12 October 2005.

Marsily G, Lavedan C, Boucher M. Interpretation of interference tests in a well field using geostatistical techniques to fit the permeability distribution in a reservoir model. In: Geostatistics for natural resources characterization. Dordrecht: Marechal Reidel Publishing Company; 1984. p. 831–49.

McFarland JW, Lasdon L, Loose V. Development planning and management of petroleum reservoirs using tank models and nonlinear programming. Oper Res. 1984;32(2):270–89.

McKie CJN, Rojas EA, Quintero NM. Economic benefits from automated optimization of high pressure gas usage in an oil production system. Paper SPE 67187 presented at SPE production and operations symposium, Oklahoma City, Oklahoma; 24–27 March 2001.

Mochizuki S, Saputelli LA, Kabir CS, et al. Real-time optimization: classification and assessment. SPE Prod Oper. 2006;21(4):455–66 (paper SPE 90213).

Moridis GJ, Reagan MT, Kuzma HA, et al. SeTES: a self-teaching expert system for the analysis, design, and prediction of gas production from unconventional gas resources. Comput Geosci. 2013;58:100–15.

Nævdal G, Brouwer DR, Jansen JD. Waterflooding using closed-loop Control. Comput Geosci. 2006;10(1):37–60.

Nævdal G, Mannseth T, Vefring E. Near well reservoir monitoring through ensemble Kalman filter. Paper SPE 75235 presented at SPE/DOE improved oil recovery symposium, Tulsa, Oklahoma; 13–17 April 2002.

Negrete E, Datta-Gupta A, Choe J. Streamline-assisted ensemble kalman filter for rapid and continuous reservoir model updating. SPE Reserv Eval Eng. 2008;11(6):1046–60 (paper SPE 104255).

Olatunji SO, Selamat A, Abdulraheem A. Modeling the permeability of carbonate reservoir using type-2 fuzzy logic systems. Comput Ind. 2011;62(2):147–63.

Oliver DS. A comparison of the value of interference and well-test data for mapping permeability and porosity. In Situ. 1996a;20(1):41–59.

Oliver DS. Multiple realizations of the permeability field from well test data. SPE J. 1996b;1(2):145–54 (paper SPE 27970).

Oliver DS, Chen Y. Recent progress on reservoir history matching: a review. Comput Geosci. 2011;15(1):185–221.

Oliver DS, Reynolds AC, Liu N. Inverse theory for petroleum reservoir characterization and history matching. Cambridge: Cambridge University Press; 2008.

Onwunalu J, Durlofsky L. Application of a particle swarm optimization algorithm for determining optimum well location and type. Comput Geosci. 2009;14(1):183–98.

Ouenes A, Brefort B, Meunier G et al. A new algorithm for automatic history matching: application of simulated annealing method (SAM) to reservoir inverse modeling. SPE general paper. (Paper SPE 26297); 1993.

Peters E, Chen Y, Leeuwenburgh O, et al. Extended Brugge benchmark case for history matching and water flooding optimization. Comput Geosci. 2013;50:16–24.

Peters L, Arts RJ, Brouwer GK, et al. Results of the Brugge benchmark study for flooding optimization and history matching. SPE Reserv Eval Eng. 2010;13(3):391–405 (Paper SPE 119094).

Powell MJD. Developments of NEWUOA for minimization without derivatives. IMA J Numer Anal. 2008;28(4):649–64.

Qian X. Simulated annealing and conditional simulation. Master Thesis in Mathematics. New Mexico Tech, Socorro; 1993.

Ramgulam A, Ertekin T, Flemings PB. Utilization of artificial neural networks in the optimization of history matching. Paper SPE 107468 presented at Latin American & caribbean petroleum engineering conference, Buenos Aires, Argentina; 15–18 April 2007.

Ramirez WF, Fathi Z, Cagnol JL. Optimal injection policies for enhanced oil recovery: part 1 theory and computational strategies. SPE J. 1984;24(3):328–32 (Paper SPE 11285).

Reynolds AC, Zafari M, Li G. Iterative forms of the ensemble Kalman filter. Presented at the proceedings of 10th European conference on the mathematics of oil recovery, Amsterdam; 4–7 September 2006.

Rodrigues JRP. Calculating derivatives for automatic history matching. Comput Geosci. 2006;10(1):119–36.

Rosenwald GW, Green DW. A method for determining the optimum location of wells in a reservoir using mixed-integer programming. SPE J. 1974;14(1):44–54 (paper SPE 3981).

Saputelli LA, Nikolaou M, Economides MJ. Real-time reservoir management: a multi-scale adaptive optimization and control approach. Comput Geosci. 2005;10(1):61–96.

Saputelli L, Malki H, Canelon J et al. A critical overview of artificial neural network applications in the context of continuous oil field optimization. Paper SPE 77703 presented at SPE annual technical conference and exhibition, San Antonio, Texas; 29 September–2 October 2002.

Sarma P, Chen WH, Durlofsky LJ, et al. Production optimization with adjoint models under nonlinear control-state path inequality constraints. SPE Reserv Eval Eng. 2008a;11(2):326–39 (paper SPE 99959).

Sarma P, Durlofsky LJ, Aziz K. Kernel principal component analysis for efficient differentiable parameterization of multipoint geostatistics. Math Geosci. 2008b;40:3–32.

Sarma P, Durlofsky LJ, Aziz K et al. A new approach to automatic history matching using kernel PCA. Paper SPE 106176 presented at SPE reservoir simulation symposium, Houston, Texas, USA; 26–28 February 2007.

Sarma P, Durlofsky L, Aziz K. Efficient closed-loop production optimization under uncertainty. Paper SPE 94241 presented at SPE Europec/EAGE annual conference, Madrid, Spain; 13–16 June 2005.

Sequeira SE, Graells M, Luis P. Real-time evolution for on-line optimization of continuous processes. Ind Eng Chem Res. 2002;41(7):1815–25.

Spall JC. Implementation of the simultaneous perturbation algorithm for stochastic optimization. IEEE Trans Aerosp Electron Syst. 1998;34(3):817–23.

Sudaryanto B, Yortsos YC. Optimization of fluid front dynamics in porous media using rate control. I. equal mobility fluids. Phys. Fluids. 2000;12(7):1656–70.

Tan TB, Kalogerakis N. A fully implicit three-dimensional three-phase simulator with automatic history-matching capability. Paper SPE 21205 presented at SPE symposium on reservoir simulation, Anaheim, California; 17–20 February 1991.

Tarantola A. Inverse problem theory and methods for model parameter estimation. Philadelphia: SIAM; 2005.

Tavakoli R, Reynolds AC. History matching with parameterization based on the SVD of a dimensionless sensitivity matrix. SPE J. 2010;15(2):495–508 (paper SPE 118952).

Tavakoli R, Reynolds AC. Monte Carlo simulation of permeability fields and reservoir performance predictions with SVD parameterization in RML compared with EnKF. Comput Geosci. 2011;15(1):99–116.

Tavallali MS, Karimi IA, Teo KM, et al. Optimal producer well placement and production planning in an oil reservoir. Comput Chem Eng. 2013;55:109–25.

Tillier E, Veiga SD, Derfoul R. Appropriate formulation of the objective function for the history matching of seismic attributes. Comput Geosci. 2013;51:64–73.

Tokuda N, Takahashi S, Watanabe M. Application of genetic algorithm to history matching for core flooding. Paper SPE 88621 presented at SPE Asia pacific oil and gas conference and exhibition, Perth, Australia; 18–20 October 2004.

Van Essen GM, Van den Hof PMJ, Jansen JD. Hierarchical long-term and short-term production optimization. SPE J. 2011;16(1):191–9 (paper SPE 124332).

Wackowski RK, Stevens CE, Masoner LO et al. Applying rigorous decision analysis methodology to optimization of a tertiary recovery project: Rangely Weber Sand Unit, Colorado. Paper SPE 24234 presented at oil and gas economics, finance and management conference, London, United Kingdom 28–29 April 1992.

Wang C, Li G, Reynolds AC. Production optimization in closed-loop reservoir management. SPE J. 2009;14(3):506–23 (paper SPE 109805).

Wang Y, Li M. Reservoir history matching and inversion using an iterative ensemble Kalman filter with covariance localization. Petrol Sci. 2011;8(3):316–27.

Wasserman ML, Emanuel AS. History matching three-dimensional models using optical control theory. J Can Petrol Technol. 1976;15(4):70–7.

Wattenbarger RA. Maximizing seasonal withdrawals from gas storage reservoirs. J Petrol Technol. 1970;22(8):994–8.

Wen X, Chen W. Real time reservoir model updating using the ensemble Kalman filter with confirming opinion. SPE J. 2006;11(4):431–42.

Wu SH, Zhang CS, Li QY, et al. A multilevel preconditioner and its shared memory implementation for new generation reservoir simulator. Petrol Sci. 2014;11(4):540–9.

Wu W. Optimum design of field-scale chemical flooding using reservoir simulation. PhD Dissertation. University of Texas, Austin. 1996.

Wu Z, Reynolds AC, Oliver DS. Conditioning geostatistical models to two-phase production data. SPE J. 1999;4(2):142–55 (paper SPE 56855).

Yang PH, Watson AT. Automatic history matching with variable-metric methods. SPE Reserv Eng. 1988;3(3):995–1001 (paper SPE 16977).

Zafari M, Reynolds AC. Assessing the uncertainty in reservoir description and performance predictions with the ensemble Kalman filter. SPE J. 2007;12(3):382–91 (paper SPE 95750).

Zandvliet MJ, Bosgra OH, Van den Hof PMJ, et al. Bang-bang control and singular arcs in reservoir flooding. J Petrol Sci Eng. 2007;58(1–2):186–200.

Zhang F, Reynolds AC, Oliver DS. Evaluation of the reduction in uncertainty obtained by conditioning a 3D stochastic channel to multiwell pressure data. Math Geol. 2002;34(6):715–42.

Zhang F, Skjervheim JA, Reynolds AC, et al. Automatic history matching in a Bayesian framework: example applications. SPE Reserv Eval Eng. 2005;8(3):214–23 (paper SPE 84461).

Zhao H, Chen C, Li G et al. Maximization of a dynamic quadratic interpolation model for production optimization. Paper SPE 141317 presented at SPE reservoir simulation symposium, the woodlands, Texas, USA; 21–23 February 2011.

Zhou K, Hou J, Zhang X, et al. Optimal control of polymer flooding based on simultaneous perturbation stochastic approximation method guided by finite difference gradient. Comput Chem Eng. 2013;55:40–9.

15

Preparation of furfural and reaction kinetics of xylose dehydration to furfural in high-temperature water

De-Run Hua[1,2] · Yu-Long Wu[2,3] · Yun-Feng Liu[2] · Yu Chen[2] · Ming-De Yang[1] ·
Xin-Ning Lu[2] · Jian Li[2]

Abstract Factors influencing dehydration of xylose to furfural, such as catalyst and extract agents, were investigated. Results indicated that high-temperature water may substitute for solid and liquid acid as a catalyst, and ethyl butyrate improved furfural yield for the high distribution coefficient. A furfural yield of 75 % was obtained at 200 °C for 3 h in ethyl butyrate/water. The reaction kinetics of xylose dehydration to furfural was investigated and it was found that the reaction order was 0.5, and the activation energy was 68.5 kJ/mol. The rate constant k showed a clear agreement with the Arrhenius law from 160 to 200 °C.

Keywords Xylose · Furfural · SO_3H-SBA-15 · High-temperature water

1 Introduction

Furfural, which is readily obtained from renewable biomass, is a key biomass-derived chemical that can be used to replace petrochemicals (Dias et al. 2005a; Lichtenthaler 1998). In addition, furfural is a common industrial solvent and an intermediate for preparation of fine-chemical products, resins/plastics, and biofuel (Climent et al. 2011; Li et al. 2014).

Furfural is conventionally produced by hydrolysis of a variety of biomass sources including almond shell, sugarcane, and corn with the aid of liquid acid catalysts at high temperatures (200–250 °C) (Kim et al. 2011). However, liquid acid catalysts are toxic and corrosive, and their applications can generate large amounts of toxic wastes. In point of the principles of green chemistry, the replacement of liquid acid catalysts with solid acid catalysts, such as zeolites (O'Neill et al. 2009) and acid salts (Dias et al. 2006), can overcome these drawbacks (Agirrezabal-Telleria et al. 2012a, b; Antunes et al. 2012). Additionally, solvent, which can rapidly and continuously remove furfural from an aqueous phase, is an important factor in increasing furfural yield. Zhang et al. (2012) reported that 1-butanol from biomass-based carbohydrates can be used as a renewable extraction solvent in a biphasic system on a MCM-41 catalyst, and 1-butanol can obviously increase the yield of furfural. A biphasic reactor system can easily separate furfural from an aqueous phase (Chheda et al. 2007). So researchers are paying more attention to green process for furfural preparation, and more rational catalytic systems have been developed recently. High-temperature water (HTW) was applied in the catalytic conversion of renewable lignocellulosic biomass to furfural (Akiya and Savage 2002). HTW can partly replace acid catalyst to produce high ionic products (Jing and LÜ 2007). Recently, kinetic studies of furfural formation have been conducted by using xylose or hemicellulose as a starting material on heterogeneous catalysts (Dias et al. 2005b; O'Neill et al. 2009), such as mineral acids (Marcotullio and De Jong 2010; Morinelly et al. 2009). Additionally, some methods to slow or inhibit side reactions (in Scheme 1) are studied by addition of extraction agent during reaction. As far as we know, biphasic systems are now

✉ Yu-Long Wu
wylong@tsinghua.edu.cn

[1] Chemical Institute of Chemical Industry, Gannan Normal University, Jiangxi 341000, China

[2] Institutes of Nuclear and New Energy Technology, Tsinghua University, Beijing 100084, China

[3] Beijing Engineering Research Center for Biofuels, Tsinghua University, Beijing 100084, China

Edited by Xiu-Qin Zhu

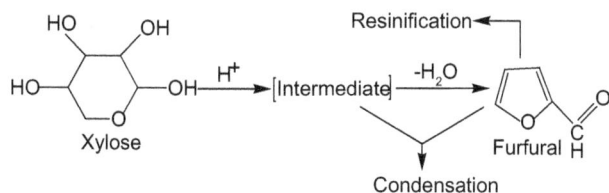

Scheme 1 Side reactions in the reaction of xylose to furfural

applied, such as toluene/water (Shi et al. 2011a), dioxane/ water (Chheda and Dumesic 2007), methyl isobutyl ketone/ water (Moreau et al. 1996), and SBP/NaCl-DMSO (Li et al. 2015), but these organic solvents are toxic and from petrochemical sources; therefore, it is important to search for environment-friendly renewable solvents for preparation of furfural.

In this study, ethyl butyrate was selected and used as extraction agent to extract furfural from the aqueous phase. Then the effect of reaction conditions, such as temperature and catalyst, on xylose conversion and furfural yield was studied. The objective of the work was to find a green process for furfural preparation. In addition, the kinetic model of xylose dehydration to furfural was investigated in ethyl butyrate solvent.

2 Experimental

2.1 Preparation of catalyst

Organosulfonic acid-functionalized mesoporous silica (SO$_3$H-SBA-15) was synthesized (Erdem et al. 2013; Hua et al. 2013; Shi et al. 2011b), and Tetraethoxysilane (TEOS) (Aldrich Co.) and copolymer Pluronic P123 (Aldrich Co.) were used as silica source and structure directing agent, respectively. Firstly, P123 (1.0 g, MW = 5800) was dissolved in 100 mL ethanol solution, and 1 mL TEOS and 0.4 g HCl (38 wt.%) were added to the solution, then 0.15 mL 3-mercaptopropyltrimethoxysilane (MPTMS) and 1 mL H$_2$O$_2$ (30 wt.%) were added and stirred at room temperature forwere added and stirred at room temperature for 20 h. After that, the mixture was evaporated using a rotary evaporator at 40 °C for 10 h, and then dried at 60 °C for 12 h, washed with ethanol, and dried at 60 °C again. At last, the template was removed from the as-synthesized material by washing with ethanol under reflux for 24 h. The sample was denoted as SBA-15–SO$_3$H.

2.2 Analysis

All the extraction agents were from J&K Scientific Ltd. (China) and used without further purification. A 50-mL self-made autoclave with a stirrer was used. In a typical

procedure, D-xylose (0.75 g, Sigma–Aldrich Co., 99 %), extraction agent (17.5 mL), and H$_2$O (7.5 mL) were poured into the reactor (autoclave), and the reactor was heated to temperature and kept at the temperature for the desired time. Then the autoclave was cooled to room temperature. Products existed in aqueous and organic phases. The aqueous phase was examined using a high-performance liquid chromatograph (Shimadzu LC-20AD, Aminex HPX-87H column 300 × 7.8 (i.d.) mm, Japan) with 0.5 M H$_2$SO$_4$ as the mobile phase. The organic phase was determined using a gas chromatograph (Aigent 6820 and PEG-20 M, 30 m × 0.32 mm × 0.25 μm, USA).

The distribution coefficient of furfural between the organic and aqueous phases was determined as follows: 50 mL furfural solution (2 wt.%) and 50 mL extraction agent were fed into a 150-mL separatory funnel, and the mixture was vigorously shaken at room temperature, then stood for 7 h. The mixture was partitioned between an organic and aqueous phase, and the furfural masses in organic and aqueous phases were determined. The distribution coefficient (K) can be calculated by Eq. (1).

$$K = \frac{W_O}{W_A}, \tag{1}$$

where W_o and W_A stand for furfural mass in organic and in aqueous phases, respectively.

The conversion of D-xylose (X), the selectivity of furfural (S), and the yield of furfural (Y) are calculated as follows.

$$X(\%) = \frac{C^s_{xylose} - C^e_{xylose}}{C^s_{xylose}} \times 100 \tag{2}$$

$$S(\%) = \frac{M_{furfural}}{M} \times 100 \tag{3}$$

$$Y(\%) = X \times S \times 100, \tag{4}$$

where $C^s_{xylose}, C^e_{xylose}, M_{furfural}$, and M stand for initial concentration of xylose, final concentration of xylose, mass of furfural, and mass of all the products, respectively.

3 Results and discussion

Table 1 shows the performance of HTW in preparation of furfural from xylose. The furfural yield of 50 % in HTW was attributed to the high value of K_w (ion product of H$_2$O) of HTW (Akiya and Savage 2002). The reason for the yield of no more than 50 % in HTW was attributed to the condensation reaction between xylose in the long residence time in HTW. The furfural yield was above 70 % in the biphasic system HTW/toluene, because furfural was immediately removed by toluene from the aqueous phase. Furfural selectivity was greatly improved as some side reactions were minimized at a high-volume ratio of

Table 1 Comparison of HTW and conventional catalysts

Catalysts	Extraction agent	Temperature, °C	Reaction time, h	Yield of furfural, wt.%
4.5 wt.% H_2SO_4	Toluene	170	1.2	65
SO_3H-SBA-15	Toluene	160	4	70
HTW	No	200	3	50
HTW	Toluene	200	3	75

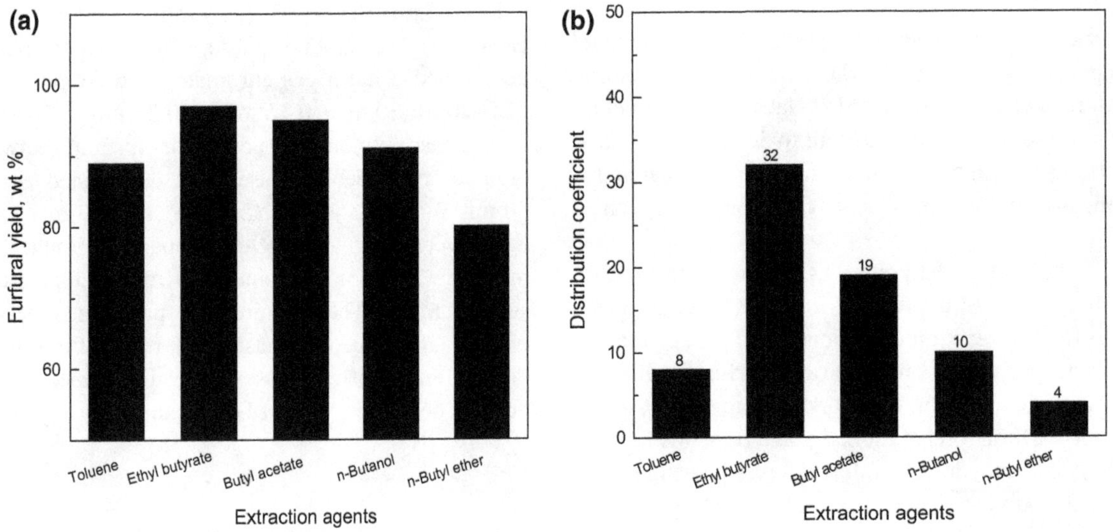

Fig. 1 Effect of extraction agents on furfural yield. Reaction conditions: $T = 200$ °C, $t = 3$ h

toluene/water. HTW as a catalyst overcomes the shortcoming of liquid and solid acid catalysts.

In this work, ethyl butyrate, butyl acetate, n-butanol, n-butyl ether, and toluene solvents were used as extraction agents. Figure 1a shows the effect of extraction agent on furfural yield. The furfural yield was the lowest in n-butyl ether/HTW system, whereas the furfural yield was above 80 % in other four extraction agents. The furfural yield was the highest in ethyl butyrate/HTW system, which was related to the distribution coefficients of furfural in extraction agents (in Fig. 1b). The distribution coefficient (K) in ethyl butyrate was the highest, so the solubility of furfural was the highest in ethyl butyrate. Ethyl butyrate can remove furfural from an aqueous phase; hence, the furfural concentration in the aqueous phase was the lowest.

HTW has a potential as a catalyst because of its high ionic product (Promdej and Matsumura 2011), which is affected by temperature. Figure 2a shows the effect of temperature on furfural yield, furfural selectivity, and xylose conversion. The results show that the xylose conversion increased with temperature from 160 to 200 °C. Xylose conversion and furfural yield were 23 and 14 wt% at 160 °C, 95 and 75 wt% at 200 °C, respectively. This is attributed to the high ionic product of HTW at high temperature (Akiya and Savage 2002). At temperatures above

200 °C, the xylose conversion changed slightly, but the furfural yield began to decrease due to side reactions. Figure 2b shows the effect of residence time on yield, furfural selectivity, and xylose conversion. Furfural yield increased and reached a maximum at 3 h, and then the furfural yield decreased above 3 h. Furfural selectivity decreased slightly with increasing of residence time.

In the initial stage of the reaction (1–3 h), when furfural was removed from the aqueous phase to an organic phase by an extraction agent, side reactions in aqueous phase were minimized, so the furfural selectivity and yield were high. Furfural selectivity and furfural yield decreased above 3 h, and this was attributed to the longer residence time (O'Neill et al. 2009; Sievers et al. 2009).

Figure 3 shows the effect of xylose concentration on furfural yield, furfural selectivity, and xylose conversion. At a concentration of xylose higher than 10 wt.%, the xylose conversion, furfural selectivity, and furfural yield decreased with increasing xylose concentration. This was attributed to two reasons: (1) the low ratio of $[H]^+/[xylose]$ and (2) the increase of side reaction rates.

The pathway of xylose dehydration was simplified for convenient discussion (see Scheme 2). The following kinetic model was applied to describe the dehydration of xylose to furfural in ethyl butyrate/HTW.

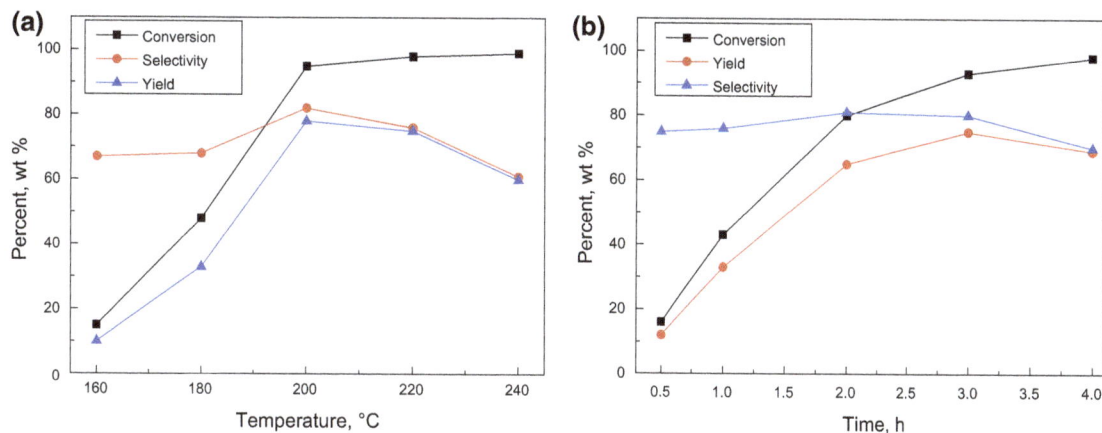

Fig. 2 Effect of temperature and time on the hydrolysis of xylose to furfural. Reaction conditions: **a** $C_{\text{xylose}} = 10$ wt.%, $t = 3$ h, **b** $C_{\text{xylose}} = 10$ wt.%, $T = 200$ °C

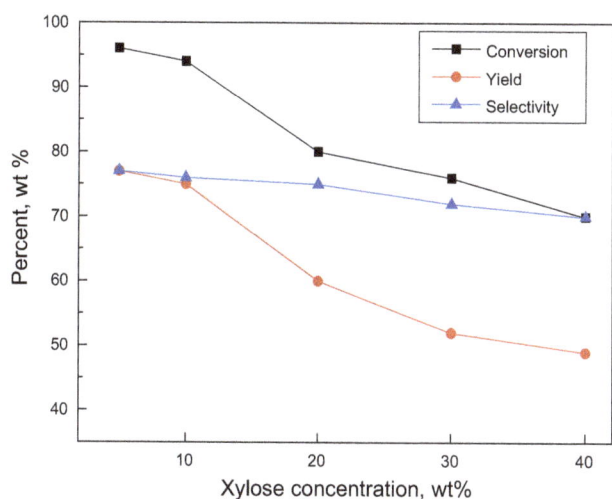

Fig. 3 Effect of concentration of xylose. Reaction conditions: $T = 200$ °C, $t = 3$ h

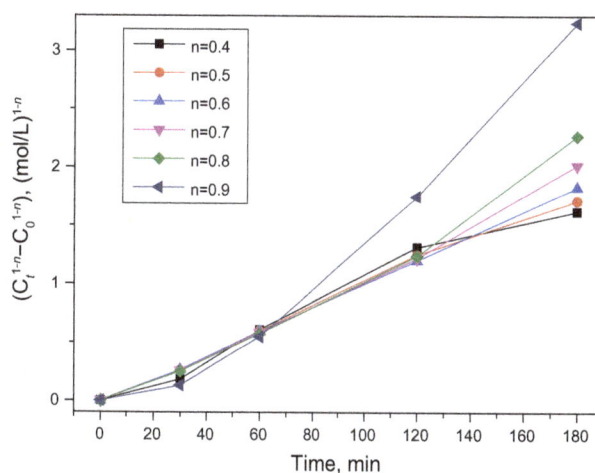

Fig. 4 Kinetic order of the dehydration of xylose to furfural. Reaction condition: $C_o = 10$ wt.%, $T = 200$ °C

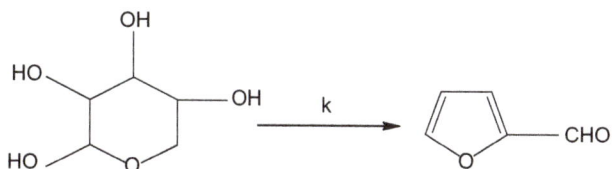

Scheme 2 Pathway of xylose hydrolysis

The above reaction was assumed to be an irreversible n-order reaction. The chemical reaction rate equation can be expressed in Eq. (5).

$$r = kC^n \tag{5}$$

$$r = -\frac{dC}{dt} = kC^n \tag{6}$$

$$\frac{(C_t^{1-n} - C_o^{1-n})}{n-1} = kt \, (n \neq 1), \tag{7}$$

where k, C, and n are the kinetic rate constant, xylose concentration, and the reaction order, respectively.

Jing and LÜ (2007) reported that the dehydration of xylose in HTLW (High-temperature liquid water) followed first-order kinetics. The result of this work (see Fig. 4) was different from their result. The trial method was used in this work to compute the kinetic order, and the kinetic order of the decomposition of xylose in ethyl butyrate/HTW was 0.5.

Figure 2a shows the conversion as the function of time at temperatures 160, 180, 190, and 200 °C. The dehydration rate constant (k) is obtained by Eq. (7). The kinetic parameter of xylose decomposition was estimated and the results are listed in Table 2. The reaction rate constant increased with temperature. The coefficient of determination (R^2 value) was 99.89 %.

We calculated the apparent activation energy and pre-exponential factor (Table 2) via Arrhenius plots (see

Table 2 Rate constants and activation energy of xylose dehydration

T, °C	k, (mol dm^{-3})$^{0.5}$ min^{-1}	E, kJ mol	k_o, (mol dm^{-3})$^{0.5}$ min^{-1}
160	1.75×10^{-3}	68.5	1.82×10^5
180	3.79×10^{-3}		
190	5.67×10^{-3}		
200	1.56×10^{-2}		

T temperature, k kinetic rate constant, E activation energy, k_o pre-exponential factor

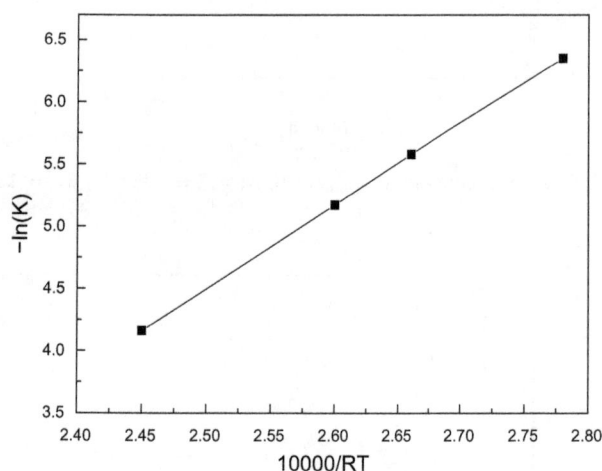

Fig. 5 Arrhenius plots of xylose dehydration

Fig. 5). The estimated activation energy was distinct from the reported values (123.27 kJ/mol) (Jing and LÜ 2007) in the high-temperature water system. The extraction agent affected values of the activation energy and the pre-exponential factor.

4 Conclusions

Factors influencing dehydration of xylose to furfural were investigated. Results indicated that HTW may be used to catalyze dehydration of xylose to furfural, and ethyl butyrate had the highest extraction efficiency and improved furfural yield. The effect of temperature on furfural yield was attributed to the change of water properties at high temperatures. The furfural yield was 75 % under optimal conditions. The rate constant k showed a clear agreement with the Arrhenius law from 160 to 200 °C. The activation energy for the reaction was 68.5 kJ/mol, and the pre-exponential factor was 1.82×10^5 (mol dm^{-3})$^{0.5}$ min^{-1}.

Acknowledgments This project was supported by the National Natural Science Foundation of China (No. 21376136,No. 21176142, No. 21376140, No. 21176142, and No. 21466001) and Program for Changjiang Scholars and Innovative Research Team in University (IRT13026).

References

Agirrezabal-Telleria I, Requies J, Guemez MB, et al. Dehydration of D-xylose to furfural using selective and hydrothermally stable arenesulfonic SBA-15 catalysts. Appl Catal B. 2012a;145:34–42.

Agirrezabal-Telleria I, Requies J, Guemez MB, et al. Pore size tuning of functionalized SBA-15 catalysts for the selective production of furfural from xylose. Appl Catal B. 2012b;115–116:169–78.

Akiya N, Savage PE. Roles of water for chemical reactions in high-temperature water. Chem Rev. 2002;102(8):2725–50.

Antunes MM, Lima S, Fernandes A, et al. Aqueous-phase dehydration of xylose to furfural in the presence of MCM-22 and ITQ-2 solid acid catalysts. Appl Catal A. 2012;417–418:243–52.

Chheda JN, Dumesic JA. An overview of dehydration, aldol-condensation and hydrogenation processes for production of liquid alkanes from biomass-derived carbohydrates. Catal Today. 2007;123(1–4):59–70.

Chheda JN, Roman-Leshkov Y, Dumesic JA. Production of 5-hydroxymethylfurfural and furfural by dehydration of biomass-derived mono- and poly-saccharides. Green Chem. 2007;9(4):342–50.

Climent MJ, Corma A, Iborra S. Converting carbohydrates to bulk chemicals and fine chemicals over heterogeneous catalysts. Green Chem. 2011;13(3):520–40.

Dias AS, Lima S, Pillinger M, et al. Acidic cesium salts of 12-tungstophosphoric acid as catalysts for the dehydration of xylose into furfural. Carbohydr Res. 2006;341(18):2946–53.

Dias AS, Pillinger M, Valente AA. Dehydration of xylose into furfural over micro-mesoporous sulfonic acid catalysts. J Catal. 2005a;229(2):414–23.

Dias AS, Pillinger M, Valente AA. Liquid phase dehydration of D-xylose in the presence of Keggin-type heteropolyacids. Appl Catal A. 2005b;285(1–2):126–31.

Erdem B, Erdem S, Öksüzoğlu RM, et al. High-surface-area SBA-15–SO₃H with enhanced catalytic activity by the addition of poly(ethylene glycol). J Porous Mater. 2013;20(5):1041–9.

Hua D, Li P, Wu Y, et al. Preparation of solid acid catalyst packing AAO/SBA-15-SO₃H and application for dehydration of xylose to furfural. J Ind Eng Chem. 2013;19(4):1395–9.

Jing Q, LÜ X. Kinetics of non-catalyzed decomposition of D-xylose in high temperature liquid water. Chin J Chem Eng. 2007;15(5):666–9.

Kim SB, Lee MR, Park ED, et al. Kinetic study of the dehydration of D-xylose in high temperature water. React Kinet Mech Catal. 2011;103(2):267–77.

Li G, Li N, Yang J, et al. Synthesis of renewable diesel range alkanes by hydrodeoxygenation of furans over Ni/Hβ under mild condition. Green Chem. 2014;16:594–9.

Li H, Ren J, Zhong L, et al. Production of furfural from xylose, water-insoluble hemicelluloses and water-soluble fraction of corncob via a tin-loaded montmorillonite solid acid catalyst. Bioresour Technol. 2015;176:242–8.

Lichtenthaler FW. Towards improving the utility of ketoses as organic raw materials. Carbohydr Res. 1998;313(2):69–89.

Marcotullio G, De Jong W. Chloride ions enhance furfural formation from D-xylose in dilute aqueous acidic solutions. Green Chem. 2010;12(10):1739–46.

Moreau C, Durand R, Razigade S, et al. Dehydration of fructose to 5-hydroxymethylfurfural over H-mordenites. Appl Catal A. 1996;145(1–2):211–24.

Morinelly JE, Jensen JR, Browne M, et al. Kinetic characterization of xylose monomer and oligomer concentrations during dilute acid pretreatment of lignocellulosic biomass from forests and switchgrass. Ind Eng Chem Res. 2009;48(22):9877–84.

O'Neill R, Ahmad MN, Vanoye L, et al. Kinetics of aqueous phase dehydration of xylose into furfural catalyzed by ZSM-5 zeolite. Ind Eng Chem Res. 2009;48(9):4300–6.

Promdej C, Matsumura Y. Temperature effect on hydrothermal decomposition of glucose in sub- and supercritical water. Ind Eng Chem Res. 2011;50(14):8492–7.

Shi X, Wu Y, Li P, et al. Catalytic conversion of xylose to furfural over the solid acid $SO_4^{2-}/ZrO_2-Al_2O_3/SBA-15$ catalysts. Carbohydr Res. 2011a;346(4):480–7.

Shi X, Wu Y, Yi H, et al. Selective preparation of furfural from xylose over sulfonic acid functionalized mesoporous SBA-15 materials. Energies. 2011b;4(4):669–84.

Sievers C, Musin I, Marzialetti T, et al. Acid-catalyzed conversion of sugars and furfurals in an ionic-liquid phase. Chem Sus Chem. 2009;2(7):665–71.

Zhang J, Zhuang J, Lin L, et al. Conversion of D-xylose into furfural with mesoporous molecular sieve MCM-41 as catalyst and butanol as the extraction phase. Biomass Bioenergy. 2012;39:73–7.

16

Reservoir characteristics, formation mechanisms and petroleum exploration potential of volcanic rocks in China

Zhi-Guo Mao · Ru-Kai Zhu · Jing-Lan Luo · Jing-Hong Wang · Zhan-Hai Du · Ling Su · Shao-Min Zhang

Abstract Characterized by complex lithology and strong heterogeneity, volcanic reservoirs in China developed three reservoir space types: primary pores, secondary pores and fractures. The formation of reservoir space went through the cooling and solidification stage (including blast fragmentation, crystallization differentiation and solidification) and the epidiagenesis stage (including metasomatism, filling, weathering and leaching, formation fluid dissolution and tectonism). Primary pores were formed at the solidification stage, which laid the foundation for the development and transformation of effective reservoirs. Secondary pores were formed at the epidiagenesis stage, with key factors as weathering and leaching, formation fluid dissolution and tectonism. In China, Mesozoic–Cenozoic volcanic rocks developed in the Songliao Basin and Bohai Bay Basin in the east and Late Paleozoic volcanic rocks developed in the Junggar Basin, Santanghu Basin and Tarim Basin in the west. There are primary volcanic reservoirs and secondary volcanic reservoirs in these volcanic rocks, which have good accumulation conditions and great exploration potential.

Keywords Volcanic reservoirs · Diagenesis · Formation mechanism · Hydrocarbon exploration

Z.-G. Mao (✉) · R.-K. Zhu · J.-H. Wang · L. Su · S.-M. Zhang
Research Institute of Petroleum Exploration & Development,
PetroChina, Beijing 100083, China
e-mail: maozhiguo@petrochina.com.cn

Z.-G. Mao · R.-K. Zhu · J.-H. Wang · L. Su · S.-M. Zhang
State Key Laboratory of Enhanced Oil Recovery,
Beijing 100083, China

J.-L. Luo
Northwest University, Xi'an 710069, Shaanxi, China

J.-L. Luo
State Key Laboratory of Continental Dynamics,
Xi'an 710069, Shaanxi, China

Z.-H. Du
International Iraq FZE, PetroChina, Beijing 100034, China

S.-M. Zhang
China University of Petroleum, Qingdao 266580, Shandong,
China

Edited by Jie Hao

1 Introduction

As sites for hydrocarbon accumulation, reservoirs are very important for research on petroliferous basins. In recent years, AAPG conferences have chosen reservoir studies as a topic and one of the major trends in studies of petroliferous basins. For a long time, research on reservoirs mostly focused on rocks related to sedimentary processes. Volcanic rocks, seldom regarded as reservoirs, have not been looked into sufficiently (Rohrman 2007; Lenhardt and Götz 2011).

Extensively distributed in a number of petroliferous basins around the world, volcanic rocks are one type of hydrocarbon-bearing rocks, and they may form hydrocarbon reservoirs (Bashari 2000; Chen et al. 2014). Since the first discovery of hydrocarbon reservoirs in volcanic rocks in the San Juan Basin, California in 1887, exploration history has extended for 120 years. Up to now, over 300 hydrocarbon reservoirs or oil/gas shows related to volcanic rocks have been identified around the world. Among them, 169 hydrocarbon reservoirs have proved reserves (Zou et al. 2008). These volcanic reservoirs have the following characteristics: (1) formed mainly in the Mesozoic–

Cenozoic continental margin settings; (2) lithological, dominated by basalt reservoirs (accounting for 32 %) and andesite reservoirs (accounting for 17 %); (3) reservoir spaces are mainly primary or secondary pores, and the common-developed fractures improve reservoir properties; and (4) generally small-scale reservoirs, but can also be high-production wells.

With progress in hydrocarbon exploration around the world and more and more discoveries of volcanic reservoirs, volcanic rocks have attracted the interest of scholars and the petroleum industry as a new domain for hydrocarbon exploration (Zhao et al. 2008; Zou et al. 2008). In the late 1990s, volcanic rock reservoir geology emerged as an important marginal discipline (Zhu et al. 2010a, b), which studies the macro distribution, internal structures, reservoir parameter distribution and pore structures in volcanic rocks, together with dynamic changes of reservoir parameters during development of fields in volcanic rocks, for the purpose of guiding exploration and development of oil and gas fields.

At the same time, petroleum geology theories and techniques related to volcanic rocks also have experienced rapid development (Sruoga and Rubinstein, 2007; Farooqui et al. 2009; Zhang et al. 2011; Zou et al. 2012a, b; Dong et al. 2013; Du et al. 2013). In the south Nagaoka gas field in Niigata, Japan, gamma ray, compensated formation density and compensated neutron logs were used together to successfully identify reservoirs in Miocene "Green tuff"—rhyolite volcanic rocks in the Qigu Formation (Yagi et al. 2009). Generally speaking, research on volcanic reservoirs is more difficult than clastic and carbonate reservoirs. Consequently, it emphasizes the comprehensive application of multiple disciplines, as petroleum geology, volcanology, petrology and reservoir physics theories, together with technologies as gravity-magnetic survey, electric prospecting, seismic, logging, mathematic geology and computer technology (Pan et al. 2008; Zhu et al. 2010a, b; Chen et al. 2014).

2 Volcanic reservoir characteristics in China

Volcanic rocks, the product of a series of volcanic activities after cooling, solidification and consolidation, are significantly different from sedimentary rocks in formation conditions, development environment and distribution pattern. Consequently, characteristics of hydrocarbon reservoirs in volcanic rocks are quite different from those in sedimentary rocks.

2.1 Lithological features of volcanic reservoirs in China

Sedimentary basins in China developed under various regional tectonic backgrounds (Lü et al. 2004; Luo et al.

2012; Zhao et al. 2009; Xie et al. 2010) have a variety of reservoir volcanic rocks (Mao et al. 2010). Among them, lava rocks mainly include basalt, andesite, dacite, rhyolite, trachyte, etc.; pyroclastic rocks mainly include agglomerate, volcanic breccia, tuff and welded pyroclastic rocks. Mesozoic volcanic reservoirs in eastern China mostly generated in the Late Jurassic to Early Cretaceous, with basic to acidic rocks, but mostly are acidic (Fig. 1); Cenozoic volcanic reservoirs in eastern China mainly include those in the Jiangling Sag of the Jianghan Basin, Jiyang Sag and the eastern part of the Liaohe Sag, and lithologically, there are acidic to basic rocks, but mostly are meso-basic rocks (Zhao et al. 2008). Volcanic rocks in western China are dominated by meso-basic rocks, such as the Permian Formation in the Tarim Basin, Carboniferous Batamayineishan Formation in the Junggar Basin and Permian Jiamuhe Formation in the Tuha Basin and the Carboniferous-Permian system in the Santanghu Basin (Fig. 1) (Zhou et al. 2010; Zhang et al. 2013a, b).

2.2 Types and features of reservoir space in volcanic rocks in China

Compared with sedimentary rocks, volcanic rocks are much more complicated in reservoir space types and features. According to observations and research on large quantities of cores and thin sections, volcanic reservoir spaces can be classified into three major categories: primary pores, secondary pores and fractures (Table 1). Primary pores are predominantly composed of pores formed during the eruption of volcanic material, residual pores incompletely filled by amygdaloidal bodies, intercrystalline micro-pores and pores among volcanic breccia (Fig. 2a, b). Secondary pores mainly include volcanic glass devitrification pores, together with various mineral and particle dissolution pores and cavities (Fig. 2c, d). Fractures can be formed by the following three reasons: bursting fractures and contraction fractures formed due to volcanism and cooling (Fig. 2a); structural fractures formed by volcanic rocks deformation and slippage induced by tectonic stresses (Fig. 2e); and dissolution fractures formed by the dissolution of formation fluids during weathering, leaching and burial processes (Fig. 2f).

Typical samples of volcanic reservoir rocks from the Yingcheng Formation in the Songliao Basin (110 samples from 49 wells) and the northern Xinjiang Carboniferous (53 samples from 15 wells) were observed by optical microscopy, combined with quantitative analyses. The results show that reservoir spaces in both areas are dominated by air pores, dissolution pores and micro-fractures. At the same time, they have significant differences in percentages. Volcanic rocks of the Yingcheng Formation in the Songliao Basin are dominated by air pores (38 %) and

Fig. 1 Lithology of volcanic rocks and evolution of petroliferous basins of China

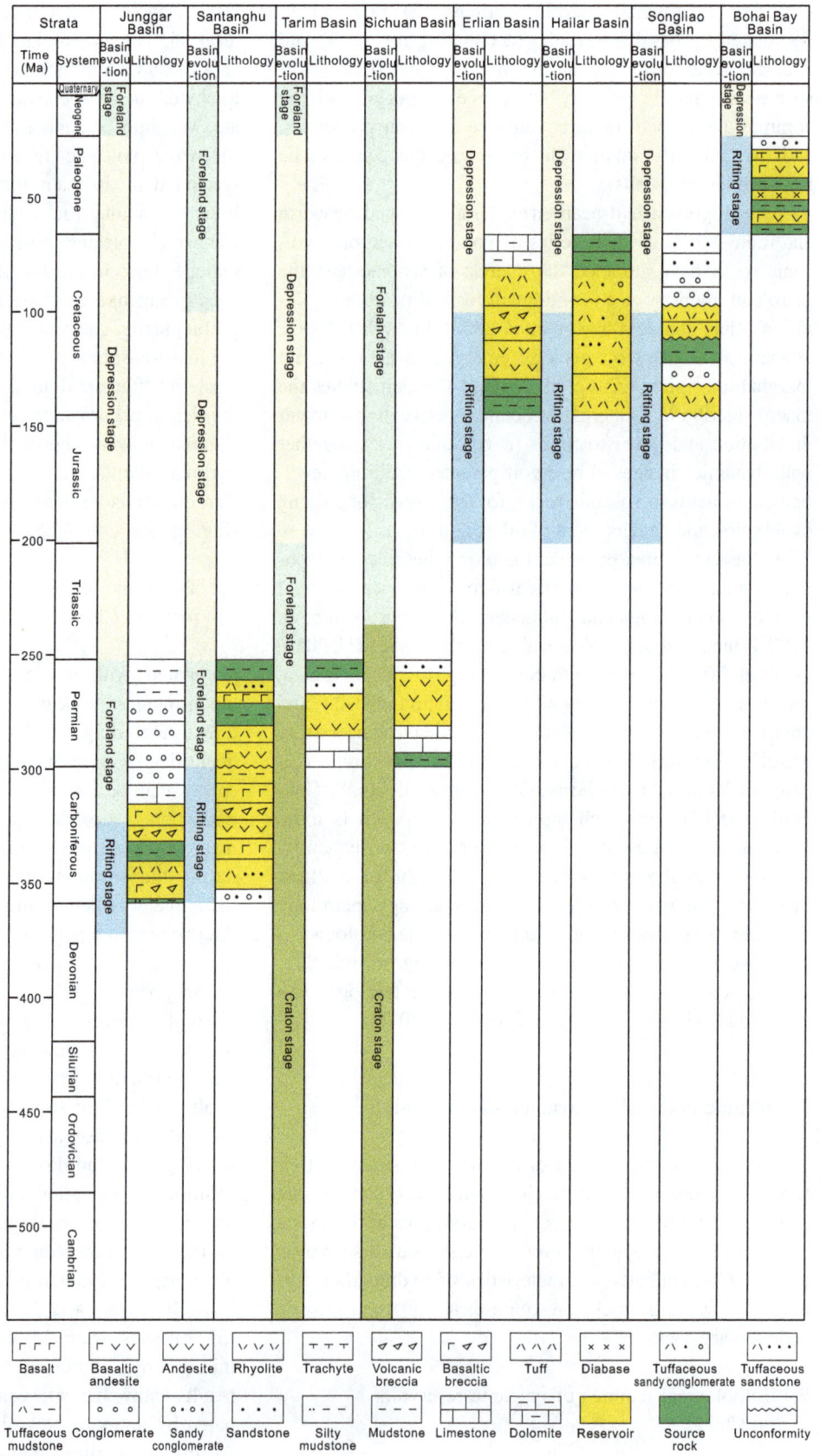

Table 1 Classification and features of reservoir space of volcanic rocks in China

Classification		Origin	Features	Corresponding lithological features	Oiliness
Primary pores	Primary air pores	Formed due to expansion of gas during diagenetic process	Mostly distributed at bottom and top of lava-flow layers in different sizes and shapes	Volcanic breccia, lava	Good in those connected with pores and fractures
	Residual air pores	Pores formed due to incomplete filling of pores by secondary minerals	Also known as semi-filling pores	Basalt, volcanic breccia	Good in those connected with pores and fractures
	Intergranular pores	Residual pores formed after diagenetic compaction between clastic particles	Mostly in pyroclastic rocks	Volcanic breccia, agglomerate, volcanic sedimentary rocks	Good
	Intercrystal, inner-crystal pores	Pores in framework of rock-forming minerals, augite, plagioclase and phenocryst minerals, mostly with cleavage. They are inner-crystal pores themselves	Mostly distributed in central parts of lava-flow layers with minor pores	Lava, pyroclastic rocks	Good
Secondary pores	Devitrification pores	Formed by vitric matter after devitrification	Micro-pores, but with favorable connectivity	Spherulitic rhyolite, ignimbrite	Fairly good reservoir space
	Phenocryst dissolution pores	Phenocryst may generate pores due to dissolution of fluids. Such dissolutions may develop along cleavage faces	Irregular in pore shape, mostly in bay shape and dominated by inner-crystal pores	Andesite	One of the most important reservoir space
	Dissolution pores in amygdaloidal bodies	Dissolution pores generated due to alternation and dissolution of filling materials in air pores	Irregular in pore shape with poor connectivity	Lava	Favorable hydrocarbon-bearing features
	Dissolution pores in matrix	Devitrification of vitric matter in matrix or dissolution of microcrystal feldspar	Fine pores, mostly dissolution pores with certain connectivity	Various lava and melted tuff	Favorable reservoir formations
	Dissolution pores among breccia	Formed due to weathering, leaching, dissolution and other epigenetic actions	Developed along fractures, cracked clastics and structural highs	Basalt, andesite, breccia	Good hydrocarbon-bearing features
Fractures	Solidification contraction fractures	Micro contraction fractures formed during solidification and crystallization of lava	Columnar jointing and split in open or facial configuration with minor dislocations	Volcanic breccia, andesite, trachyte	Generally good
	Explosion fractures	Self-cracking or concealed eruption	Restorable	Self-cracking breccia, lava and secondary volcanic rocks	Good
	Structural fractures	Micro-fractures generated under tectonic stress	Developed near faults, flat and straight, mostly high angle fractures	Basalt, andesite	Related to timing of tectonic activities
	Weathering fractures	Usually connected with dissolution pores, fractures and structural fractures to cut rocks into debris of various sizes	Connected with dissolution pores, fractures and structural fractures	Pyroclastic rocks, volcanic breccia	Good
	Dissolution fractures	Weathering and leaching, dissolution by formation fluid	Dissolution and extension of original fractures	Amygdaloidal andesite, volcanic breccia	Fairly good hydrocarbon-bearing features

Modified after Du et al. (2013) and Zou et al. (2012a, b)

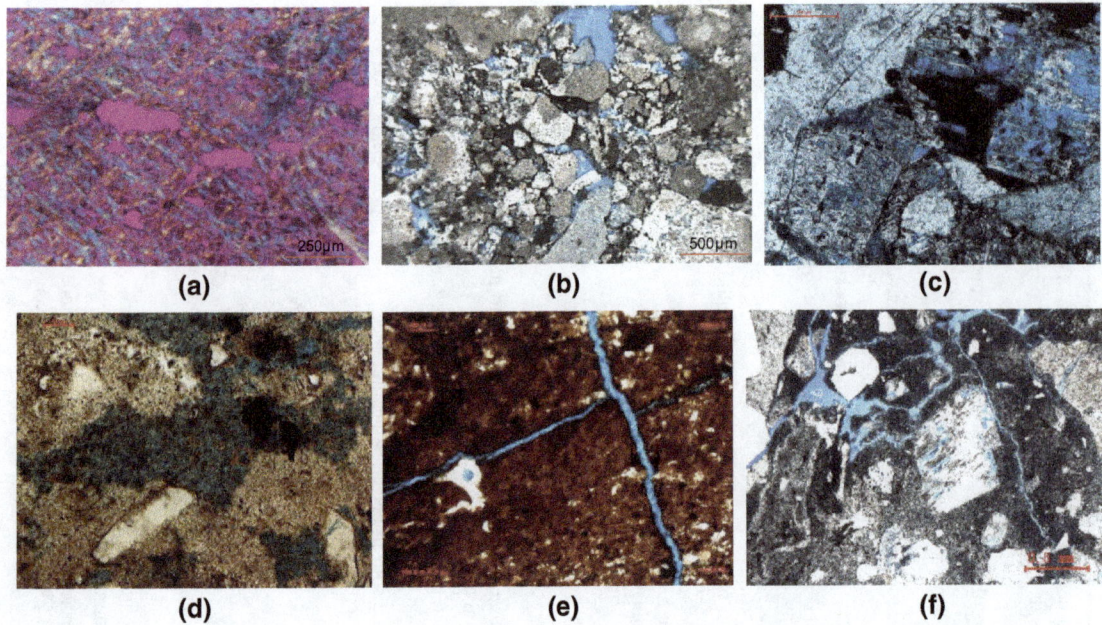

Fig. 2 Types and microscope photos of reservoir spaces in volcanic rocks. **a** Rhyolitic ignimbrite, air pores and contraction fractures, well Madong-1, 4266.85 m, ×2.5. **b** Volcanic breccia, intergranular pores, well Dixi-5, 3649.19 m, ×4 (+). **c** Lava, feldspar phenocryst, dissolved pores, well Huang-95, 2908.00 m, ×5 (−). **d** Quartz porphyry, matrix dissolution pores, well Mana-1, 5166.0 m, ×10 (−). **e** Tuff, structural micro-fractures, well Niudong-9-8, 1524.75 m, ×10 (−). **f** Tuff, weathering and leaching dissolution pores and fractures, well Di-403, 3818.55 m, ×4

Fig. 3 Pore types and average contents in typical volcanic reservoirs of China

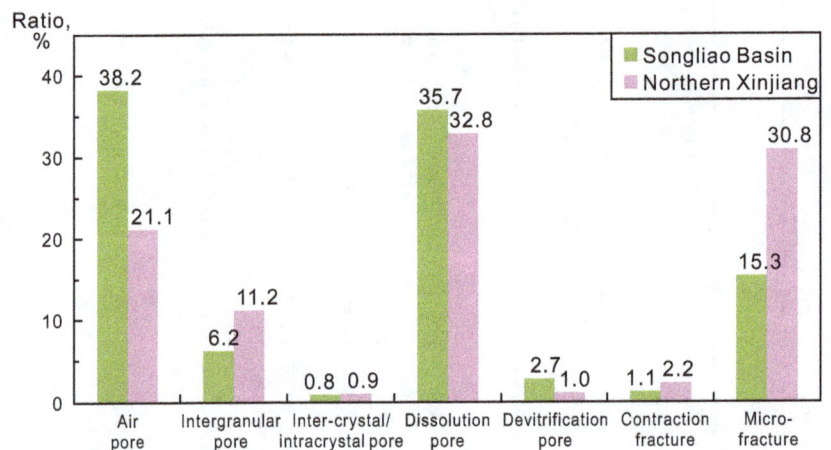

dissolution pores (36 %) (Fig. 3), whereas Carboniferous volcanic rocks in the Northern Xinjiang have mainly dissolution pores (33 %) and micro-fractures (31 %) (Fig. 3).

2.3 Physical properties of volcanic rocks in China

There are a variety of reservoir spaces in volcanic reservoirs. Different reservoir spaces may combine with each other to form pore-fracture dual-media reservoirs which have complicated pore structures, leading to widely variable physical properties and severe heterogeneity. Physical properties of volcanic rock samples (112) from the Junggar, Santanghu, Songliao and other basins show that pyroclastic rocks, lava rocks and hypabyssal intrusive rocks can serve as effective reservoirs, mostly with medium to high porosity, low to medium permeability, high heterogeneity, and poor correlation between porosity and permeability. Porosity may be over 30 %, with 10 % on average, whereas permeability may display intense heterogeneity with the maximum over $1,000 \times 10^{-3} \ \mu m^2$. But on the whole, the permeability is quite low, mostly less than $1 \times 10^{-3} \ \mu m^2$ and a large portion may be less than $0.01 \times 10^{-3} \ \mu m^2$ (Fig. 4).

Fig. 4 Correlation between porosity and permeability of volcanic rocks (978 core samples, reservoir classification according to SY/T 6285-2011)

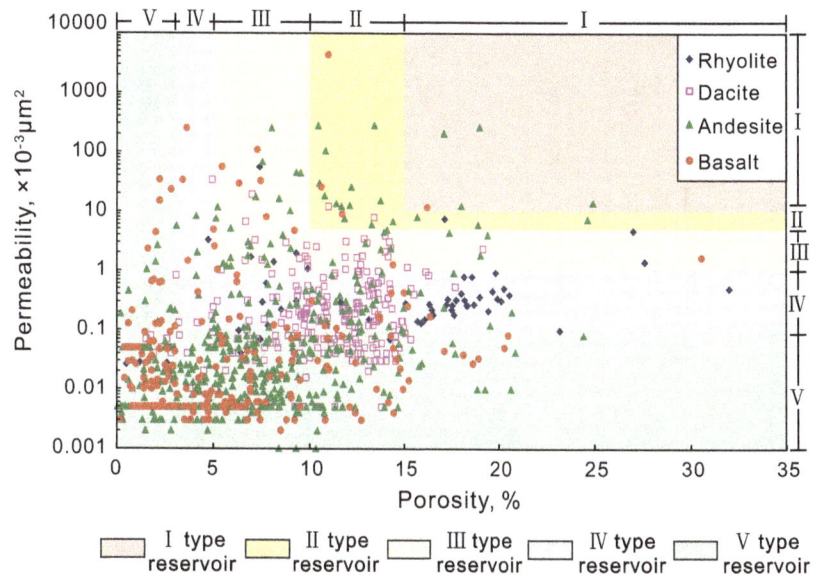

3 Origin and controlling factors of volcanic reservoirs in China

Compared with clastic reservoirs, volcanic reservoirs are more complex in pore types. Intercrystal pores, contraction pores and other primary pores formed after cooling of igneous rocks are not much affected by compaction, but complicated raw minerals and their combination may experience dramatic changes due to hydrothermal fluids during tectonic evolution and diagenetic burial. These changes may affect diagenesis routes of volcanic rocks and diagenetic products, resulting in different pores and throats occurrence and storage properties. In the end, such differences may directly affect the formation and evolution of high-quality reservoirs in volcanic rocks (Surour and Moufti 2013; Borgia et al. 2014; Caricchi et al. 2014). Therefore, it can be seen that volcanic reservoirs are the combined product of volcanism, tectonization, diagenesis, fluids, epidiagenesis, burial transformation and many other factors.

3.1 Diagenesis evolution of volcanic reservoirs in China

Due to their special features, volcanic reservoirs have obvious stages and periods in diagenesis and evolution (Sruoga and Rubinstein 2007). Diagenesis is distinctive in different stages and periods. Consequently, it is possible to divide the diagenesis processes of volcanic rocks into several stages and further into different periods (Table 2).

Since different diageneses may evolve constantly with changes in diagenetic environments, some diagenesis may happen during different diagenetic stages. In addition, different sedimentary basins have various burial–tectonic–thermal evolution history, so volcanic rocks in different basins would experience quite different diageneses at different stages. Therefore, different volcanic reservoir types have their unique diagenesis sequence and diagenetic stage.

Research on volcanic reservoir diagenesis of different periods and types in different basins shows that there are two major diagenesis sequences in volcanic rocks in China: eruption—burial diagenesis sequence and eruption—weathering—burial diagenesis sequence.

(1) Eruption—burial diagenesis sequence of volcanic rocks in the Yingcheng Formation, Songliao Basin. Eruption—burial diagenesis sequence includes volcanic eruption, cooling, solidification and consolidation, and subsidence and burial by sediments. Reservoir spaces are dominated by primary pores generated during eruption, cooling, solidification and consolidation of volcanic rocks, and these are primary volcanic reservoirs. The Cretaceous Yingcheng volcanic reservoir in the Songliao Basin is a typical reservoir of this type (Jin et al. 2010; Cai et al. 2012).

After eruption in the Early Cretaceous, the Yingcheng Formation volcanic rocks in the Songliao Basin were formed after experiencing dissolution, cooling, crystallization and some other consolidation (approximately 156–125 Ma) (Sun et al. 2008; Meng et al. 2010; Shao et al. 2013; Xiang et al. 2013; Zhang et al. 2013a, b). Then the formation was covered directly by overlying sedimentary strata after short periods of hydrothermal fluids and weathering and leaching between eruption intermissions. During these periods, the formation experienced no less than two stages of large-scale

Table 2 Diagenetic stages of volcanic rocks

Stage	Period	Diagenesis	Mechanisms of diagenesis	Diagenesis markers	Pore types
Cooling and solidification stage	Volcanic active period	Explosion and cracking	Eruption, cracking and explosion of volcano	Pyroclastic rocks of various compositions and particle sizes	Air pores, intergranular pores, explosion fractures, contraction fractures, intercrystalline pores
	Solidification period	Crystalline differentiation	Differentiation, separation and crystallization of lava	Volcanic rocks with different crystals and mineral compositions	
		Solidification	Solidification and contraction	Volcanic rocks contraction fractures	
Re-construction stage	Hydrothermal period	Alteration	Changes in temperatures in uprising hot fluids in deep layers	Chlorite and zeolite	Intercrystalline micro-pores in clay minerals, pores in amygdaloidal bodies, residual pores, dissolution pores, dissolution fractures
		Filling	Crystallization and precipitation of minerals carried out by volcanic hydrothermal solution	Chlorite, zeolite filling	
		Dissolution	Dissolution, alternation of volcanic hydrothermal solution	Chlorite, zeolite solution pores	
	Weathering & leaching period	Weathering, breaking	Thermal expansion of rocks	Weathering fractures	Weathering fractures
		Leaching dissolution	Leaching and dissolution of rocks	Intergranular, inner-granular solution holes, dissolution fractures	Dissolution pores, dissolution fractures
	Burial period	Compaction	Compaction	Intergranular, intercrystalline contacts and modification of debris	
		Tectonization	Tectonic stress	High angle fractures, near horizontal fractures, reticular fractures	Structural fractures
		Dissolution	Dissolution by formation water and organic acid	Huge amounts of secondary solution holes	Matrix solution holes, phenocryst solution holes, intergranular dissolution pores, dissolution fractures
		Alteration	Increase of formation temperatures and pressures, together with activities of formation fluids	Zeolite, chlorite, clay and other associated minerals	
		Filling and cementation	Formation fluid dissolution, mineral precipitation	Filling and cementation of zeolite, chlorite, clay and other minerals	
		Devitrification	Increase of formation temperatures and pressures	Felsitic texture, cryptocrystalline texture after devitrification	

(a)

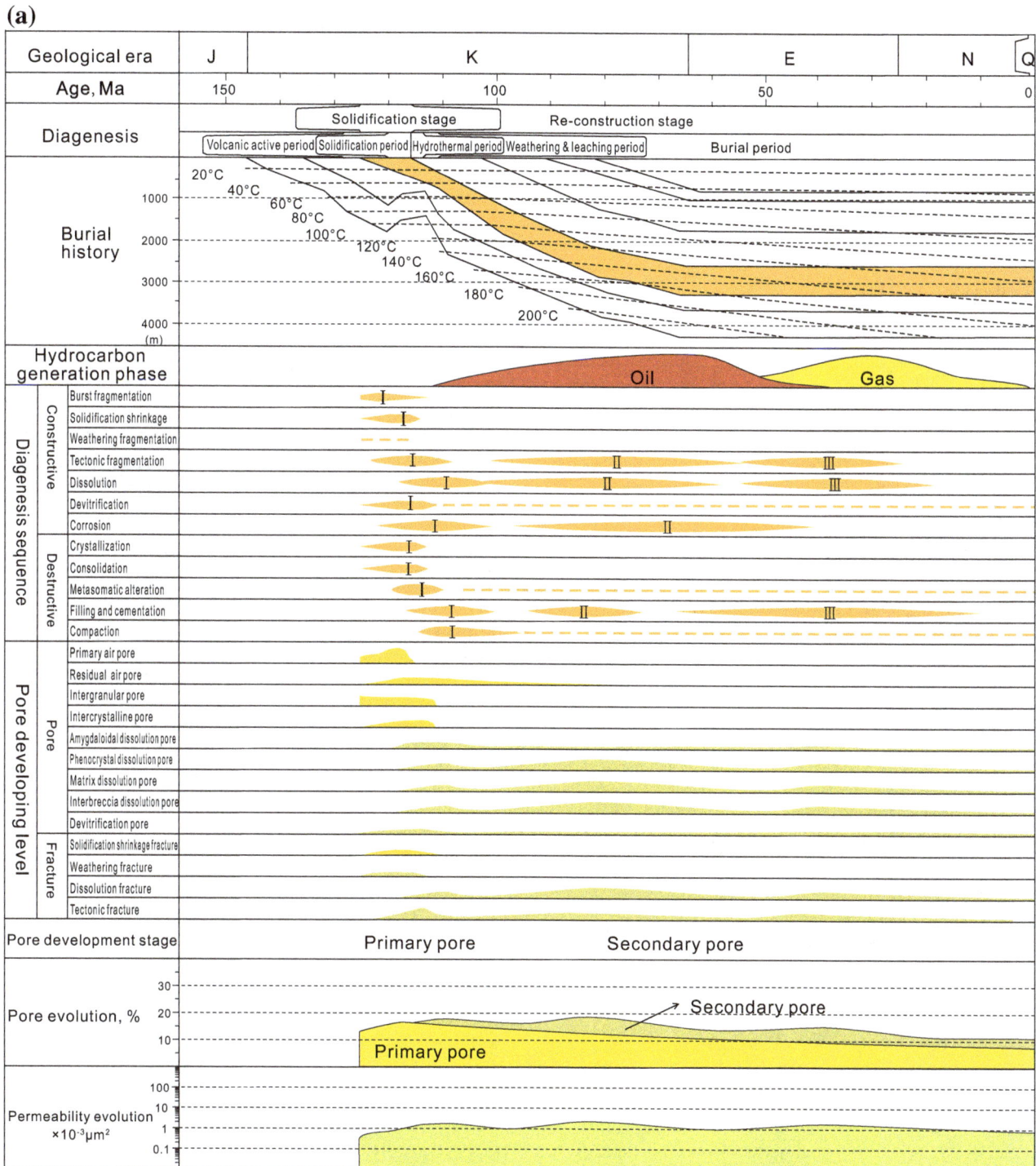

Fig. 5 Diagenesis and porosity evolution of typical volcanic reservoirs in China. **a** Cretaceous volcanic rocks of the Yingcheng Formation in the Songliao Basin. **b** Carboniferous volcanic rocks in the Junggar Basin

secondary dissolution, two stages of hydrocarbon charge, three stages of fracture generation, 2–3 stages of siliceous cementation, three stages of carbonate cementation, 2–3 stages of chlorite cementation and formation of zeolite, albite, fluorite and other secondary minerals (Qin et al. 2010; Sun et al. 2012; Luo et al. 2013) (Fig. 5a).

(2) Eruption—weathering—burial diagenesis sequence of Carboniferous—Permian volcanic rocks in northern Xinjiang.

(b)

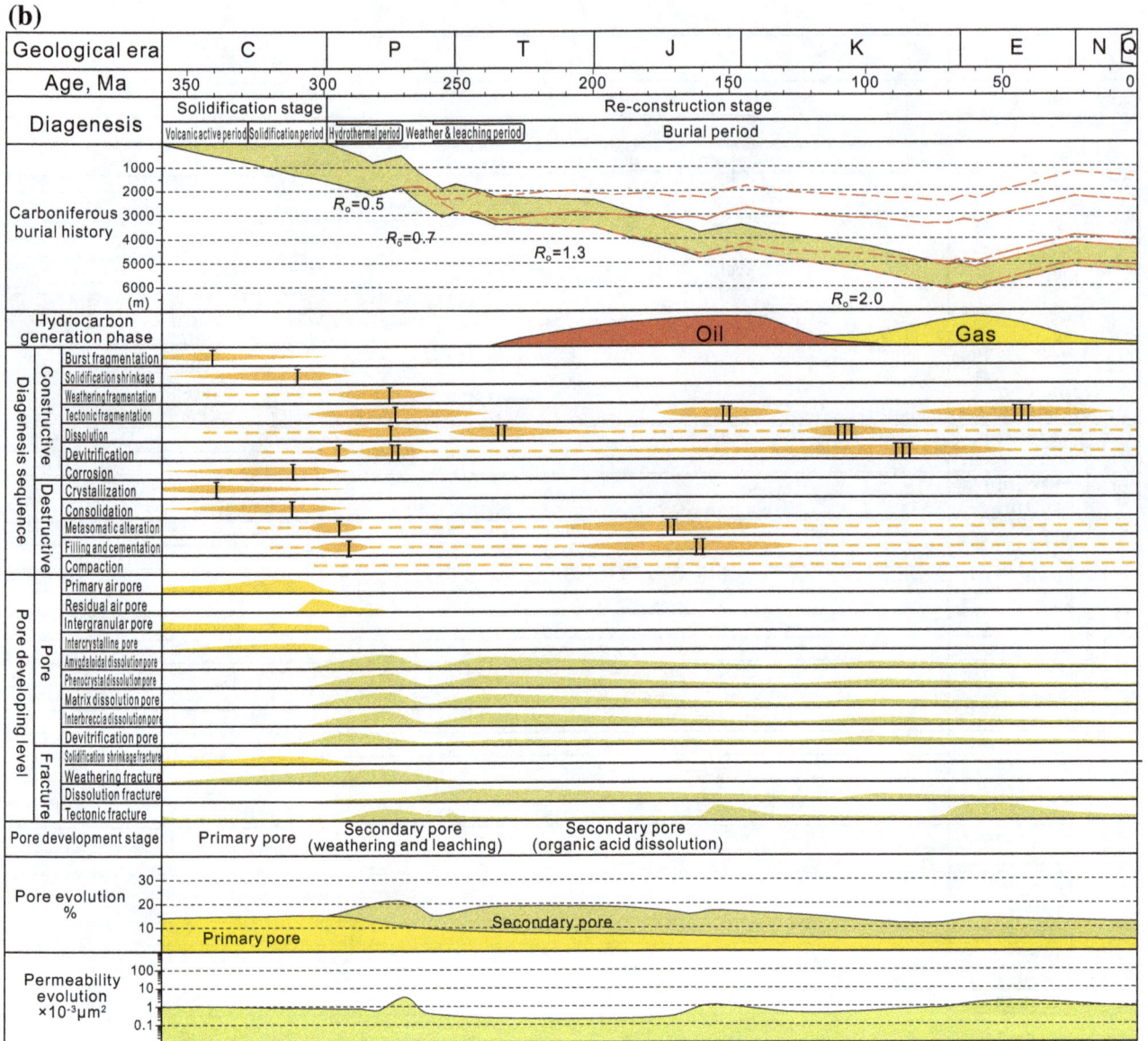

Fig. 5 continued

The eruption—weathering—burial diagenesis sequence involves the uplift and exposure of consolidated volcanic rocks at first, then prolonged weathering and leaching and reformation next, and lastly burial by sediments. Reservoir spaces are mainly secondary pores and fractures from weathering, leaching and reformation, and these form secondary volcanic reservoirs. The Carboniferous—Permian volcanic rock reservoirs in the Junggar Basin and Santanghu Basin are representatives of this type (Hou et al. 2012, 2013; Zou et al. 2012a, b).

After eruption in the Carboniferous, the Batamayinei-shan Formation volcanic reservoir in the Junggar Basin formed through dissolution, cooling, crystallization,

welding and some other consolidation (approximately 359–320 Ma). After a short period of corrosion by hydrothermal fluids and filling of pores, volcanic rocks in this area had been uplifted and exposed till the Early Triassic (approximately 246 Ma), subject to prolonged weathering and leaching in hypergene environments, which resulted in fragmentation of rocks and secondary dissolution of minerals, and large numbers of weathering fractures and secondary pores formed (Mo and Lian 2010). After that, these formations were covered and buried by sediments. In later stages of burial, the entire region experienced several major tectonic movements, which played constructive roles in volcanic reservoir development. Generally speaking, Carboniferous volcanic rocks in the Junggar Basin experienced no less than two stages of large-scale secondary dissolution, 2–3 stages of hydrocarbon-filling events, three stages of fracture formation, one stage of siliceous cementation, two stages of carbonate cementation, two stages of chlorite cementation and three stages of anhydrite/gypsum cementation (Fig. 5b).

3.2 Origin of volcanic reservoirs

Reservoir spaces in volcanic rocks experienced extremely complicated processes of formation, development, blocking and re-construction. Different evolution processes may involve different diagenesis types, which may play dual roles of destruction and improvement to reservoirs at the same time. By a number of means, research shows that volcanic reservoirs were the product of combined effect of long-term multiple diagenesis. Complicated in formation and evolution processes, there are multiple stages of filling and dissolution. At the same time, diagenesis processes in different zones and different reservoirs also vary significantly.

(1) Development of reservoir spaces during cooling and consolidation.

Primary pores mainly form at this stage. After volcanic lava is ejected onto the surface, pores are formed by the escape of a large amount of volatile gases. Crystallization of volcanic lava resulted in the forming of small intercrystalline pores between phenocrysts, between microcrystallites, and between phenocrysts and microcrystallites. Pores among volcanic breccias were formed under volcanic eruption and explosion. Solidification of volcanic lava produced contraction fractures and other primary reservoir pores. These primary pores lay a solid foundation for the development of reservoirs at later stages. In vertical profile, lithological combinations with large amounts of lava and tuff or volcanic breccia, lava and pyroclastics can be observed, which reflect volcanic activities.

(2) Development of reservoir spaces during the epigenetic alteration stage.

Tectonic movements, weathering, leaching and fluid actions are the major geological actions affecting the development of reservoir spaces.

During periods of hydrothermal activities, alteration may occur in many rock-forming minerals. For example, augite and amphiboles may convert into chlorite; basic plagioclase may convert into kaolinite, sericite and chlorite; iddingsitization of olivine; chlorite may convert into zeolite, carbonate or other minerals; as well as carbonation and laumontitization of tuff matrix. During such alteration and conversion of minerals, hydrothermal fluids may carry large amounts of secondary minerals, e.g. chlorite, zeolite, calcite and quartz. Under suitable conditions, these minerals would crystallize, precipitate and fill up reservoir spaces, reducing volcanic reservoir capacities significantly (Yu et al. 2014). At the same time, since these minerals are mostly soluble minerals, they provide necessary materials for the later dissolution.

During weathering and leaching periods, weathering and leaching facilitate the formation of large numbers of dissolution pores in the surface and upper sections of volcanic rocks, which may connect primary reservoir spaces and enhance reservoir properties in volcanic rocks significantly. Core analysis data show that the Carboniferous basalts in the Ludong area of the Junggar Basin without weathering have an average original porosity of 7.6 %, basalts with weak weathering have an average porosity of 8.7 % and basalts with strong weathering have an average porosity of 15.3 %. Therefore, it can be seen that weathering and leaching are the major forces for the formation of secondary pores in this area.

During burial stages, volcanic rocks were covered and compacted by sedimentary rocks. Various pores and fractures generated by lava eruption, invasion, cooling, solidification, and tectonic activities and dissolution in later stages may connect pores that are originally isolated. At the same time, they may communicate with formation water in sedimentary formations to form more secondary dissolution pores and fractures. Volcanic rocks formed during early cooling, solidification and consolidation may experience prolonged dissolution by the formation of water and organic acids during their deep burial. These are the main causes for the formation of secondary reservoir spaces in volcanic rocks. Intermediate to basic basalt and andesite mainly experience re-dissolution of early fillings and alternation products. For example, partial dissolution of chlorite and calcite can be often observed in Well Shinan-3, Well Shinan-4 and Well Madong-2, but dacite may predominantly experience dissolution of amphiboles, feldspar phenocrysts and matrix. The Carboniferous dacite in the Shixi-1 well block in the Shixi Oilfield experienced the dissolution of plagioclase phenocrysts, and dissolution pores or expansion dissolution fractures can be found in the matrix. As for alkaline or strongly alkaline trachyte and phonolite, their alkalinity makes them even more sensitive to acid environments. As soon as the media environment changed from alkalinity to acidity, large amounts of alkaline feldspar phenocrysts and matrix may undergo dissolution, e.g. large amounts of dissolution pores can be observed in tephritic phonolite and trachyte andesite cores in Well Xiayan-2 and Well Shidong-8. In fact, acidic fluids can be further classified as inorganic and organic acids. They may participate in reactions individually or jointly in different areas. Areas with volcano eruption and in vicinity of major faults may be dominated by inorganic acids, whereas dissolution induced by organic acids may be stronger in areas near oil sources. In addition, occurrence of dissolution over a large area may be closely related to the development of faults. As for various minerals, especially those most affected by dissolution, such as feldspar, they may have significantly different dissolution mechanisms for organic acid dissolution and for inorganic acid dissolution.

Above research conclusions show that diagenesis is a double-edged sword for volcanic reservoirs, the matching of filling and dissolution in different diagenetic stages, together with intensities of diagenesis processes directly affect the quality of reservoir reformation at later stages.

3.3 Controlling factors for the development of volcanic reservoirs in China

Formation, preservation and reformation of reservoir spaces in volcanic rocks during the evolution process are highly complicated. Primary pores and fractures are mostly controlled by original eruption states, namely lithology of volcanic rocks. Under the same tectonic stress, the development and preservation of structural fractures are also controlled by the original eruption states. Volcanic rocks formed after volcano eruption, cooling, solidification, compaction and consolidation contain disconnected primary pores without permeability. Only after various geological transformations in later stages, they can have reservoir capacities. On the whole, volcanism, tectonic movements, weathering, leaching and fluids are key factors and geological actions for the formation and development of reservoir spaces in volcanic rocks.

Differences in diagenesis sequences of volcanic reservoirs in eastern and western China lead to the differences of reservoir types. Primary reservoir formations in volcanic rocks are represented by the Yingcheng Formation in the Songliao Basin in eastern China. Secondary weathering reservoir formations in volcanic rocks are represented by Carboniferous formations in northern Xinjiang in western China. Different factors control the development of these two types of volcanic rocks.

Volcanic rocks in the Songliao Basin experienced relatively short weathering or no weathering at all. As a result, volcanic structures are completely preserved. Volcanic rocks have relatively complete facies with reservoirs predominantly developed in eruption facies. Major controlling factors for the storage capacity of volcanic reservoirs include lithological features, rock facies, structural fractures and dissolution by acidic fluids. To be more specific, the developments of primary pores in volcanic rocks are mainly determined by lithological features and rock facies; structural fractures may enhance the permeability of

volcanic rocks and serve as seepage channels for acidic fluids in later stages. Later dissolution by acidic fluids can enlarge primary pores, whereas weathering and leaching can facilitate the formation of secondary pores and fractures. Favorable facies of volcanic rocks include intrusive facies, upper subfacies of effusive facies and air-fall subfacies of eruption facies, all of which develop primary pores. In later stages of volcanic activity, acidic fluids may rise along fractures and enter pores through fractures, stimulating the formation of dissolution pores, and eventually enlarging reservoir spaces. The upper subfacies of effusive facies, air-fall subfacies of eruption facies and intrusive facies near craters below the weathering crust, which have been reformed due to acidic fluid dissolution in later stages of volcanic activities below the unconformity surface (approximately 3,600–3,800 m), are favorable facies for the development of reservoir spaces.

The residual Carboniferous formations developed over the Paleozoic folding base in the Junggar Basin. These strata experienced uplifting and denudation over a period of 30–60 Ma from the late Carboniferous to the early and middle Permian. Consequently, volcanic structures are generally incompletely preserved with weathering crusts developed. Usually, effective reservoirs with a thickness of 300 m might develop under the weathering crust (Li et al. 2007; Zhu et al. 2010a, b). The Lower Carboniferous formation is preserved in areas with severe denudation on uplifting zones, such as Kelameilishan and Dixi Uplifts. The preserved strata in ancient sags are relatively young, mostly the Upper Carboniferous formations, such as Wucaiwan Sag and Mahu Sag. From the sag center to the surrounding areas or from the structural low parts to highs, the Carboniferous formations experienced more severe denudation with newer covering formations and better reservoir properties. High landform induced by prolonged volcanic activities facilitates weathering and denudation. Consequently, the intensities of weathering, leaching, dissolution and filling reformation during volcanic rock evolution control the effectiveness of reservoirs. Palaeohighs subject to prolonged weathering and leaching commonly contain well-developed volcanic reservoirs where hydrocarbons can accumulate.

4 Hydrocarbon exploration in volcanic rocks in sedimentary basins of China

Volcanic rocks are key components of filling series for various sedimentary basins. During the early development stage of basins, volcanic rocks were not only large in volume, but also mostly associated with rapid subsiding hydrocarbon source rocks. Therefore, they are important targets in hydrocarbon exploration (Zeng et al. 2013;

Batkhishig et al. 2014). In later burial stage, volcanic rocks, less affected by burial depths than conventional sedimentary rocks, in deep parts of basins, could be better reservoirs than conventional sedimentary rocks (Feng 2008). Accordingly, they are considered as key target layers for exploration in deep basins.

Volcanic rocks are distributed extensively in China. Three packages of favorable volcanic rocks in Carboniferous—Permian, Jurassic—Cretaceous and Paleogene formations developed in existing petroliferous basins. Volcanic rocks in eastern basins are predominantly intermediate to acidic, whereas those in western parts are intermediate to basic. These volcanic rocks cover a total area of $39 \times 10^4 \text{ km}^2$, including $5.0 \times 10^4 \text{ km}^2$ in the Songliao Basin in eastern China, $2.0 \times 10^4 \text{ km}^2$ in the Bohai Bay Basin, $6.0 \times 10^4 \text{ km}^2$ in the Junggar Basin, $1.0 \times 10^4 \text{ km}^2$ in the Santanghu Basin, $2.0 \times 10^4 \text{ km}^2$ in the Tuha Basin, $13 \times 10^4 \text{ km}^2$ in the Tarim Basin and $7.0 \times 10^4 \text{ km}^2$ in the Sichuan-Tibet area.

At present, volcanic reservoirs have been found in the Songliao, Bohai Bay, Junggar, Santanghu, Halar, Erlian and some other basins, and they are still under-explored on the whole. Preliminary studies show that the total oil resources in volcanic rocks amount to $(19–26) \times 10^8 \text{ t}$, and natural gas resources are $4.2 \times 10^{12} \text{ m}^3$ with an oil discovery rate of 19 %–25 % and a natural gas discovery rate of 2 %. The total equivalent hydrocarbon reserves reach $(52–59) \times 10^8 \text{ t}$ with a discovery rate of 6 %–7 %. With abundant remaining resources and great potential for exploration, volcanic reservoirs are important replacement domain for hydrocarbon exploration.

Novel techniques should be fully utilized in future exploration of volcanic rocks, aiming at dissolution type, fracture type secondary volcanic rocks and other favorable primary reservoir formations of pyroclastic rocks (eruption facies) and lava (effusive facies) types. Based on research of deep layers in the Songliao Basin and Carboniferous formations in the Junggar Basin, two major gas-producing areas of volcanic rocks can be constructed. Exploration of volcanic rocks in the Santanghu Basin and Bohai Bay Basin should be strengthened to bring the reserve to one hundred million tons; more exploration activities should be carried out in the Carboniferous—Permian formations in the Tuha Basin, Carboniferous basins in northern Xinjiang, Permian formations in the Tarim and Sichuan Basins, Ordos and other new areas in the hope of making new breakthroughs.

Hydrocarbon-bearing layers of volcanic rocks in eastern China are dominated by Mesozoic–Cenozoic formations, which can be classified as intra-continent rift volcanic rocks formed under an extensional environment. The volcanic rock distribution is related to major faults, whereas hydrocarbon reservoir combinations are controlled by the development of faulted basins (Yang et al. 2014). Generally speaking, volcanic rocks in eastern China have identical structural environment to hydrocarbon source rocks with their distribution ranges coinciding with each other. In this way, a self-generation and self-preservation reservoir combination is formed. Due to different basin evolutions, deep layers in the Songliao Basin are mostly dominated by gas reservoirs in volcanic rocks, whereas the Bohai Bay, Erlian and Halar Basins are dominated by oil reservoirs. The distribution areas of volcanic rocks in northern Xinjiang primarily include Carboniferous—Permian formations in the Junggar, Santanghu, Tuha and some other basins, which were generated in the Xingmeng oceanic trench. Their reservoir combinations experienced significant changes due to intense basin reformation in later stages. There are not only near-source combinations, such as Ludong-Wucaiwan area in the Junggar Basin and Malang Sag in the Santanghu Basin, but also far-source reservoir combinations, such as the northwestern edges of the Junggar Basin. The distribution of hydrocarbons in these structures is predominantly controlled by unconformities and faults.

Acknowledgments This work was sponsored by the National Key Basic Research Program of China (973 Program, 2014CB239000, 2009CB219304) and National Science and Technology Major Project (2011ZX05001). We are grateful for the help and support given to this study by Du Jinhu of PetroChina, Chen Shumin of Daqing Oilfield Company, Kuang Lichun of Xinjiang Oilfield Company, Liang Shijun of Tuha Oilfield Company, Zou Caineng of Exploration and Development Research Institute of CNPC, journal editors and anonymous reviewers.

References

Bashari A. Petrography and clay mineralogy of volcanoclastic sandstones in the Triassic Rewan Group, Bowen Basin, Australia. Pet Geosci. 2000;6:151–63.

Batkhishig B, Noriyoshi T, Bignall G. Magmatic-hydrothermal activity in the Shuteen area, South Mongolia. Econ Geol. 2014;109(7):1929–42.

Borgia A, Mazzoldi A, Brunori CA, et al. Volcanic spreading forcing and feedback in geothermal reservoir development, Amiata Volcano, Italia. J Volcanol Geoth Res. 2014;284:16–31.

Cai ZR, Huang QT, Xia B, et al. Development features of volcanic rocks of the Yingcheng Formation and their relationship with fault structure in the Xujiaweizi Fault Depression, Songliao Basin, China. Pet Sci. 2012;9(4):436–43.

Caricchi L, Biggs J, Annen C, et al. The influence of cooling, crystallisation and re-melting on the interpretation of geodetic signals in volcanic systems. Earth Planet Sci Lett. 2014;388:166–74.

Chen HQ, Hu YL, Jin JQ, et al. Fine stratigraphic division of volcanic reservoir by uniting of well data and seismic data—Taking volcanic reservoir of member one of Yingcheng Formation in

Xudong area of Songliao Basin for an example. J Earth Sci. 2014;25(2):337–47.

Dong JX, Tong M, Ran B, et al. Nonlinear percolation mechanisms in different storage-percolation modes in volcanic gas reservoirs. Pet Explor Dev. 2013;40(3):372–7 (in Chinese).

Du J, Chen H, Guo P, et al. The simulation study of full diameter cores depletion on volcanic condensate gas reservoirs. Pet Sci Technol. 2013;31(22):2388–95.

Farooqui MY, Hou HJ, Li GX, et al. Evaluating volcanic reservoirs. Oilfield Rev. 2009;21:36–47.

Feng ZQ. Volcanic rocks as prolific gas reservoir: a case study from the Qingshen gas field in the Songliao Basin, NE China. Mar Pet Geol. 2008;25(4):416–32.

Hou LH, Luo X, Wang JH, et al. Weathered volcanic crust and its petroleum geological significance: a case study of the Carboniferous volcanic crust in northern Xinjiang. Pet Explor Dev. 2013;40(3):277–86 (in Chinese).

Hou LH, Zou CN, Liu L, et al. Geologic essential elements for hydrocarbon accumulation within Carboniferous volcanic weathered crusts in northern Xinjiang, China. Acta Petrolei Sinica. 2012;33(4):533–40 (in Chinese).

Jin XH, Yan XB, Li LN. Characteristics and oil-source correlation of the source rocks in Changling fault depression, Songliao Basin. J Southwest Pet Univ. 2010;32(6):53–6 (in Chinese).

Lenhardt N, Götz AE. Volcanic settings and their reservoir potential: an outcrop analog study on the Miocene Tepoztlán Formation, Central Mexico. J Volcanol Geoth Res. 2011;204:66–75.

Li W, Zhang ZH, Yang YC, et al. Oil source of reservoirs in the hinterland of the Junggar Basin. Pet Sci. 2007;4(4):34–43.

Luo JL, Hou LH, Jiang YQ, et al. Chronology and tectonic settings of igneous rocks and origin of volcanic reservoirs in Ludong area, eastern Junggar Basin. Acta Petrolei Sinica. 2012;33(3):352–60 (in Chinese).

Luo JL, Shao HM, Yang YF, et al. Temporal and spatial evolution of burial—hydrocarbon filling—diagenetic process of deep volcanic reservoirs in the Songliao Basin. Earth Sci Front. 2013;20(5):175–87 (in Chinese).

Lü XX, Yang HJ, Xu SL, et al. Petroleum accumulation associated with volcanic activity in the Tarim Basin—Taking Tazhong-47 Oilfield as an example. Pet Sci. 2004;1(3):30–6.

Mao ZG, Zou CN, Zhu RK, et al. Geochemical characteristics and tectonic settings of Carboniferous volcanic rocks in the Junggar Basin. Acta Petrologica Sinica. 2010;26(1):207–16 (in Chinese).

Meng YL, Liang HW, Meng FJ, et al. Distribution and genesis of the anomalously high porosity zones in the middle-shallow horizons of the northern Songliao Basin. Pet Sci. 2010;7(3):302–10.

Mo BB, Lian B. Study on feldspar weathering and analysis of relevant impact factors. Earth Sci Front. 2010;17(3):281–9 (in Chinese).

Pan BZ, Xue LF, Huang BZ, et al. Evaluation of volcanic reservoirs with the "QAPM mineral model" using a genetic algorithm. Appl Geophys. 2008;5(1):1–8.

Qin LM, Zhang ZH, Wu YY, et al. Organic geochemical and fluid inclusion evidence for filling stages of natural gas and bitumen in volcanic reservoirs of the Changling faulted depression, southeastern Songliao Basin. J Earth Sci. 2010;21(3):303–20.

Rohrman M. Prospectivity of volcanic basins: trap delineation and acreage de-risking. AAPG Bull. 2007;91(6):915–39.

Shao HM, Luo JL, Yang YF, et al. Origin and quality evolution of volcanic reservoir in the Yingcheng Formation, Songliao Basin. Chin J Geol. 2013;48(4):1187–203 (in Chinese).

Sruoga P, Rubinstein N. Processes controlling porosity and permeability in volcanic reservoirs from the Austral and Neuquén basins, Argentina. AAPG Bull. 2007;91(1):115–29.

Sun SM, Wu XS, Liu HT, et al. Genetic models of structural traps related to normal faults in the Putaohua oilfield, Songliao Basin. Pet Sci. 2008;5(4):302–7.

Sun YH, Kang L, Bai HF, et al. Fault systems and their control of deep gas accumulations in Xujiaweizi Area. Acta Geologica Sinica. 2012;86(6):1546–58.

Surour AA, Moufti AMB. Opaque mineralogy as a tracer of magmatic history of volcanic rocks: an example from the Neogene-Quaternary Harrat Rahat Intercontinental Volcanic Field, North Western Saudi Arabia. Acta Geologica Sinica. 2013;87(5):1281–305.

Xiang CF, Danišík M, Feng ZH. Transient fluid flow in the Binbei district of the Songliao Basin, China: evidence from apatite fission track thermochronology. Pet Sci. 2013;10(3):314–26.

Xie Q, He SL, Pu WF. The effects of temperature and acid number of crude oil on the wettability of acid volcanic reservoir rock from the Hailar Oilfield. Pet Sci. 2010;7(1):93–9.

Yagi M, Ohguch T, Akiba F, et al. The Fukuyama volcanic rocks: submarine composite volcano in the Late Miocene to Early Pliocene Akita-Yamagata back-arc basin, northeast Honshu, Japan. Sed Geol. 2009;220(4):243–55.

Yang RF, Wang YC, Cao J. Cretaceous source rocks and associated oil and gas resources in the world and China: a review. Pet Sci. 2014;11(3):331–45.

Yu ZC, Liu L, Qu XY, et al. Melt and fluid inclusion evidence for a genetic relationship between magmatism in the Shuangliao volcanic field and inorganic CO_2 gas reservoirs in the southern Songliao Basin, China. Int Geol Rev. 2014;56(9):1122–37.

Zeng HS, Li JK, Huo QL. A review of alkane gas geochemistry in the Xujiaweizi fault-depression, Songliao Basin. Mar Pet Geol. 2013;43:284–96.

Zhang K, Marfurt K, Wan Z, et al. Seismic attribute illumination of an igneous reservoir in China. Lead Edge. 2011;30(3):266–70.

Zhang M, Li HB, Wang X. Geochemical characteristics and grouping of the crude oils in the Lishu fault depression, Songliao Basin, NE China. J Petrol Sci Eng. 2013a;110:32–9.

Zhang YY, Pe-Piper G, Piper DJW, et al. Early Carboniferous collision of the Kalamaili orogenic belt, North Xinjiang, and its implications: evidence from molasse deposits. Geol Soc Am Bull. 2013b;125(5–6):932–44.

Zhao WZ, Zou CN, Feng ZQ, et al. Geological features and evaluation techniques of deep-seated volcanic gas reservoirs, Songliao Basin. Pet Explor Dev. 2008;35(2):129–42.

Zhao WZ, Zou CN, Li JZ, et al. Comparative study on volcanic hydrocarbon accumulations in western and eastern China and its significance. Pet Explor Dev. 2009;36(1):1–11.

Zhou XY, Pang XQ, Li QM, et al. Advances and problems in hydrocarbon exploration in the Tazhong area, Tarim Basin. Pet Sci. 2010;7(2):164–78.

Zhu RK, Mao ZG, Guo HL, et al. Volcanic oil and gas reservoir geology: thinking and forecast. Lithologic Reservoirs. 2010a;22(2):7–13 (in Chinese).

Zhu XM, Zhu SF, Xian BZ, et al. Reservoir differences and formation mechanisms in the Ke-Bai overthrust belt, northwestern margin of the Junggar Basin, China. Pet Sci. 2010b;7(1):40–8.

Zou CN, Guo QL, Wang JH, et al. A fractal model for hydrocarbon resource assessment with an application to the natural gas play of volcanic reservoirs in Songliao Basin, China. Bull Can Pet Geol. 2012a;60(3):166–85.

Zou CN, Hou LH, Tao SZ, et al. Hydrocarbon accumulation mechanism and structure of large-scale volcanic weathering crust of the Carboniferous in northern Xinjiang, China. Sci China Earth Sci. 2012b;55(2):221–35.

Zou CN, Zhao WZ, Jia CZ, et al. Formation and distribution of volcanic hydrocarbon reservoirs in sedimentary basins of China. Pet Explor Dev. 2008;35(3):257–71.

17

Research into polymer injection timing for Bohai heavy oil reservoirs

Lei-Ting Shi · Shi-Jie Zhu · Jian Zhang ·
Song-Xia Wang · Xin-Sheng Xue · Wei Zhou ·
Zhong-Bin Ye

Abstract Polymer flooding has been proven to effectively improve oil recovery in the Bohai Oil Field. However, due to high oil viscosity and significant formation heterogeneity, it is necessary to further improve the displacement effectiveness of polymer flooding in heavy oil reservoirs in the service life of offshore platforms. In this paper, the effects of the water/oil mobility ratio in heavy oil reservoirs and the dimensionless oil productivity index on polymer flooding effectiveness were studied utilizing relative permeability curves. The results showed that when the water saturation was less than the value, where the water/oil mobility ratio was equal to 1, polymer flooding could effectively control the increase of fractional water flow, which meant that the upper limit of water/oil ratio suitable for polymer flooding should be the value when the water/oil mobility ratio was equal to 1. Mean while, by injecting a certain volume of water to create water channels in the reservoir, the polymer flooding would be the most effective in improving sweep efficiency, and lower the fractional flow of water to the value corresponding to $\Delta J_{\max}$. Considering the service life of the platform and the polymer mobility control capacity, the best polymer injection timing for heavy oil reservoirs was optimized. It has been tested for reservoirs with crude oil viscosity of 123 and 70 mPa s, the optimum polymer flooding effectiveness could be obtained when the polymer floods were initiated at the time when the fractional flow of water were 10 % and 25 %, respectively. The injection timing range for polymer flooding was also theoretically analyzed for the Bohai Oil Field utilizing relative permeability curves, which provided methods for improving polymer flooding effectiveness.

Keywords Heavy oil reservoir · Mobility ratio · Polymer injection timing · Injection timing range · Timing optimization

L.-T. Shi (✉) · J. Zhang · X.-S. Xue · W. Zhou
State Key Laboratory of Offshore Oil Exploitation,
Beijing 100027, China
e-mail: flygoslt@126.com

L.-T. Shi · S.-J. Zhu · Z.-B. Ye
State Key Laboratory of Oil and Gas Reservoir Geology
and Exploitation, Southwest Petroleum University,
Chengdu 610500, Sichuan, China

S.-X. Wang
Research Institute of Exploration and Development,
PetroChina Southwest Oil & Gas Field Company,
Chengdu 610041, Sichuan, China

Edited by Yan-Hua Sun

1 Introduction

Heavy oils account for about 69 percent of the total of about 17.8×10^8 m^3 original oil in place discovered in offshore reservoirs in the Bohai Oil Field. For water flooding in heavy oil reservoirs, the swept volume by water during a pilot flood turns out to be very limited due to an excessively high water/oil mobility ratio and serious fingering. The average oil recovery of the overall development program is 20.2 percent in the Bohai Oil Field (Zhou 2009). Therefore, specific measures should be taken for most conventional heavy oil reservoirs to improve oil displacement efficiency during different oil production stages (Levitt et al. 2013). As for the Bohai Oil Field even if the oil recovery is improved by only 1 %, it will in effect result in obtaining the equivalent of another major oil field of hundreds of millions of tonnes without any exploration and development investment. Among different EOR techniques, polymer flooding, as a relatively mature technique (Chang 2011; Abu-shiekah et al. 2012; Delamaide et al. 2013), has been proven to be very effective in increasing oil recovery and decreasing water cuts in different scales of on-site polymer flood pilots conducted in the Bohai Oil Field (Ye et al. 2010; Zhang et al. 2011a, b; Kang and Zhang 2013; Shi et al. 2013).

In the course of water flooding in conventional heavy oil reservoirs, with an increase in water saturation, the water/oil mobility ratio will increase as well, resulting in a much higher increase in the water cut in heavy oil reservoirs than in light oil reservoirs (Kumar et al. 2005; Asghari and Nakutnyy 2008; Aktas et al. 2008; Mogbo 2011; Morelato et al. 2011; Liu et al. 2012). Influenced by such factors as reservoir heterogeneity, the polymer flood was initiated when the water cut of the produced fluids reached 60 % for the SZ36-1 oil field, and the oil recovery factor increased by only 5–7 % due to polymer floods (Zhang et al. 2007, 2009, 2013; Jiang et al. 2010; Kang et al. 2011), which was lower than that in the onshore light crude reservoirs (Asghari and Nakutnyy 2008). Meanwhile, the largest difference between onshore and offshore oil production is that the design service life of an offshore platform is less than 30 years (Rivas and Gathier 2013). Therefore, it is extremely important for offshore oil fields to select an optimum injection timing to initiate polymer floods in heavy oil reservoirs (Zhang et al. 2013) for better ultimate oil recovery.

Field tests show that the earlier the polymer flooding is performed, the better the water/oil mobility control is maintained with polymer solutions. This means that the injection timing of polymer solutions has a significant influence on the ultimate recovery factor (Ma 1995; Hu 2004; Alzayer and Sohrabi 2013). Therefore, appropriate polymer injection timing would helpful to effectively enhance oil recovery, lower the risks in polymer flooding in heavy oil reservoirs, and displace more oil during the service life of offshore platforms. In this paper, the effect of the water/oil mobility ratio on the efficiency of polymer flooding is studied at different fractional flows of water in heavy oil reservoirs using relative permeability curves, and then the polymer injection timing is optimized.

2 Experimental

2.1 Materials

Hydrophobically associating polyacrylamide (HAP2010, industrial purity) was commercially available, with a relative molecular weight of 1.0×10^7, degree of hydrolysis of 20 percent, and a hydrophobic group content of 1.0 percent. Injection water used in all tests had a total salinity of

9,047 mg/L, in which the mass concentrations of Na^+/K^+, Ca^{2+}, Mg^{2+}, SO_4^{2-}, HCO_3^-, and Cl^- were 2,552, 569, 229, 37, 191, and 5,471 mg/L, respectively. The types of oils used, namely oil A and B, were the mixtures of diesel and dehydrated crude from the Bohai Oil Field, with viscosities of 70 and 123 mPa s, respectively, at 65 °C. The oil viscosity was measured with a DV-IIIBrook-field viscometer.

2.2 Experimental procedures and data processing method

Natural cores were taken from the Bohai oil reservoirs, with a dimension of 2.5 cm in diameter and 7 cm in length. These cores had a gas permeability of about $2,000 \times 10^{-3}$ μm^2 and porosity of about 30 %.

Water–oil/polymer solution–oil relative permeability curves were measured by an unsteady state method according to oil and natural gas industrial specifications SY/T 5435-2007, P.R. China. The experimental procedures are as follows (Crotti and Rosbaco 1988; Bakhitov et al. 1980): (a) The core was cleaned with solvent and dried with hot nitrogen and then evacuated; (b) Permeability was measured with gas; (c) The core was saturated with injection water, then the core porosity and permeability to water were calculated; (d) Oil was injected into the core at a rate of 0.1 mL/min up to 3 PV (pore volume) in order to reach the irreducible water saturation. Oil was injected into the core by an ISCO Model 260D pump; (e) The oil permeability was measured at irreducible water saturation; (f) The injection water was injected into the core at a rate of 1 mL/min up to 10 PV in order to reach the irreducible oil saturation; (g) The water permeability was measured at residual oil saturation; (h) The core was cleaned with solvent and dried with hot nitrogen and evacuated. Then steps from b to e were repeated; (i) The polymer solution (polymer concentration was 1,750 mg/L) was injected into the core at a rate of 1 mL/min up to 10 PV in order to reach the residual oil saturation; and (j) The permeability to polymer solution was measured at irreducible oil saturation. The volumes of oil and water produced were measured dynamically.

The water saturation at different periods was calculated by the material balance method, and the modified "J·B·N" method (Dou et al. 2007; Shi et al. 2001; Zhou et al. 2010; Delgado et al. 2013) was used to calculate the water–oil/polymer solution–oil relative permeability. The effective viscosity μ_{eff} of the polymer solution was calculated with Eq. (1).

$$\mu_{eff} = \frac{RF}{RRF} \mu_w, \tag{1}$$

where RF is the resistance factor, dimensionless; RRF is the residual resistance factor, dimensionless; and μ_w is the viscosity of water solution, mPa s.

3 Results and discussion

3.1 Relative permeability curves during water flooding and polymer flooding

In experiments, the viscosity of the injection water and the polymer solution was 0.60 and 1.05 mPa s, respectively, at 65 °C. The relative permeability of the displaced and displacing phases (oil and water during water flooding, oil and polymer solution during polymer flooding) was measured, and the relative permeability curves are shown in Fig. 1.

In Fig. 1, K_{ro} and K_{rw} are the oil and water relative permeability, respectively, during water flooding; and K_{rpo} and K_{rp} are the oil and polymer solution relative permeability, respectively, during flooding; S_w is the water saturation.

With an increase in oil viscosity, the residual oil saturation increased in the relative permeability curves, which demonstrated that the performance of the displacing phase dropped (Jiang et al. 2008). Under the same water saturation, the relative permeability of cores to the polymer phase was far lower than that to the water phase, while the relative permeability of cores to the oil phase changed slightly; and the residual oil saturation decreased in the relative permeability curve during polymer flooding. This showed that polymer flooding could control the water/oil mobility and improve the oil displacement efficiency for heavy oil reservoirs (Shi et al. 2010).

3.2 Effect of the fractional flow of water on injection timing of polymer floods

According to the relative permeability curves during water flooding and polymer flooding, the mobility ratio (M), the fractional flow of water (f_w), and the rate of increase of fractional water flow with water saturation (df_w/dS_w) could be obtained in the fractional flow equation

(Romero-Zeron et al. 2009; Torabi et al. 2013). During water flooding, the curves of the mobility ratio versus water saturation are shown in Fig. 2, and the fractional water flow curves and the rate of increase in the fractional water flow with water saturation are demonstrated in Fig. 3

Figure 2 indicates that at a water saturation of 0.35, the water/oil mobility ratio is typically very unfavorable for oil B; and a similar tendency is present for oil A when the water saturation is about 0.38. During polymer flooding, the polymer solution/oil mobility ratio would be reduced to just slightly above 1.0. The injection of polymer solutions may increase the viscosity of the water phase and decrease the water mobility, indicating that improved sweep efficiency can be expected from polymer flooding.

As shown in Fig. 3, when the oil viscosity was relatively high, the fractional flow of water increased rapidly and reached its maximum levels at low water saturations. As a result, water channels may be formed in a short time due to serious viscous fingering, causing early water breakthrough and poor oil displacement efficiency.

The water/oil mobility ratio and the fractional flow of water increased quickly when the water saturation (S_w) increased. Therefore, oil recovery can be improved by injecting polymer solutions before the rate of increase of fractional water flow reached its maximum level. This was helpful to control mobility for stabilizing the water front, improve sweep efficiency and to increase displacement efficiency.

For oil B and A, when the water saturation was 0.186 and 0.252, respectively, the rate of increase of fractional water flow reached its maximum, as soon as the water/oil mobility ratio was higher than 1. The mobility ratio would increase dramatically when the water saturation was higher than 0.186 and 0.252. At high mobility ratios viscous fingering occurred, leading to a poor sweep efficiency (Shi et al. 2012). A frontal-advance velocity cutoff would be established for mobility control with polymer floods

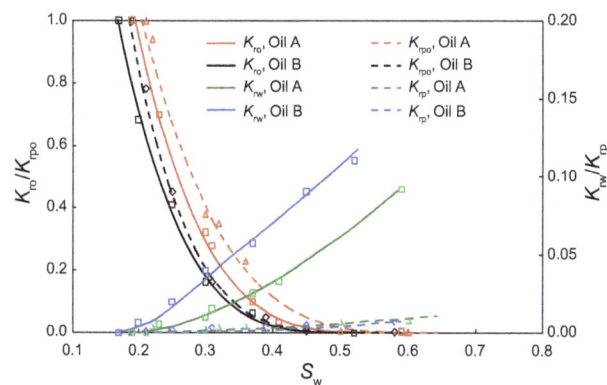

Fig. 1 Relative permeability curves during water flooding and polymer flooding

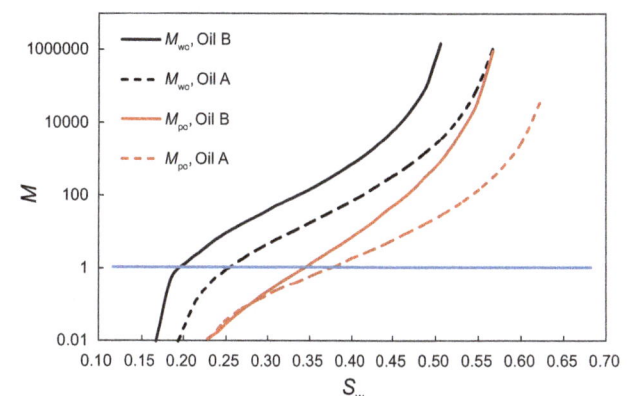

Fig. 2 Mobility ratios for crude oils of different viscosity

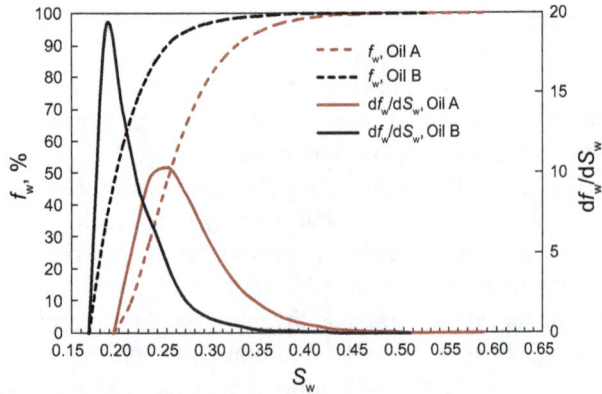

Fig. 3 The curves of fractional water flow and the rate of increase of fractional water flow

(Lee 2010), but the mobility control of polymer solution could become limited when the mobility ratio was too high during water flooding, leading to a poor polymer flood sweep efficiency in high viscosity reservoirs.

It was favorable to perform polymer flooding at relatively low water/oil mobility ratio when the polymer solution had enough ability to control flow mobility. Injection of polymer solutions would decrease the rate of increase of fractional water flow and extended the stable production period, so good displacing efficiency would be achieved. The upper water saturation limit of polymer flooding should be the saturation (S_{wu}) when the rate of increase of fractional water flow reached its maximum level. When $S_w \leq S_{wu}$, the ability of polymer floods to control flow mobility ratio has appeared to be affected significantly by the water saturation in the heavy oil reservoirs. While $S_w > S_{wu}$, polymer floods could not control flow mobility, finally resulting in poor displacement efficiency.

Thus, it was favorable to perform polymer floods before the rate of increase of fractional water flow reached its maximum level. The suitable water saturation for polymer floods should be less than 0.186 and 0.252, respectively, for oil B and A. The water saturation was 0.186 and 0.252, the rates of increase of fractional water flow was 34 % and 50 %, respectively.

3.3 Effect of dimensionless productivity index on injection timing of polymer floods

The objective of mobility control is to improve the volumetric sweep efficiency of polymer floods. The oilfield responses are observed through an increase in injection pressure and reduced water cut (Mahani et al. 2011). The polymer solution slug would improve the volumetric sweep efficiency, resulting in a decrease in water cut; on the other hand, the polymer solution slug would lower the water

injection capacity and the liquid producing capacity. The reduction of the liquid producing capacity generally took place in the initial stage of the polymer flooding process, and decreased slowly thereafter.

Figure 4 indicates that the water saturation of oil reservoirs increased during polymer flooding when the fractional flow of water was the same. Figure 5 shows that the dimensionless productivity index in the polymer flooding process was less than that in the water flooding process.

The difference between the dimensionless productivity indexes in the water flooding and polymer flooding processes are defined as follows:

$$\Delta J = J_{orw} - J_{orp}, \tag{2}$$

where J_{orw} is the dimensionless productivity index in the water flooding process; J_{orp} is the dimensionless productivity index in the polymer flooding process.

Figure 5 shows that ΔJ increased at first and then decreased as the water saturation increased. The water/oil mobility ratio was low when the water saturation was relatively low. The water front was relatively stable, injection water could connect the flow channels in oil reservoirs, and thus J_{orw} was relatively high. If the polymer solution was injected at this moment, the polymer mobility should be less than the oil mobility, resulting in an excessive pressure increase and lowering the dimensionless productivity index. This showed that the flow channels for initial water were connected together in oil reservoirs when the water saturation was relatively low. ΔJ reached its maximum level (ΔJ_{max}) as the water saturation increased, which indicated that flow channels were formed in oil reservoirs due to water injection and the oil was displaced by injection water. After that, the injection water would preferentially flow through the high permeability channels, the fractional flow of water increased rapidly, water breakthrough was very fast and the volumetric sweep efficiency was low, so it was desirable to inject polymer solution to control the flow resistance in flow channels, which ultimately enhanced the displacement efficiency.

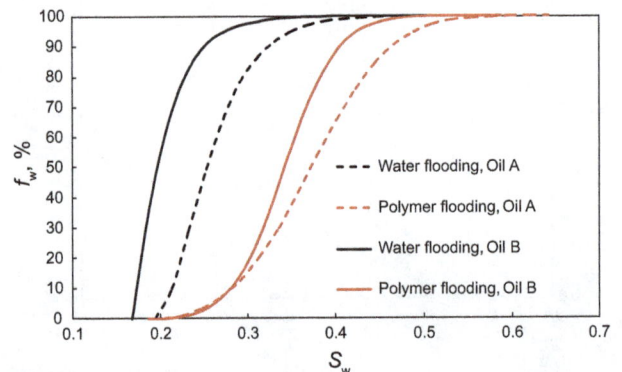

Fig. 4 The fractional water flow curves

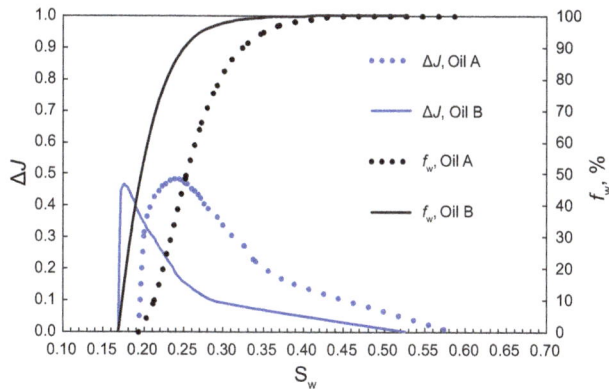

Fig. 5 Dimensionless productivity index during water flooding and polymer flooding

When $S_w \leq S_{wL}$ (where S_{wL} is the water saturation when ΔJ reached its maximum level), the displacement effect of the polymer flood was similar to water flood. The main function of polymer floods was to form flow channels. The displacement efficiency of polymer floods would be below the desired level if the polymer solution was only used to build up flow channels. S_{wL} should be the water saturation lower limit for polymer floods. A certain amount of water is recommended to be injected into heavy oil reservoirs to create or connect flow channels before polymer flooding. Then the polymer solution is injected to modify the oil/water mobility, improve the volumetric sweep efficiency, and to enhance success rate of polymer floods. Therefore, when ΔJ reaches maximum (ΔJ_{max}), the corresponding water saturation should be the lower limit of injection timing, which is also the start for polymer slug injection in heavy oil reservoirs.

When $S_w \geq S_{wL}$, the injection of a polymer solution slug would improve the volumetric sweep efficiency and oil recovery. For oil B and A, the corresponding suitable water saturation should be higher than 0.179 and 0.240, respectively, for polymer flooding. When the water saturations were 0.179 and 0.240, the rates of increase of fractional water flow were 10 % and 25 %, respectively.

By analyzing the displacement mechanism of water flooding and polymer flooding, the injection timing for heavy oil B was in the range of 10–34 % of the fractional flow of water; and the injection timing for oil A was in the range of 25–50 % of the fractional flow of water. During the service life of the offshore platform, the optimum injection timing for polymer flooding should be selected as S_{wL} in order to maximize the improvement of the oil recovery for the Bohai heavy oil reservoirs. Normally, under consideration of the mobility control ability of polymer solution, the higher the crude oil viscosity is, the earlier the polymer flooding injection timing should be. It could be realized that the polymer injection timing should be a little later if the ability of the polymer solution to control flow mobility is high enough, which also could achieve good displacement efficiency.

4 Conclusions

(1) For polymer solutions of the same mobility control ability, the polymer flood may be injected at relatively low water saturation to stabilize the water front and decrease the water cut when the oil viscosity is higher.

(2) The injection timing for polymer floods is when the upper limit of water saturation is less than S_{wu}. Before the fractional flow of water reaches the maximum value, the injection polymer solution can effectively control the increase of fractional water flow.

(3) The injection timing for polymer floods is when the lower limit of water saturation is greater than S_{wL}. A certain amount of water injected before polymer flooding may effectively connect the flow channels in heavy oil reservoirs, cut the polymer cost and ensure the displacement efficiency of polymer flooding.

(4) Influenced by the service life of offshore platforms in Bohai heavy oil reservoirs, the injection timing should be selected at S_{wL}. When the oil viscosity is 123 mPa s, the optimum injection timing for polymer floods should be 10 % of the fractional flow of water; and when the oil viscosity is 70 mPa s, the optimum injection timing for polymer floods should be 25 % of the fractional flow of water. Meanwhile, for polymer solutions with different mobility control capabilities, the optimum injection timing for polymer flooding should be varied as well.

Acknowledgments This paper was supported by Open Fund (CRI2012RCPS0152CN) of State Key Laboratory of Offshore Oil Exploitation and the National Science and Technology Major Project (2011ZX05024-004-01).

References

Abu-shiekah IM, Nieuwenhuijs RA, Ross RW, et al. Developing a layered heterogeneous Precambrian reservoir by polymer flooding. SPE EOR conference at oil and gas West Asia, 16–18 Apr 2012, Muscat, Oman. (SPE 154465).

Aktas F, Clemens T, Castanier LM, et al. Viscous oil displacement with aqueous associative polymers. SPE symposium on improved oil recovery, 20–23 Apr 2008, Tulsa, Oklahoma, USA. (SPE 113264-MS).

Alzayer A, Sohrabi M. Numerical simulation of improved heavy oil recovery by low-salinity water injection and polymer flooding. SPE Saudi Arabia section technical symposium and exhibition, 19–22 May 2013, Khobar, Saudi Arabia. (SPE 165287).

Asghari K, Nakutnyy P. Experimental results of polymer flooding of heavy oil reservoirs. Canadian international petroleum conference, 17–19 Jun 2008, Calgary, Alberta. (PETSOC-2008-189).

Bakhitov GG, Ogandzhanyants VG, Polishchuk AM. Experimental investigation into the influence of polymer additives in water on the relative permeabilities of porous media. Fluid Dyn. 1980;15(4):611–5.

Chang HG. Advances of polymer flood in heavy oil recovery. SPE heavy oil conference and exhibition, 12–14 Dec 2011, Kuwait City, Kuwait. (SPE 150384).

Crotti MA, Rosbaco JA. Relative permeability curves: the influence of flow direction and heterogeneities. SPE improved oil recovery symposium held in Tulsa, Oklahoma, 19–22 Apr 1988. (SPE 39657).

Delamaide E, Zaitoun A, Renard G, et al. Pelican lake field: first successful application of polymer flooding in a heavy oil reservoir. SPE enhanced oil recovery conference, 2–4 Jul 2013, Kuala Lumpur, Malaysia. (SPE 165234).

Delgado DE, Vittoratos E, Kovscek AR. Optimal voidage replacement ratio for viscous and heavy oil water floods. SPE western regional & AAPG pacific section meeting 2013 joint technical conference, 19–25 Apr 2013, Monterey, California, USA. (SPE 165349).

Dou HE, Chen CC, Chang YW, et al. A new method of calculating water invasion for heavy oil reservoir by steam injection. Production and operations symposium, 31 Mar–3 Apr 2007, Oklahoma City, Oklahoma, USA. (SPE 106233).

Hu FZ. Polymer flooding production engineering. Beijing: Petroleum Industry Press; 2004. pp. 127–30.

Jiang RZ, Liu XT, Zhao W, et al. Study of a quantitative method for mobility design in polymer flooding. Pet Geol Recover Effic. 2010;17(5):39–41 (in Chinese).

Jiang WD, Lu GX, Ren YB. Study of oil and water relative permeability and water flooding efficiency in heavy oil reservoirs. Pet Geol Oilfield Dev Daqing. 2008;27(5):50–3 (in Chinese).

Kang XD, Zhang J, Sun FJ et al. A review of polymer EOR on offshore heavy oil field in Bohai Bay, China. SPE enhanced oil recovery conference, 19–21 Jul 2011, Kuala Lumpur, Malaysia (SPE 144932).

Kang XD, Zhang J. Offshore heavy oil polymer flooding test in JZW area. SPE heavy oil conference, 11–13 Jun, 2013, Calgary, Alberta, Canada. (SPE 165473).

Kumar M, Hoang V, Satik C, et al. High mobility ratio water flood performance prediction: challenges and new insights. SPE international improved oil recovery conference in Asia Pacific, 5–6 Dec 2005, Kuala Lumpur, Malaysia. (SPE 97671).

Lee KS. Effects of polymer adsorption on the oil recovery during polymer flooding processes. Pet Sci Technol. 2010;4(28):351–9.

Levitt D, Jouenne S, Igor B, et al. Polymer flooding of heavy oil under adverse mobility conditions. SPE enhanced oil recovery conference, 02–04 Jul, 2013, Kuala Lumpur, Malaysia. (SPE 165267).

Liu C, Liao XW, Zhang YL. Field application of polymer microspheres flooding: a pilot test in offshore heavy oil reservoir. SPE Annual Technical Conference and Exhibition, 8-10 Oct 2012, San Antonio, Texas, USA. (SPE 158293).

Ma SY. The pragmatic engineering method of polymer flood. Beijing: Petroleum Industry Press; 1995. pp. 42–4 (in Chinese).

Mahani H, Sorop TG, van den Hoek PJ, et al. Injection fall-off analysis of polymer flooding EOR. SPE reservoir characterisation and simulation conference and exhibition, 9–11 Oct 2011, Abu Dhabi, UAE. (SPE 145125).

Mogbo OC. Polymer flood simulation in a heavy oil field: offshore Niger-delta experience. SPE enhanced oil recovery conference, 19–21 Jul 2011, Kuala Lumpur, Malaysia. (SPE 145027).

Morelato P, Rodrigues L, Romero OJ. Effect of polymer injection on the mobility ratio and oil recovery. SPE heavy oil conference and exhibition, 12–14 Dec 2011, Kuwait City, Kuwait. (SPE 148875).

Rivas C, Gathier F. C-EOR projects-offshore challenges. The twenty-third international offshore and polar engineering conference, 30 June–5 Jul, 2013, Anchorage, Alaska. (ISOPE-I-13-188).

Romero-Zeron LB, Li L, Ongsurakul S, et al. Visualization of waterflooding through unconsolidated porous media using magnetic resonance imaging. Pet Sci Technol. 2009;17(27): 1993–2009.

Shi JP, Li FQ, Cao WZ. Measurement of relative permeability curve of polymer flooding with non-steady state method. Daqing Pet Geol Dev. 2001;20(5):53–5 (in Chinese).

Shi LT, Chen L, Ye ZB, et al. Effect of polymer solution structure on displacement efficiency. Pet Sci. 2012;9(2):230–5.

Shi LT, Ye ZB, Zhang Z, et al. Necessity and feasibility of improving the residual resistance factor of polymer flooding in heavy oil reservoirs. Pet Sci. 2010;7(02):251–6.

Shi LT, Li C, Zhu SS. Study on properties of branched hydrophobically modified polyacrylamide for polymer flooding. J Chem. 2013;2013:1–5 Article ID 675826.

Torabi F, Zarivnyy O, Mosavat N. Developing new Corey-based water/oil relative permeability correlations for heavy oil systems. 2013 SPE Heavy Oil Conference, 11–13 Jun 2013, Calgary, Alberta, Canada. (SPE 165445).

Ye ZB, He EQ, Xie SY. The mechanism study of disproportionate permeability reduction by hydrophobically associating water-soluble polymer gel. J Petrol Sci Eng. 2010;72(1–2):64–6.

Zhang FJ, Jiang W, Sun FJ, et al. Key technology research and field tests of offshore viscous polymer flooding. Eng Sci. 2011a;13(5):28–33 (in Chinese).

Zhang JP, Sun FJ, An GR. Study of an incremental law of water cut and decline law in a water drive oilfield. Pet Geol Recover Effic. 2011b;18(6):82–5 (in Chinese).

Zhang XS, Sun FJ, Feng GZ, et al. Research into the factors influencing polymer flooding and field testing in Bohai heavy oil fields. China Offshore Oil Gas. 2007;19(01):30–4 (in Chinese).

Zhang XS, Wang HJ, Tang EG, et al. Research into reservoir potentials and polymer flooding feasibility for EOR technology in the Bohai offshore oilfield. Pet Geol Recover Effic. 2009;16(05):56–9 (in Chinese).

Zhang XS, Tang EG, Xie XG, et al. Study of the characteristics and development patterns of early polymer flooding in offshore oilfields. J Oil Gas Technol. 2013;35(07):123–6 (in Chinese).

Zhou FJ, Zhang LF, Wang HJ, et al. Experimental study into relative permeability curves of polymer flooding. Pet Geol Eng. 2010;24(6):117–9 (in Chinese).

Zhou SW. Exploration and practice of offshore oilfield effective development technology. Eng Sci. 2009;11(10):55–9 (in Chinese).

18

Positive carbon isotope excursions: global correlation and genesis in the Middle–Upper Ordovician in the northern Tarim Basin, Northwest China

Cun-Ge Liu[1,2] · Li-Xin Qi[2] · Yong-Li Liu[2] · Ming-Xia Luo[2] · Xiao-Ming Shao[2] · Peng Luo[2] · Zhi-Li Zhang[3]

Abstract Stable carbon isotope ratio ($\delta^{13}C_{carb}$) analysis has been widely applied to the study of the inter-continental or global marine carbonate correlation. Large-scale Cambrian–Ordovician carbonate platforms were developed in the Tarim Basin. But research on fluctuation characteristics and global correlation of $\delta^{13}C_{carb}$ is still weak. Based on conodont biostratigraphy and whole-rock $\delta^{13}C_{carb}$ data in the Tahe oil–gas field of the northern Tarim Basin, the global correlation and genesis of positive carbon isotope excursions in the Darriwilian—Early Katian was examined. Three positive excursions were identified in the Tahe oil–gas field including the middle Darriwilian carbon isotope excursion (MDICE), the Guttenberg carbon isotope excursion (GICE), and a positive excursion within the *Pygodus anserinus* conodont zone which is named the Early Sandbian carbon isotope excursion (ESICE) in this paper. Furthermore, these positive excursions had no direct relation with sea level fluctuations. MDICE and GICE could be globally correlated. The Middle–Upper Ordovician Saergan Formation source rocks of the Kalpin outcrops were in accordance with the geological time of MDICE and ESICE. GICE had close relationship with the source rock of the Lianglitag Formation in the basin.

Massive organic carbon burial was an important factor controlling the genesis of these positive excursions.

Keywords Tarim Basin · Tahe oil–gas field · Middle–Upper Ordovician · Carbon isotope · Darriwilian · Sandbian · Katian source rock

1 Introduction

The stable carbon isotope ratio in dissolved inorganic carbon ($\delta^{13}C_{carb}$) reflects the initial isotope composition in the original seawater (Saltzman 2005; Ainsaar et al. 2010; Munnecke et al. 2011). Diagenesis to a certain degree cannot change the isotope fluctuation trend (Saltzman 2005). Paleoceanographic environment change results in the death of many organisms and the burial of ^{12}C in them, which can enrich the ^{13}C in the sea water and hence increase the $\delta^{13}C_{carb}$ values of carbonate rocks (Sial et al. 2013). $\delta^{13}C_{carb}$ fluctuation is taken as an indicator of environmental change (Ainsaar et al. 2010). During the past decades, $\delta^{13}C_{carb}$ had been widely used in studies of regional or global correlation on marine carbonate strata and paleoceanographic environment (Zhang et al. 2010a; Ainsaar et al. 2010; Fan et al. 2011).

The Ordovician is a special period in geological history in which some significant events happened, such as the Great Ordovician Biodiversification Event, Late Ordovician glaciation and one mass extinction (Trotter et al. 2008; Monnecke et al. 2010; Zhang et al. 2010b; Thompson et al. 2012). At present, three short-lived $\delta^{13}C_{carb}$-positive excursions have been globally identified in the Ordovician (Zhang et al. 2010a; Sial et al. 2013), including the middle Darriwilian carbon isotope excursion (MDICE), Early Katian Guttenberg carbon isotope excursion (GICE), and

✉ Cun-Ge Liu
liucunge@163.com

[1] State Key Laboratory of Oil and Gas Reservoir Geology and Exploitation, Chengdu University of Technology, Chengdu 610059, Sichuan, China

[2] Northwest Oilfield Company, China Petroleum & Chemical Corporation, Urumqi 830011, Xinjiang, China

[3] Institute of Petroleum Exploration and Development, China Petroleum & Chemical Corporation, Beijing 100083, China

Edited by Jie Hao

the Hirnantian carbon isotope excursion (HICE). It is generally considered that the MDICE and GICE are closely related to the burial of abundant organic matter (Rosenau et al. 2012; Pancost et al. 2013; Sial et al. 2013), and Hirnantian glaciation and biotic extinction events are closely related to HICE (Zhang et al. 2010a; Monnecke et al. 2011).

In 1990s, the $\delta^{13}C_{carb}$ negative excursions were used to investigate Cambrian-Ordovician sea level changes in the Kalpin outcrops of the Tarim Basin (Wang and Yang 1994). In addition, Wang (2000) employed $\delta^{13}C_{carb}$-positive excursions as the proof of marine source rock developing in the Middle–Upper Ordovician. In recent years, some scholars have been engaged in the study of carbon isotope stratigraphy at Cambrian-Ordovician outcrops in the Kalpin and Bachu regions (Jing et al. 2008; Hu et al. 2010; Wang et al. 2011; Zhao 2015). Zhang et al. (2014) discussed $\delta^{13}C_{carb}$ curve features of the Ordovician using analyses of cores and cuttings from the Bachu and Tazhong uplifts (Fig. 1a) and concluded the existence of two $\delta^{13}C_{carb}$-positive excursions in the Dapingian–Darriwilian and Sandbian–Early Katian.

In this paper, whole-rock carbon and oxygen isotope and conodont analyses were undertaken on the cores from two wells of the Tahe oil–gas field, Akekule uplift (Fig. 1b). On the basis of conodont biostratigraphy, the identified $\delta^{13}C_{carb}$-positive excursions were compared with those found in the Tarim Basin and other regions of the world. The genetic relationships of $\delta^{13}C_{carb}$-positive excursions with sea level fluctuation, source rocks, and kerogen thermal evolution were also investigated.

2 Geological setting

The Tarim plate had broken from Gondwana and moved as an independent plate during Cambrian. The plate moved from north to south to the equator in the Middle Ordovician (Wang et al. 2013a; Torsvik and Cocks 2013) and was close to the northwestern margin of Gondwana. Large carbonate platforms had been developed in the western part of the plate during the Cambrian-Middle Ordovician (Wu et al. 2012). The Tarim plate had moved northward in the Late Ordovician, when three small carbonate platforms formed in the Tabei Uplift, Bachu-Tazhong Uplift, and Kalpin area in the Early Katian (Hu et al. 2010; Liu et al. 2012).

The Tahe oil–gas field is located in the Akekule uplift of the northern Tarim Basin (Fig. 1b). According to the lithology and well logging features, the Lower–Middle Ordovician can be divided into Penglaiba, Yingshan, and Yijianfang Formations, and the Upper Ordovician is composed of the Qiaerbak, Lianglitag, and Sangtamu

Formations. The Lower–Middle Ordovician exhibits platform facies, and the margins of the platforms were located in the eastern region of what is now the Tahe oilfield (Fig. 1b).

The lithology of the Penglaiba Formation is gray dolomite and sandwiched limy dolomite, characterized by restricted platform facies. The Yingshan Formation which deposited on open platforms is composed of yellowish-gray limestone and thin-bedded calcarenite with dolomitization intensified in the lower part. The Yijianfang Formation consists of yellowish-gray calcarenites sandwiched with thin-bedded limestones and deposited in a shoal environment. Deng et al. (2007) considered the Yijianfang Formation as a Highstand System Tract (HST). The conodont fauna from the Penglaiba Formation to the Yijianfang Formation has North American Midcontinent species (Zhao et al. 2006), characterized by warm water, low abundance, and slow evolution of species. It is difficult to carry out stratigraphic correlation using conodonts in these strata (Wang et al. 2007).

The Qiaerbak Formation consists of drowned platform deposits. The lithology of the lower part is grayish-green muddy limestone, which belongs to a transgressive system tract (TST). The lithology of the upper part is red-brown nodular argillaceous limestone, which belongs to an HST (Wu et al. 2012). The conodont fauna of this formation is of the North Atlantic type, which is characterized by cold water, high abundance, and more species (Zhao et al. 2006). The *Nemagraptus gracilis* graptolite zone is taken as the boundary of the Middle–Upper Ordovician globally, but this boundary is higher than that of the *P. anserinus* conodont zone at the bottom of Upper Ordovician (Chen and Wang 2003; Wang et al. 2013b). Therefore, grayish-green muddy limestones under the finger peak of gamma ray curves should partly belong to the Middle Ordovician (Fig. 2). Red-brown nodular argillaceous limestone of the upper part of the Qiaerbak Formation is a marker bed in the Tabei Uplift (Fig. 1a). The geological time of the Qiaerbak Formation is equal to that of the Kanling Formation in the Kalpin outcrop (Zhao et al. 2006). The lithology and color of the marker bed can be easily distinguished.

The Lianglitag Formation consists of limestones, calcarenites, and muddy limestone and was deposited in carbonate ramp environments containing both TST and HST sequences (Liu et al. 2012; Wu et al. 2012). The ramp margin was distributed circularly (Fig. 1b) with the thickness of strata thinning eastward and southward. The thickness of the Lianglitag Formation of the Tahe oil–gas field is between 10 and 120 m, and that of the Kalpin outcrop is 163.7 m (Hu et al. 2010). The thickness of the Lianglitag Formation of the Tazhong Uplift is between 200 and 800 m (Liu et al. 2012), and is larger than that of the Tahe oil–gas field and Kalpin region. The conodonts are

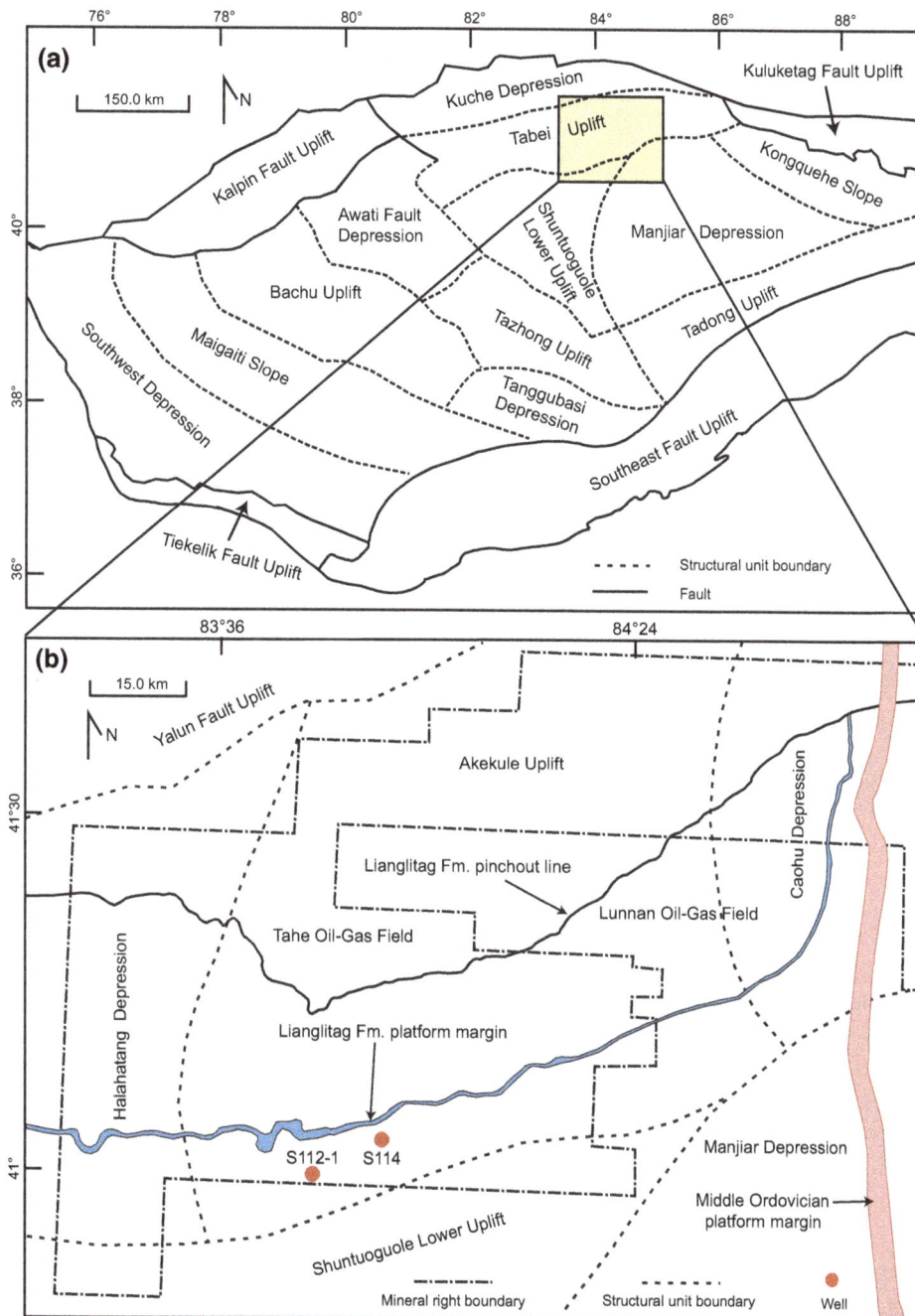

Fig. 1 Location map of the Tahe oil–gas field in the Tarim Basin. **a** The division map of structural units of the Tarim Basin. **b** Geological map of the Middle–Upper Ordovician in the Tahe oil–gas field

from the North American Midcontinent fauna. Typical conodont zones were lacking on the top of the Qiaerbak Formation and Lianglitag Formation. A *Belodina compressa* conodont zone generally developed from the upper Qiaerbak Formation to the Lianglitag Formation (Zhao et al. 2006).

The Sangtamu Formation is deep-water shelf deposits. The lower part is composed of grayish-black mudstone, and the upper part consisted of grayish-black mudstone and

thin-bedded argillaceous limestone. The *Yaoxianognathus neimengguensis* conodont zone can be found at the top of the Lianglitag Formation and developed mostly in the Sangtamu Formation (Zhao et al. 2006).

A short-term deposition hiatus occurred at the end of both the Yijianfang Formation and Lianglitag Formation (Liu et al. 2010; Chen et al. 2013) in the Akekule Uplift. The exposed area of the Yijianfang Formation appears mainly in the eastern and northern region. The

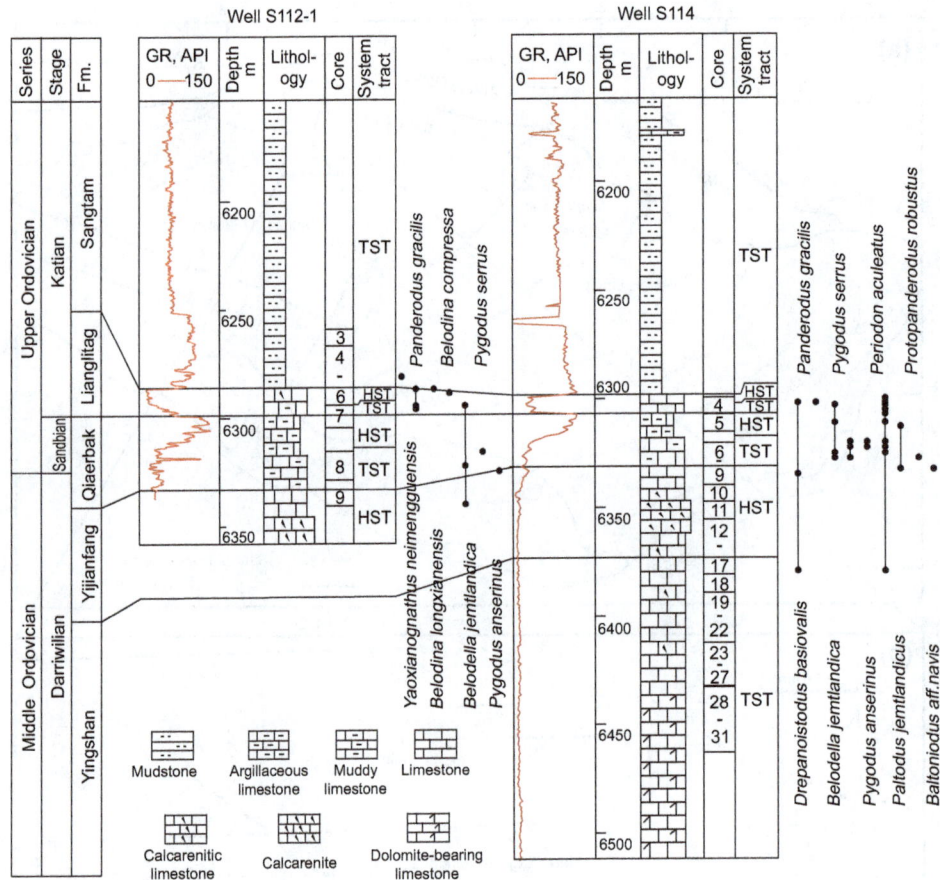

Fig. 2 Stratigraphic correlation and conodonts distribution in well S112-1 and well S114 in the Tahe oil–gas field

southwestern region was successive deposition (Liu et al. 2010). The exposed area of the Lianglitag Formation was located in the northern ramp edge. The southern region was successive deposition.

The Akekule Uplift was controlled by NW–SE extension stress in the Early Hercynian tectonic movement, so a large nose uplift plunging to the southwest was formed. The Ordovician carbonate strata were severely denuded in the northern region of the Akekule Uplift (Liu et al. 2010). The Akekule Uplift was further transformed by the Late Hercynian and Late Himalayan tectonic movements.

3 Methodology

3.1 Samples and experiments

Conodont and carbon–oxygen isotope samples for this study were collected from cores from well S112-1 and well S114 in the southern Tahe oil–gas field (Fig. 1b). The distance between the wells is 12 km. Continuous coring was not performed in the middle of the Qiaerbak Formation but separately in 11-m intervals in well S112-1 and 5-m

intervals in well S114. The total organic carbon (TOC) analysis of well S112-1 samples was also carried out. Conodont and TOC samples were analyzed by the Geological Laboratory of the Northwest Oilfield Branch Company.

Samples for carbon and oxygen isotope analysis were collected from micrites avoiding organic-rich drill cores with strong diagenesis and high argillaceous content, and from conglomerates or bands with high limy content in muddy limestones and argillaceous limestones. Samples from well S114 were collected in 2006. $\delta^{13}C_{carb}$ and $\delta^{18}O_{carb}$ analysis were measured with an MAT251 mass spectrograph in the Isotope Laboratory of Ministry of Land and Resources. Samples from well S112-1 were collected in 2010. $\delta^{13}C_{carb}$ and $\delta^{18}O_{carb}$ analysis were measured with an MAT253 mass spectrograph in the Analysis and Test Research Center of the Beijing Research Institute of Uranium Geology. The phosphate-continuous flow test method was adopted with errors less than ±0.1 ‰ Vienna Pee Dee Belemnite (VPDB) for $\delta^{13}C_{carb}$ and less than ±0.2 ‰ (VPDB) for $\delta^{18}O_{carb}$. The analytical process followed Liu et al. (2013). Analytical data are shown in Table 1 while the depth of samples were adjusted.

Table 1 Data of carbon and oxygen isotope for well S112-1 and well S114 (relative to VPDB standard)

Well	Formation	Depth, m	Core number	Lithology	$\delta^{13}C$, ‰	$\delta^{18}O$, ‰
S112-1	Sangtamu	6259.05	3 4/60	Mudstone	0.8	−6.1
		6262.17	3 21/60	Mudstone	0.9	−6.1
		6263.28	3 33/60	Mudstone	1.0	−5.9
		6264.52	3 39/60	Mudstone	1.0	−6.2
		6265.41	3 47/60	Mudstone	0.9	−5.9
		6266.51	3 57/60	Mudstone	0.9	−6.3
		6266.81	4 2/72	Mudstone	1.0	−6.0
		6269.93	4 18/72	Mudstone	1.0	−5.9
		6270.32	4 22/72	Mudstone	1.0	−6.1
		6271.73	4 34/72	Mudstone	1.0	−5.6
		6273.2	4 39/72	Mudstone	0.9	−6.3
		6273.4	4 42/72	Mudstone	0.9	−6.2
		6276.06	4 56/72	Mudstone	1.0	−6.3
		6276.26	4 59/72	Mudstone	0.9	−6.3
		6278.27	4 67/72	Mudstone	0.9	−6.2
		6280.27	5 3/8	Mudstone	1.0	−5.9
		6281.0	5 6/8	Mudstone	0.8	−6.5
		6281.7	6 7/106	Mudstone	0.6	−6.2
		6283.09	6 14/106	Mudstone	0.6	−6.5
		6284.49	6 27/106	Mudstone	0.9	−6.2
		6285.49	6 37/106	Mudstone	1.1	−6.1
	Lianglitag	6286.6	6 46/106	Limestone	0.4	−5.5
		6287.57	6 53/106	Limestone	0.7	−4.8
		6288.52	6 61/106	Limestone	0.9	−5.1
		6289.7	6 70/106	Limestone	0.9	−4.5
		6290.48	6 78/106	Limestone	1.8	−3.9
		6290.97	6 87/106	Limestone	1.7	−4.2
		6291.93	6 93/106	Limestone	1.5	−4.5
		6293.04	6 103/106	Muddy limestone	1.6	−5.0
		6294.28	7 3/57	Muddy limestone	2.0	−3.9
		6295.41	7 11/57	Muddy limestone	2.5	−4.3
		6295.84	7 14/57	Muddy limestone	2.0	−4.8
		6297.11	7 24/57	Muddy limestone	2.5	−3.8
	Qiaerbak	6298.95	7 36/57	Argillaceous limestone	1.9	−3.8
		6300.77	7 40/57	Argillaceous limestone	1.8	−4.5
		6300.97	7 43/57	Argillaceous limestone	1.8	−4.0
		6301.37	7 48/57	Argillaceous limestone	1.6	−4.8
		6303.77	7 56/57	Argillaceous limestone	1.0	−5.8
		6315.57	8 6/80	Muddy limestone	1.3	−4.8
		6317.45	8 11/80	Muddy limestone	1.1	−5.7
		6317.85	8 13/80	Muddy limestone	1.2	−4.4
		6319.47	8 20/80	Muddy limestone	0.9	−4.4

Table 1 continued

Well	Formation	Depth, m	Core number	Lithology	$\delta^{13}C$, ‰	$\delta^{18}O$, ‰
S112-1	Qiaerbak	6319.93	8 27/80	Muddy limestone	1.0	−4.6
		6320.12	8 40/80	Muddy limestone	1.3	−6.5
		6320.75	8 43/80	Muddy limestone	1.5	−4.4
		6322.62	8 45/80	Muddy limestone	0.9	−6.1
		6323.42	8 51/80	Muddy limestone	0.8	−5.6
		6324.69	8 59/80	Muddy limestone	1.0	−5.4
		6326.21	8 67/80	Muddy limestone	0.8	−6.0
		6327.28	8 73/80	Muddy limestone	0.5	−6.5
		6328.5	8 80/80	Muddy limestone	0.6	−5.5
		6329.1	9 6/47	Muddy limestone	0.6	−5.9
		6330.71	9 12/47	Muddy limestone	0.4	−6.9
		6331.39	9 16/47	Muddy limestone	0.4	−6.0
		6331.64	9 19/47	Muddy limestone	0.4	−6.2
	Yijianfang	6334.28	9 34/47	Limestone	0.5	−5.6
		6334.78	9 38/47	Limestone	0.2	−6.0
		6339.9	9 41/47	Limestone	0.3	−6.0
		6340.1	9 43/47	Limestone	0.1	−6.4
S114	Qiaerbak	6320.35	6 1/35	Muddy limestone	1.4	−4.0
		6325.0	7 14/64	Muddy limestone	1.8	−4.7
		6329.0	7 47/64	Muddy limestone	0.8	−5.3
	Yijianfang	6331.2	8 13/47	Limestone	1.4	−2.6
		6337.35	8 44/47	Limestone	0.6	−5.4
		6339.9	10 3/69	Limestone	0.8	−5.8
		6346.25	10 61/69	Limestone	1.2	−6.4
		6349.55	11 21/80	Limestone	1.4	−6.5
		6354.6	11 72/80	Limestone	0.9	−6.0
		6358.9	12 29/39	Limestone	0.8	−6.2
		6365.8	13 34/51	Limestone	0.6	−6.5
		6369.86	14 26/57	Limestone	0.4	−6.2
		6374.7	15 16/25	Limestone	0.2	−6.7
	Yingshan	6385.1	18 37/69	Limestone	−0.2	−6.6
		6391.15	19 20/23	Limestone	−0.2	−6.5
		6397.1	20 52/73	Limestone	−0.4	−6.5
		6403.5	21 42/57	Limestone	−0.3	−6.4
		6414.85	23 14/52	Limestone	−0.3	−6.3
		6419.9	24 13/15	Limestone	−0.4	−6.5
		6424.6	26 8/43	Limestone	−0.4	−6.8
		6433.15	28 4/72	Limestone	−0.6	−6.2
		6437.9	28 44/72	Limestone	−0.7	−6.0
		6442.4	29 13/75	Limestone	−0.8	−6.5
		6446.5	29 50/75	Limestone	−0.5	−6.3
		6462.8	31 64/65	Limestone	−0.8	−5.7

3.2 Analysis of diagenetic alteration

Atmospheric freshwater diagenesis had great impacts on $\delta^{13}C_{carb}$ and $\delta^{18}O_{carb}$ ratios of carbonate rocks leading to the values of $\delta^{13}C_{carb}$ and $\delta^{18}O_{carb}$ decreasing significantly (Gradstein et al. 2012). The top surface of the Yijianfang and Lianglitag Formations in the Tahe oil–gas field had been exposed for a short time and undergone atmospheric freshwater karstification. The exposed areas are located, respectively, in the northeastern and northern regions of the study area. Well S112-1 and well S114 exhibited successive deposition in the Middle–Upper Ordovician, indicating the samples from these wells were unlikely to have been altered by atmospheric freshwater. The $\delta^{13}C_{carb}$ and $\delta^{18}O_{carb}$ curves in the upper Yijiangfang Formation of well S112-1 showed positive excursion features (Fig. 3). Discordant phenomena appeared in the upper Yijiangfang Formation of well S114. These characteristics reflected that the samples were less affected by meteoric water.

Compared with $\delta^{13}C_{carb}$, $\delta^{18}O_{carb}$ is more susceptible to the influence of diagenesis (Gradstein et al. 2012; Metzger et al. 2014). The initial composition of $\delta^{13}C_{carb}$ may be obviously changed when the $\delta^{18}O_{carb}$ value is less than −10 ‰ (Derry et al. 1992). The $\delta^{18}O_{carb}$ values of well S112-1 and well S114 were both more than −10 ‰, reflecting that the $\delta^{13}C_{carb}$ value of samples from the two wells has been influenced by relatively weak diagenesis.

Correlation of $\delta^{13}C_{carb}$ and $\delta^{18}O_{carb}$ is also a method for judging diagenetic alteration of carbonate rock samples (Sial et al. 2013). It is indicated that the $\delta^{13}C_{carb}$ and $\delta^{18}O_{carb}$ of partial samples had an obvious positive association (Fig. 4), which appeared in the upper Yijianfang Formation to middle Qiaerbak Formation, upper Qiaerbak Formation to lower Lianglitag Formation, and middle–upper

Lianglitag Formation (Fig. 3). But these positive associations had no relationship with depth, but with formations (as shown in Fig. 3). The correlation of $\delta^{13}C_{carb}$ and $\delta^{18}O_{carb}$ from the Yingshan Formation to the lower Yijianfang Formation in well S114 was not good because of weak fluctuations. The coordination change phenomenon of $\delta^{13}C_{carb}$ and $\delta^{18}O_{carb}$ in Middle–Late Ordovician also occurred in South America (Sial et al. 2013). Therefore, this positive association might reflect a coordination change relationship between $\delta^{13}C_{carb}$ and $\delta^{18}O_{carb}$ in a specific sedimentary background, indicating a fluctuation trend of $\delta^{13}C_{carb}$ and $\delta^{18}O_{carb}$ of original seawater.

4 Results

4.1 Carbon isotope curve of well S114

The $\delta^{13}C_{carb}$ curve from the Yingshan Formation to the Qiaerbak Formation in well S114 showed generally positive excursion features (Fig. 3), ranging from −0.8 to 1.8 ‰. The excursion process could be divided into three intervals, where intervals A and C were positive excursions. The fluctuation range of interval A is the largest, about 2.2 ‰ among the three intervals with a maximum $\delta^{13}C_{carb}$ value of 1.4 ‰. Interval B appeared in the upper

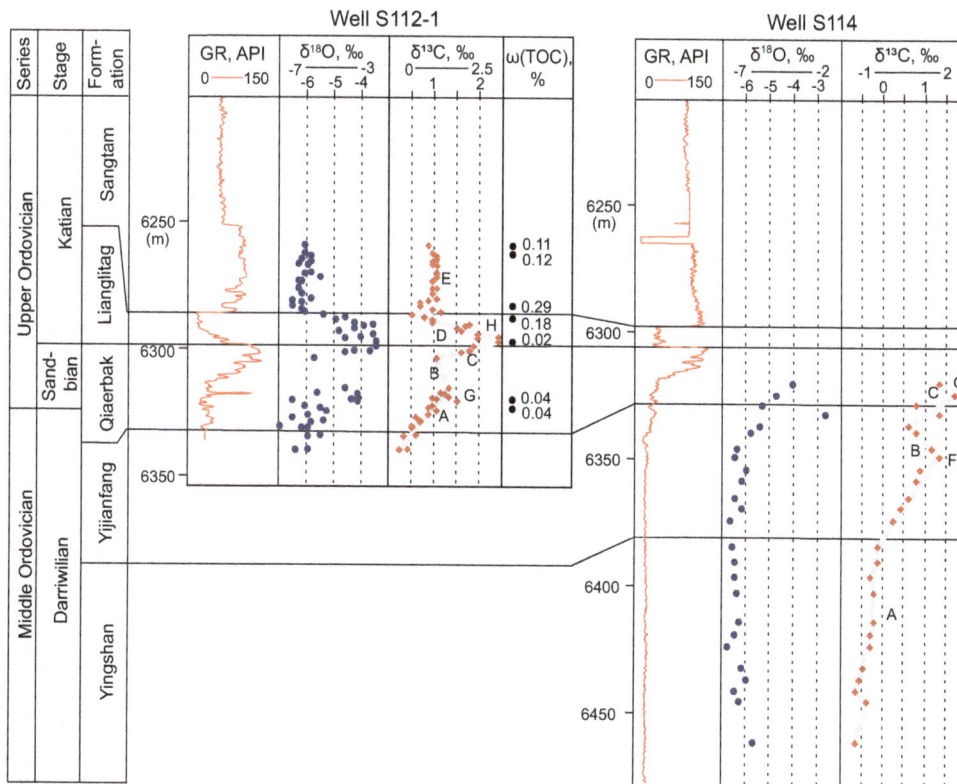

Fig. 3 The curves of $\delta^{13}C_{carb}$ and $\delta^{18}O_{carb}$ in the Middle–Upper Ordovician for well S112-1 and well S114 in the Tahe oil–gas field

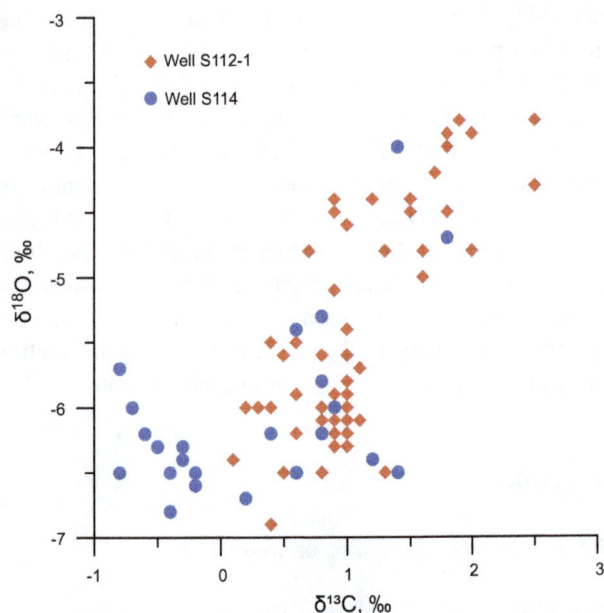

Fig. 4 Cross plot of $\delta^{13}C_{carb}$ versus $\delta^{18}O_{carb}$ for well S112-1 and well S114 in the Tahe oil–gas field

part of the Yijianfang Formation with the $\delta^{13}C_{carb}$ value moving -0.8 ‰ compared with the value below. Interval C appeared from the top part of the Yijianfang Formation to the lower part of the Qiaerbak Formation with the fluctuation range reaching 1.2 ‰.

4.2 Carbon isotope curve of Well S112-1

The $\delta^{13}C_{carb}$ curve of well S112-1 fluctuates sharply from a minimum value of 0.1 ‰ at the top of the Yijianfang Formation to a peak value of 2.5 ‰ in the Lianglitag Formation. The excursion process in this well can be divided into five intervals. Interval A is a positive excursion, where the $\delta^{13}C_{carb}$ values shift from 0.1 to 1.3 ‰ from the top of the Yijianfang Formation to the lower Qiaerbak Formation. Interval B is a negative excursion, where the fluctuation range of $\delta^{13}C_{carb}$ values in the middle of the Qiaerbak Formation is at least -0.3 ‰. This negative excursion is estimated from the difference between upper and lower values, because of non-coring in this well. But it is confirmed by other wells drilled in the western Tabei Uplift (Bao et al. 2006). Interval C is a positive excursion, and the $\delta^{13}C_{carb}$ values fluctuate up to 1.4 ‰ from the top of the Qiaerbak Formation to the bottom of the Lianglitag Formation. Interval D is a negative excursion, where $\delta^{13}C_{carb}$ values decrease rapidly in the upper part of the Lianglitag Formation and fluctuate up to 2.1 ‰. Interval E appears in the lower Sangtamu Formation, where the $\delta^{13}C_{carb}$ values with positive excursion quickly move to near 1 ‰ and narrowly fluctuate around 1 ‰.

From the characteristics of $\delta^{13}C_{carb}$ curves in Fig. 3, there are mainly three $\delta^{13}C_{carb}$-positive excursions from F

to H occurring in well S114 and well S112-1, developing in the Yijianfang Formation, Upper Yijianfang Formation to lower Qiaerbak Formation, and upper Qiaerbak Formation to the middle Lianglitag Formation, respectively.

5 Discussion

Well S112-1 and well S114 are relatively close to each other and their sedimentary environment is also similar. Therefore, taking the top of the Yijianfang Formation as the boundary, $\delta^{13}C_{carb}$, $\delta^{18}O_{carb}$, and logging curves of the two wells were put together to contrast with the Ordovician $\delta^{13}C_{carb}$ curve of the Kalpin outcrop (Hu et al. 2010) and the generalized curve of the world (Albanesi et al. 2013) (Figs. 5, 6).

5.1 Positive carbon isotope excursions and sea level change

A $\delta^{13}C_{carb}$ excursion in a stratigraphic succession may simply record the lateral movement of an isotopically unique water body during a sea level change. Considering the asynchronization of sea level change, $\delta^{13}C_{carb}$ excursions resulting from sea level change have few contributions to global comparison (Edwards and Saltzman 2014). Therefore, the relationship between three $\delta^{13}C_{carb}$-positive excursions shown in Figs. 3 and 5 with sea level changes is very important.

According to research on Ordovician sequence stratigraphy in the Tahe oil–gas field and Tarim Basin and our results shown in Fig. 2 (Deng et al. 2007; Wu et al. 2012; Liu et al. 2012), the positive excursion A occurred in the HST of the Yijianfang Formation, the positive excursion B in the TST of the Qiaerbak Formation, and the positive excursion C in the HST of the Qiaerbak Formation and the TST of the Lianglitag Formation, which indicated that three positive excursions had no direct relationships with sea level fluctuation, and could be correlated regionally or globally.

5.2 Correlation of $\delta^{13}C_{carb}$-positive excursions in the Tarim Basin and the world

The *Pygodus serra* conodont zone in the Darriwilian, the *P. anserinus* conodont zone in the Sandbian, and the *Belodina compressa* conodont zone in the Early Katian (shown in Fig. 2) of the Tahe oil–gas field in the northern Tarim Basin can be compared or partly compared with that of the Argentine Precordillera (Sial et al. 2013; Albanesi et al. 2013; Edwards and Saltzman 2014), Oklahoma and Virginia, USA (Young et al. 2008), Cincinnati region, USA (Bergström et al. 2010), Sweden (Wu et al. 2015), and

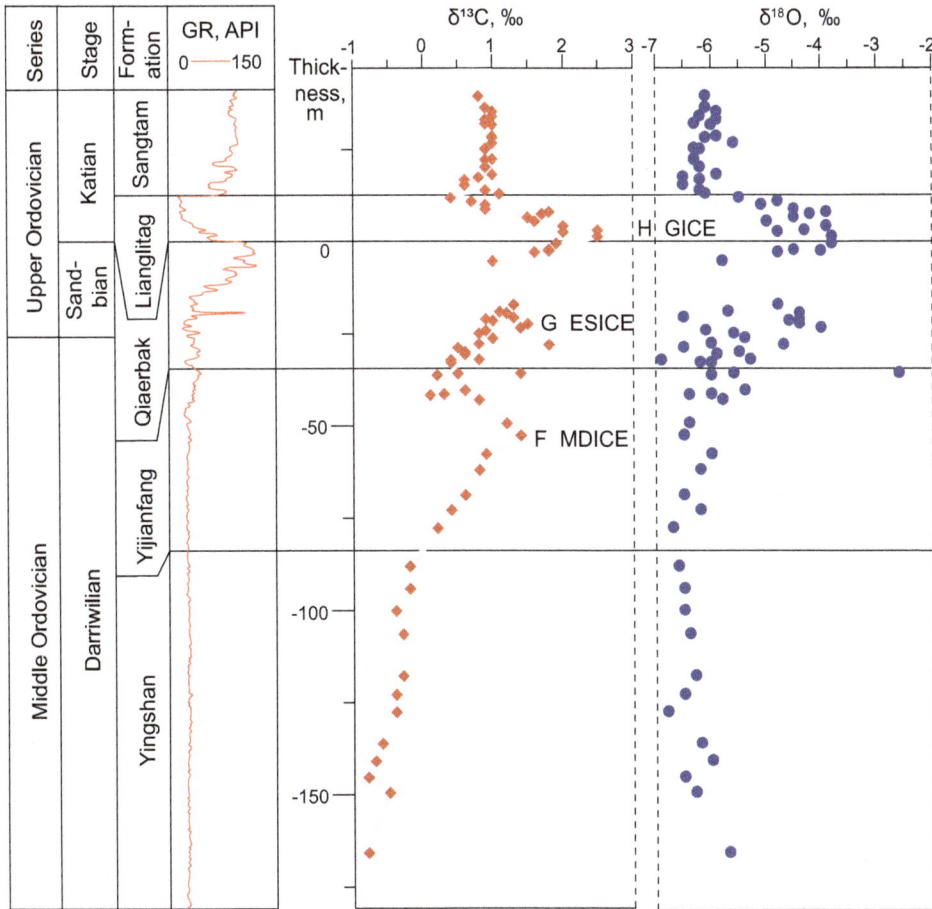

Fig. 5 The curves of $\delta^{13}C_{carb}$ and $\delta^{18}O_{carb}$ of the Darriwilian—Early Katian in the Tahe oil–gas field

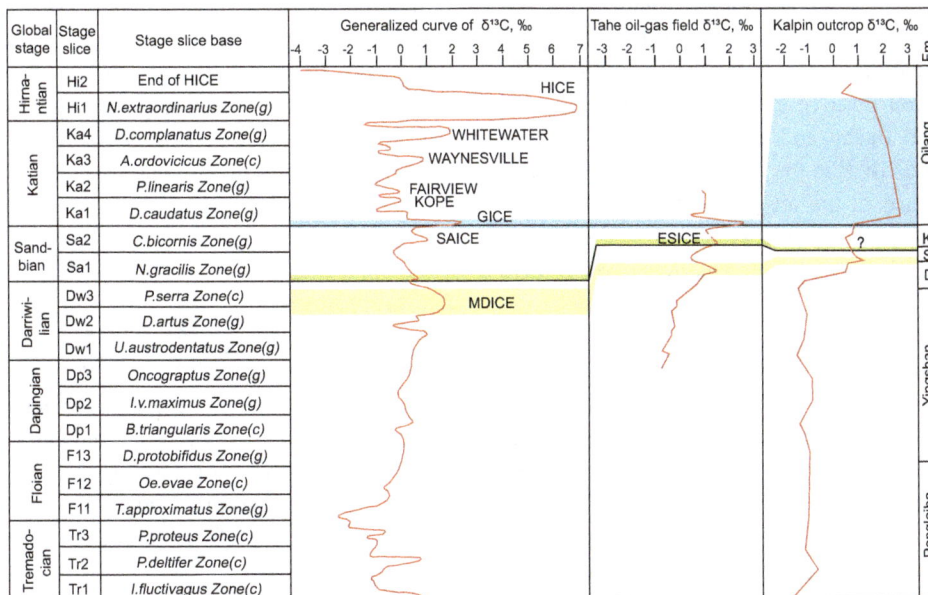

Fig. 6 The $\delta^{13}C_{carb}$ curve of the Darriwilian—Early Katian in the Tahe oil–gas field contrasting with the Ordovician $\delta^{13}C_{carb}$ curves of the Kalpin outcrop and the world. *g* graptolite zonation, *c* conodont zonation, *D* Dawangou Formation, *S* Saergan Formation, and *K* Kanling Formation

Baltoscandia (Ainsaar et al. 2010), which provides a basis for time limits on $\delta^{13}C_{carb}$-positive excursions.

According to Figs. 5 and 6, the positive excursion F in the Middle Ordovician Yijianfang Formation corresponds with the MDICE of the global Ordovician $\delta^{13}C$ curve, with a peak value generally less than 2 ‰ in South China, North America, South America, and Europe. Positive excursion H is consistent with the GICE. Maximum $\delta^{13}C_{carb}$ values are generally less than 3 ‰. However, it is slightly greater than 3 ‰ in Pennsylvania (Pancost et al. 2013) and Virginia (Young et al. 2008), USA. Evidence of the GICE from the upper Sandbian can also be found in Oklahoma, USA (Rosenau et al. 2012).

Positive excursion G between the MDICE and GICE in the Tahe oil–gas field was developed in *P. anserinus* conodont zone of the Early Sandbian. The fluctuation range was more than the corresponding positive excursion in the global Ordovician curve (Fig. 6). This excursion is currently not officially named, so in this paper, it is named the ESICE. Sandbian carbon isotope excursion (SAICE), also called the Spechts Ferry Excursion, in the Late Sandbian close to the GICE was not found in the Tahe oilfield (Fig. 6). Whether it can be globally correlated or not is controversial (Sial et al. 2013).

Chen et al. (2012) argues the age of the Middle–Upper Ordovician Saergan Formation black shale in the Kalpin outcrop is from the Middle Darriwilian to the Early Sandbian from graptolite data. According to Fig. 6, the MDICE of the Yijianfang Formation in the Tahe oil–gas field is equivalent to the positive excursion in the bottom of the Saergan Formation. But the features of the ESICE in the Saergan Formation are not clear, which might be caused by sparse sampling points. The positive excursion of the Qilang Formation corresponded with the GICE. The Lianglitag Formation positive excursions from drill cores and cuttings in the Bachu and Tazhong Uplifts were in accordance with the GICE in the Tahe oil–gas field (Zhang et al. 2014). Because of the absence of the Darriwilian–Sandbian in the Bachu and Tazhong Uplifts, the ESICE and MDICE might be developed in areas with consecutive strata.

5.3 Genesis of positive carbon isotope excursion

According to outcrop and drilling data from the Tarim Basin, Ordovician source rocks consist of the Middle–Lower Ordovician Heituwa Formation in the Manjiaer Depression (Fig. 1a), the Middle–Upper Ordovician Saergan Formation in the Kalpin outcrop and the Upper Ordovician Lianglitag Formation in the basin (Wang 2000; Zhang et al. 2006; Gao et al. 2007; Zhao et al. 2008). The thickness of the Heituwa Formation source rock can reach a maximum of 56 m with an average TOC of 3.1 % (Zhao et al. 2008). There are no published carbon isotope data from cores in the Heituwa Formation. Chen et al. (2012) believed the age of the Heituwa Formation in the Kuruktag outcrop should be Middle Tremadocian—Early Darriwilian. The genesis of this source rock may be related to the increase in organic matter productivity and organic carbon burial which was caused by a gradual decline in sea surface temperature during the Early to Middle Ordovician (Trotter et al. 2008).

The thickness of the Saergan Formation source rocks in the Kapin outcrop is 13.8 m with an average TOC of black shale of 1.92 % (Gu et al. 2012). The age of the Saergan Formation corresponds with the MDICE and ESICE (shown in Fig. 6). The Upper Ordovician Lianglitag Formation source rocks had been characterized by drilling in the Tazhong and Tabei Uplifts (Gao et al. 2007; Zhao et al. 2008), with an average TOC of 1.1 % and 0.5 %, respectively (Zhao et al. 2008). The age of Lianglitag Formation source rocks corresponds with the GICE. The Shuntuoguole low uplift and Awati fault depression, between the Tabei and Tazhong Uplifts, were located in a low area during the deposition of the Lianglitag Formation, and may be favorable for source rock development corresponding with the GICE (Fig. 1a).

Carbon isotope reversal in kerogen and soluble organic components of Cambrian and Ordovician carbonate rocks in the Tarim Basin is common, which may be controlled by thermal evolution or parent materials (Zhang et al. 2006; Liu et al. 2013a). Enrichment of ^{13}C in residues of kerogen during thermal evolution can make the $\delta^{13}C$ of source rock heavier. The TOC value of the Qiaerbak to Sangtamu Formations in the Upper Ordovician of well S112-1 is low (Fig. 3), which indicates that the source rock is not developed. Therefore, the genesis of the MDICE, ESICE, and GICE was not related to the kerogen thermal evolution, but to the $\delta^{13}C_{carb}$ value increase in carbonate rocks by massive amounts of organic matter buried during the deposition period.

6 Conclusions

(1) There are three $\delta^{13}C_{carb}$-positive excursions identified in the Middle–Upper Ordovician (Darriwilian—Early Katian) in the Tahe oil–gas field according to the conodont biostratigraphy. The positive excursions have no direct relation with sea level fluctuations. The MDICE and GICE could be globally correlated.

(2) The $\delta^{13}C_{carb}$-positive excursion within the *P. anserinus* conodont zone developed in the Early Sandbian was named the ESICE in this paper. The fluctuation range of the ESICE is less than that of the MDICE

and GICE. Previous studies have paid little attention on this excursion. Whether the ESICE can be applied to global correlation or not needs further investigation.

(3) The genesis of the Saergan Formation marine source rocks of the Middle–Upper Ordovician in the Kalpin outcrop was in accord with the MDICE and ESICE. But the GICE had strong ties to the source rock of the Lianglitag Formation in the basin. These relationships indicated that massive organic carbon burial is an important factor controlling the genesis of the $\delta^{13}C_{carb}$-positive excursion.

Acknowledgments The authors are grateful to three anonymous reviewers for their constructive reviews. This work was supported by the National Key Scientific Project of China (No. 2011ZX05005-004; 2016ZX05005-002) and the National Basic Research Program of China (973 Program) (No. 2012CB214806).

References

Ainsaar L, Kaljo D, Martma T, et al. Middle and Upper Ordovician carbon isotope chemostratigraphy in Baltoscandia: a correlation standard and clues to environmental history. Palaeogeog Palaeoclimatol Palaeoecol. 2010;294(3–4):189–201.

Albanesi GL, Bergström SM, Schmitz B, et al. Darriwilian (Middle Ordovician) $\delta^{13}C_{carb}$ chemostratigraphy in the Precordillera of Argentina: documentation of the Middle Darriwilian isotope carbon excursion (MDICE) and its use for intercontinental correlation. Palaeogeog Palaeoclimatol Palaeoecol. 2013;389(4):48–63.

Bao ZD, Jin ZJ, Sun LD, et al. Sea-level fluctuation of the Tarim area in the Early Paleozoic respondence from geochemistry and karst. Acta Geol Sin. 2006;80(3):366–73 (**in Chinese**).

Bergström SM, Young S, Schmitz B. Katian (Upper Ordovician) $\delta^{13}C$ chemostratigraphy and sequence stratigraphy in the United States and Baltoscandia: a regional comparison. Palaeogeog Palaeoclimatol Palaeoecol. 2010;296(3–4):217–34.

Chen X, Bergström SM, Zhang YD, et al. A regional tectonic event of Katian (Late Ordovician) age across three major blocks of China. Chin Sci Bull. 2013;58(34):4292–9.

Chen X, Wang ZH. Global auxiliary stratotype section of the Upper Ordovician in China. J Stratig. 2003;27(3):264–6 (**in Chinese**).

Chen X, Zhang YD, Li Y, et al. Biostratigraphic correlation of the Ordovician black shales in Tarim Basin and its peripheral regions. Chin Sci Bull. 2012;55(8):1230–7.

Deng XJ, Li GR, Xu GS, et al. Sequence stratigraphic study, prediction and quality evaluation of reservoir for the Ordovician Yijianfang Formation of the Southern part of Tahe Oilfield. Acta Sedmentol Sin. 2007;25(3):392–400 (**in Chinese**).

Derry LA, Kaufman AJ, Jacobsen SB. Sedimentary cycling and environmental change in the Late Proterozoic: evidence from stable and radiogenic isotopes. Geochim Cosmochim Acta. 1992;56(3):1317–29.

Edwards CT, Saltzman MR. Carbon isotope ($\delta^{13}C_{carb}$) stratigraphy of the Lower–Middle Ordovician (Tremadocian–Darriwilian) in The Great Basin, Western United States: implications for global correlation. Palaeogeog Palaeoclimatol Palaeoecol. 2014;399:1–20.

Fan R, Deng SH, Zhang XL. Significant carbon isotope excursions in the Cambrian and their implications for global correlations. Sci China Earth Sci. 2011;54(11):1686–95.

Gao ZY, Zhang SC, Zhang XY, et al. Relations between spatial distribution and sequence types of the Cambrian-Ordovician marine source rocks in Tarim Basin. Chin Sci Bull. 2007;52(S1):92–102.

Gradstein FM, Ogg JG, Schmitz MD, et al. The geologic time scale. London: Elsevier Science Ltd; 2012.

Gu Y, Zhao YQ, Jia CS, et al. Analysis of hydrocarbon resource potential in Awati Depression of Tarim Basin. Pet Geol Exp. 2012;34(3):257–66 (**in Chinese**).

Hu MY, Qian Y, Hu ZG, et al. Carbon isotopic and element geochemical responses of carbonate rocks and Ordovician sequence stratigraphy in Keping area. Tarim Basin. Acta Pet Mineral. 2010;29(2):199–205 (**in Chinese**).

Jing XC, Deng SH, Zhao ZJ, et al. Carbon isotope composition and correlation across the Cambrian-Ordovician boundary in Kepin region of the Tarim Basin. China. Sci China Earth Sci. 2008;51(9):1317–29.

Liu CG, Li T, Lv HT, et al. Stratigraphic division of Middle-Upper Ordovician and characteristics of the first episode karstification of Middle Caledonian in Akekule uplift, Xinjiang, China. J Chengdu Univ Technol Sci Technol Ed. 2010;37(1):55–63 **in Chinese**.

Liu H, Liao ZW, Zhang HZ, et al. Review of the study on stable carbon isotope reversal between kerogen and its evolution products: implication for the research of the marine oil reservoirs in the Tarim Basin, NW China. Bull Mineral Pet Geochem. 2013a;32(4):497–502 (**in Chinese**).

Liu HB, Jin GS, Li JJ, et al. Determination of stable isotope composition in uranium geological samples. World Nucl Geosci. 2013b;30(3):174–9 (**in Chinese**).

Liu JQ, Li Z, Huang JC, et al. Distinct sedimentary environments and their influences on carbonate reservoir evolution of the Lianglitag Formation in the Tarim Basin, Northwest China. Sci China Earth Sci. 2012;55(10):1641–55.

Metzger JG, Fike DA, Smith LB. Applying carbon-isotope stratigraphy using well cuttings for high-resolution chemostratigraphic correlation of the subsurface. AAPG Bull. 2014;98(8):1551–76.

Munnecke A, Calner M, Harper DAT, et al. Ordovician and Silurian sea-water chemistry, sea level, and climate: a synopsis. Palaeogeog Palaeoclimatol Palaeoecol. 2010;296(3–4):389–413.

Munnecke A, Zhang YD, Liu X, et al. Stable carbon isotope stratigraphy in the Ordovician of South China. Palaeogeog Palaeoclimatol Palaeoecol. 2011;307(1–4):17–43.

Pancost RD, Freeman KH, Herrmann AD, et al. Reconstructing Late Ordovician carbon cycle variations. Geochim Cosmochim Acta. 2013;105:433–54.

Rosenau NA, Herrmann AD, Leslie SA. Conodont apatite $\delta^{18}O$ values from a platform margin setting, Oklahoma, USA: implications for initiation of Late Ordovician icehouse conditions. Palaeogeog Palaeoclimatol Palaeoecol. 2012;315–316:172–80.

Saltzman MR. Phosphorus, nitrogen, and the redox evolution of the Paleozoic oceans. Geology. 2005;33(7):573–6.

Sial AN, Peralta S, Gaucher C, et al. High-resolution stable isotope stratigraphy of the Upper Cambrian and Ordovician in the Argentine Precordillera: carbon isotope excursions and correlations. Gondwana Res. 2013;24(1):330–48.

Thompson CK, Kah LC. Sulfur isotope evidence for widespread euxinia and a fluctuating oxycline in Early to Middle Ordovician greenhouse oceans. Palaeogeog Palaeoclimatol Palaeoecol. 2012;313–314:189–214.

Torsvik TH, Cocks LRM. Gondwana from top to base in space and time. Gondwana Res. 2013;24(3–4):999–1030.

Trotter JA, Williams IS, Barnes CR, et al. Did cooling oceans trigger Ordovician biodiversification? Evid Conodont Thermom Sci. 2008;321(5888):550–4.

Wang DR. Macro-evidence of carbonate isotopes for the Middle–Upper Ordovician source rocks in the Tarim Basin. Geol Rev. 2000;46(3):328–34 (in Chinese).

Wang HH, Li JH, Yang JY, et al. Paleo-plate reconstruction and drift path of Tarim Block from Neoproterozic to Early Palaeozoic. Adv Earth Sci. 2013a;28(6):637–47 (in Chinese).

Wang XL, Hu WX, Li Q, et al. Negative carbon isotope excursion on the Cambrian series 2–series 3 boundary for Penglaiba section in Tarim Basin and its significances. Geol Rev. 2011;57(1):16–23 (in Chinese).

Wang ZH, Qi YP, Bergström SM. Ordovician conodonts of the Tarim region, Xinjiang, China: occurrence and use as palaeoenvironment indicators. J Asian Earth Sci. 2007;29(5–6):832–43.

Wang ZH, Wu RC, Bergström SM. Ordovician conodonts from the Lunnan area of Northwestern Taklimakan desert, Xinjiang, China, with remarks on the evolution of pygodus. Acta Palaeontol Sin. 2013b;54(2):408–23 in Chinese.

Wang ZZ, Yang JD. Features of the carbon isotope changes in the Early Palaeozoic rocks of the Kalpin area, Xinjiang and their significance. J Stratig. 1994;18(1):45–52 (in Chinese).

Wu RC, Calner M, Lehnert O, et al. Lower–Middle Ordovician 13C chemostratigraphy of western Baltica (Jämtland, Sweden). Palaeoworld. 2015;24(1–2):110–22.

Wu XN, Shou JF, Zhang HL, et al. Characteristics of the petroleum system in Cambrian and Ordovician sequence frameworks of the Tarim Basin and its exploration significance. Acta Pet Sin. 2012;33(2):225–31 (in Chinese).

Young SA, Saltzman MR, Bergström SM, et al. Paired $\delta^{13}C_{carb}$ and $\delta^{13}C_{org}$ records of Upper Ordovician (Sandbian-Katian) carbonates in North America and China: implications for paleoceanographic change. Palaeogeog Palaeoclimatol Palaeoecol. 2008;270(1–2):166–78.

Zhang YD, Cheng JF, Munnecke A, et al. Carbon isotope development in the Ordovician of the Yangtze Gorges region (South China) and its implication for stratigraphic correlation and paleoenvironmental change. J Earth Sci. 2010a;21(S1):70–4.

Zhang YD, Zhan RB, Fan JX, et al. Principal aspects of the Ordovician biotic radiation. Sci China Earth Sci. 2010b;53(3):282–394.

Zhang ZL, Li HL, Tan GH, et al. Carbon isotope chemostratigraphy of the Ordovician system in Central Uplift of the Tarim Basin. J Stratig. 2014;38(2):181–9 (in Chinese).

Zhang ZN, Liu WH, Zheng JJ, et al. Carbon isotopic reversed distribution of the soluble organic components for the Cambrian and Ordovician carbonate rocks in Tabei and Tazhong areas, Tarim Basin. J Mineral Pet. 2006;26(4):69–74 (in Chinese).

Zhao MJ, Wang ZM, Pan WQ, et al. Lower Palaeozoic source rocks in Manjiaer Sag, Tarim Basin. Pet Explor Dev. 2008;34(4):417–23 (in Chinese).

Zhao ZJ. Indicators of global sea-level change and research methods of marine tectonic sequences: taking Ordovician of the Tarim Basin as an example. Acta Pet Sin. 2015;36(3):262–73 (in Chinese).

Zhao ZJ, Zhao ZX, Huang ZB. Ordovician conodont zones and sedimentary sequences of the Tarim Basin, Xinjiang, NW China. J Stratig. 2006;30(3):193–203 (in Chinese).

19

Modification of waste polyacrylonitrile fiber and its application as a filtrate reducer for drilling

Pei-Zhi Yu[1]

Abstract Cationic polymer fluid loss additive (CPFL) was prepared by using the reaction of 2,3-epoxypropyltrimethyl ammonium chloride (EPTMAC) (as cationic reagent) with the amide group in the molecular structure of the sodium salt of partially hydrolyzed polyacrylonitrile fibers (HPAN-Na). The chemical reaction was determined by studying the infrared absorption peaks of the materials and the products. The results proved that the cationic groups of EPTMAC were successfully grafted onto the HPAN molecular chain. The composition of the molecular chain of the product CPFL was determined by investigation and calculation of the elemental analysis results of the grafted HPAN and the final reaction product CPFL. The drilling fluid performance was evaluated, and the result showed that when the cation content was more than 0.27 mmol/g, the drilling fluid would have good resistance to fluid loss and to pollution from calcium chloride.

Keywords Acrylic fiber · Cationic agents · Fluid loss agent · Spectroscopy

1 Introduction

In oil-drilling operations, a filtrate reducer is the most commonly used drilling fluid additive. Its main functions are reducing drilling fluid filtration by forming thin and dense filter cakes on the wellbore, thus reducing the mutual permeability between the mud filtrate and wellbore, minimizing wellbore instability, and ensuring the stability of drilling fluid performance (Zhang and Deng 1998). Partially hydrolyzed sodium polyacrylonitrile with an anionic hydration group, polyacrylonitrile calcium salts, and polyacrylonitrile ammonium salts are widely used as filtrate reducers, which meet the requirements of well drilling operations below 120 °C (Cui and Zhu 1999; Ahmed et al. 2013). The application of an allyl copolymer with cationic groups in the drilling fluid has been reported. This mainly revolves around its application to cationic and zwitterionic drilling fluid systems (Zhang and Deng 1998; Warren et al. 2000). Hydrophobically modified polyacrylamide as a sealing agent has been used in water-based drilling fluid. This can effectively improve the temperature and salt resistance of water-based drilling fluid system (Xie et al. 2013). Lubricants based on water-soluble polyacrylamide have also been used in drilling, particularly in low-density drilling fluid systems (Wang 2013). In recent years, standards of drilling fluid systems have become higher with the increase in deep and complex wells. Therefore, a filtrate reducer containing cationic groups is believed to have a promising application (Zlotnikov et al. 2006). There have been reports about acrylic waste fiber that can enable the drilling fluid loss agent to be cationically grafted under alkaline conditions (Yu et al. 2004; Celik et al. 2012). Still missing are the molecular structure of the cationic agent and the cationic degree of the final products. The characterization of the structure of such reaction products is yet to be reported. The application of the products and their mechanisms of activity in the drilling fluid should be further investigated.

✉ Pei-Zhi Yu
yupz@cugb.edu.cn

[1] School of Engineering and Technology, China University of Geosciences (Beijing), Beijing 100083, China

Edited by Xiu-Qin Zhu

2 Synthesis of cationic fluid loss additive

2.1 Materials and instruments

Industrial by-product waste polyacrylonitrile fiber was provided by the Anqing Chemical Fiber Factory (China). AR NaOH, AR 2,3-epoxypropyltrimethyl ammonium chloride (EPTMAC), and CR Ethanol were provided by Sinopharm Chemical Reagent Beijing Co., Ltd.

Instrumentation included a reflux reaction device; API (American Petroleum Institute) filtration press; high-speed mixer; a Magna-IR 750 Fourier transform infrared spectrometer (Thermo Nicolet Corporation, USA); and a Vario EL element analyzer (Elementar Analysensysteme GmbH, Germany).

2.2 Chemical reaction principle

Cationic fluid loss additive was prepared by reactions of EPTMAC with amide groups, catalyzed by alkali (Tang et al. 1995; Ma and Zhao 1995). The principle of the chemical reaction is shown in Fig. 1.

2.3 Synthesis of cationic fluid loss additive

A total of 15 g of partially hydrolyzed polyacrylonitrile sodium salt (HPAN) and 150 mL of distilled water were added to a 250 mL three-necked flask, mixed, and then heated to 70–80 °C in a water bath. After that, a cationic

additive 2,3-epoxypropyltrimethyl ammonium chloride solution with a mass ratio of 20 %–50 % based on HPAN was added and stirred at constant temperature for three hours.

The reaction product was washed three times with ethanol to obtain viscous products which were dried at 105 ± 5 °C in an oven, and then pulverized to obtain cationic product CPFL with a cationic degree of 0.2–0.5 m-mol/g.

3 Structural characterization of cationic fluid loss additives

Acrylic waste fiber, hydrolysis products HPAN, and cationic product CPFL (with a cationic degree of 0.49 mmol/g determined by chemical titration) were tested by Fourier transform infrared spectroscopy. The graft substrate HPAN and cationic fluid loss additive CPFL were characterized by elemental analysis to determine the composition of their molecular chains.

3.1 IR spectrum

Figure 2 shows the infrared spectrum of acrylic waste fiber PAN, partially hydrolyzed polyacrylonitrile sodium HPAN, and fluid loss agent CPFL.

In the infrared spectrum of acrylic waste fiber PAN, the stretching vibration peak of nitrile group was at

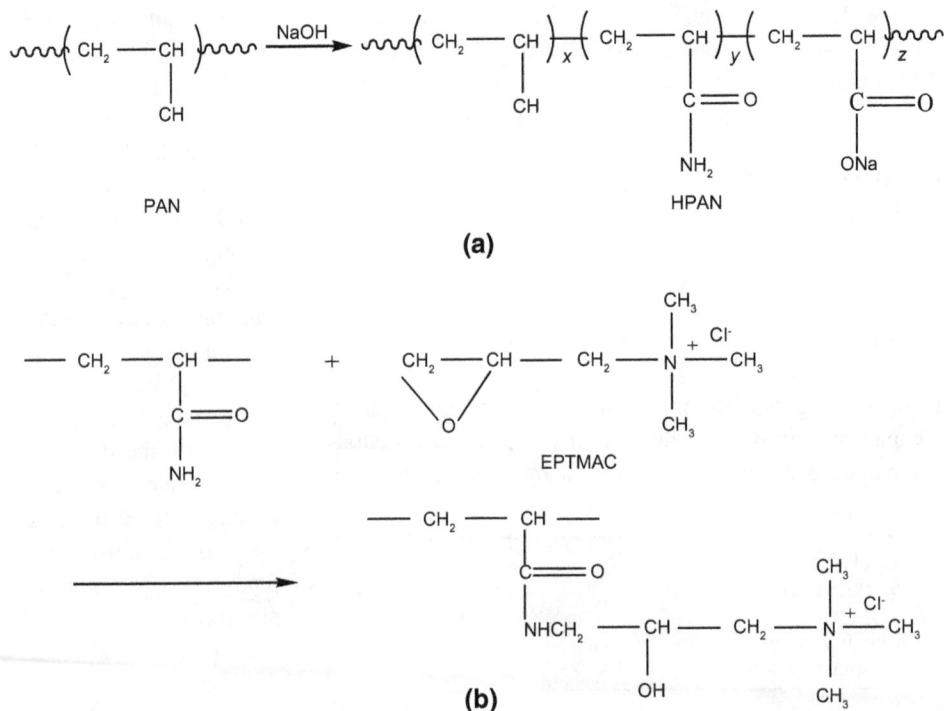

Fig. 1 Schematic diagram of the fluid loss agent made by using cationic acrylic fiber waste a Hydrolysis of acrylic waste fiber, b Cationic agent

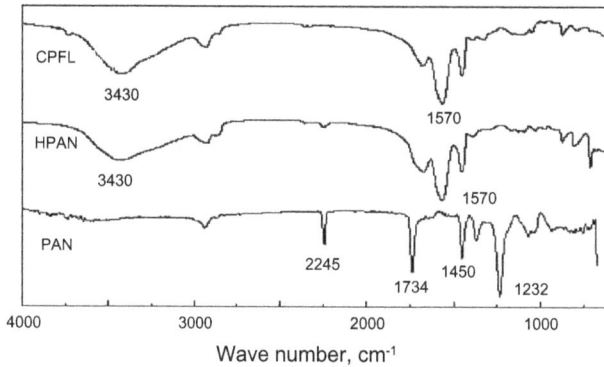

Fig. 2 FTIR of raw materials and products

$2245\ cm^{-1}$. The peak at $1450\ cm^{-1}$ was assigned to the variable angle vibration of methylene. These were the characteristic absorption peaks of polyacrylonitrile and there was a strong absorption peak at $1734\ cm^{-1}$, which was the unique stretching vibration absorption peak of carbonyl groups. It indicated that the raw material contained acrylate, the copolymer of acrylonitrile and acrylic-ester.

In the infrared spectrum of partially hydrolyzed sodium polyacrylonitrile (HPAN), which was boiled for two hours for base-catalyzed hydrolysis at atmospheric pressure, the absorption peak of nitrile group at $2245\ cm^{-1}$ became very weak, indicating that the vast majority of the nitrile groups were hydrolyzed to amide and carboxyl groups. The amide carbonyl stretching vibration peak was at $1675\ cm^{-1}$; the N–H stretching vibration absorption peak appeared at $3431\ cm^{-1}$ and the characteristic peaks of –COO– appeared at 1570 and $1400\ cm^{-1}$.

The infrared spectrum of cationic filtrate reducer CPFL (Fig. 2) showed that nitrile peak was very weak after the HPAN cationic reaction, indicating that nitrile was hydrolyzed in the cationic process. A pair of peaks appeared at 1415 and $1326\ cm^{-1}$, which are the characteristic absorption peaks of secondary alcohol, indicating that the epoxy ring was opened, and cationic groups were successfully grafted to HPAN chains.

3.2 Elemental analysis

The elemental analysis results of partially hydrolyzed polyacrylonitrile sodium (HPAN) and its cationic products CPFL are shown in Table 1.

Table 1 Elemental analysis results of HPAN and CPFL

Element	HPAN	CPFL
C	36.69	38.05
N	1.98	2.49
O	30.22	29.54

According to the structural formula of the hydrolysis product of acrylic fiber and its cationic products (Li and Zhang 1999), the elemental concentration of C, N, and O can be calculated with the following equations:

Cationic product CPFL:

$$C(\%) = 12 \times (3x + 3y + 9z)$$

$$N(\%) = 14 \times (x + 2z)$$

$$O(\%) = 16 \times (x + 2y + 2z)$$

$$\text{Deduction: } x = \frac{O}{16} - \frac{C}{18} + \frac{N}{7}$$

$$y = \frac{O}{32} - \frac{N}{28}$$

$$z = \frac{C}{36} - \frac{O}{32} - \frac{N}{28}$$

With $C = 38.05$, $N = 2.49$, and $O = 29.54$, we obtained $x = 0.088$, $y = 0.834$, and $z = 0.045$. The molar ratio of the three structural units was 9.10/86.2/4.65. The molar masses for each of the structure units were 71, 94, and 222.5 g/mol, respectively.

Among them, the x, y, and z represent the number of chains in three different structure units of HPAN molecular chain, as shown in Fig. 1.

Cationic degree of cationic polymer filtrate reducer CPFL:

$$\frac{0.045 \times 10^3}{71 \times 0.088 + 94 \times 0.834 + 222.5 \times 0.045}$$
$$= 0.475\ (\text{mmol/g})$$

The calculation from elemental analysis agreed well with the titration data of 0.49 mmol/g. The result of the elemental analysis further demonstrated that the cationic reagent was successfully grafted onto the molecular chain of the partially hydrolyzed polyacrylonitrile sodium salt.

4 Evaluation of drilling fluid performance

4.1 Influence of cationic degree on drilling fluid filtration

The base drilling fluid was prepared as follows: 100 g of bentonite and 8 g of anhydrous Na_2CO_3 were added to 1000 mL of water, and stirred for about 20 min with a high-speed mixer, and then placed under a sealed condition for about 24 h at room temperature. The property of the drilling fluid was then determined according to the API standard.

CPFL sample (with a concentration of 0.5 or 1.0 wt%) with different cationic degrees was added to the base drilling fluid. The drilling fluid property was tested after

Table 2 Influence of cationic degree on drilling fluid filtration

Concentration of CPFL (wt%)	Cationic degree, mmol/g	AV, mPa s	PV, mPa s	YP, Pa	FL, mL
0.5	0.00	6.5	6.5	0.0	6.9
	0.27	8.8	7.5	1.3	5.2
	0.47	9.0	7.0	2.0	5.5
	0.54	9.0	7.0	2.0	6.0
	1.03	11.0	6.0	5.0	9.0
1.0	0.00	12.0	11.0	1.0	4.6
	0.27	17.0	14.0	3.0	3.2
	0.54	15.5	12.0	3.5	4.0

AV apparent viscosity, *PV* plastic viscosity, *YP* yield point, *FL* filtration

20 min of stirring at a high rate. The results are shown in Table 2.

Table 2 shows that the drilling fluid filtration decreased with the increase in cationic degree. The CPFL product had a good filter reducing effect in the cationic degree ranging from 0.27 to 0.54 mmol/g. Drilling fluid filtration achieved a minimum at a cationic degree of 0.27 mmol/g.

4.2 Influence of cationic degree on drilling fluid salt resistance

A total of 1.0 % of CPFL sample with different cationic degrees was added to the base drilling mud containing 10 % of bentonite and stirred for 20 min, and then contaminated using 15 % of NaCl, after that $CaCl_2$ with different mass fractions of 1 %, 2 % or 3 % was added. After 20 min of stirring, the performance of the mixed drilling fluid system was evaluated, and the influence of the cationic degree on the performance of contaminated drilling fluids was measured as shown in Fig. 3.

The results showed that for the drilling fluid system contaminated by NaCl and $CaCl_2$ salts, the API filtration

loss would increase as the $CaCl_2$ concentration increased. When cationic degree reached 0.15 mmol/g, the drilling fluid filtration loss had a sharp decline; at the cationic degree of 0.27 mmol/g, it had its best fluid loss effect.

4.3 Analysis of action mechanism of cationic filtration reducer

The electrostatic attraction between anionic groups and cationic groups was enhanced by the addition of cations to the molecular chain of partially hydrolyzed sodium polyacrylonitrile, which in turn further enhances the electrostatic attraction between filtration reducer cationic groups and clay particle anionic groups. Therefore, a network structure was formed to a certain extent, and could improve the polymer's resistance to salt pollution and its ability to confine free water. The filtration-reducing effect significantly improved (Wang and Zhang 1995). But when the cationic degree is too big, with the enhancement of the inhibitory effect of cationic quaternary ammonium, flocculation capacity is too strong, the solid particles in the system become excessively big, the number of colloidal microparticles reduces, the density of filter cake becomes small, so the permeability increases, resulting in greater filtration.

5 Conclusion

Partially hydrolyzed sodium polyacrylonitrile was prepared through the hydrolytic reaction of acrylic fiber under alkaline conditions, which grafted cationic monomer to synthesize cation-modified polyacrylonitrile. The synthetic modification process is simple, easy to control, and can synthesize good drilling fluid loss additive cationic products. The characterization of the product molecular structure and the drilling fluid application performance indicated that if the cationic reaction and the cationic degree could be properly controlled, it would be applicable as

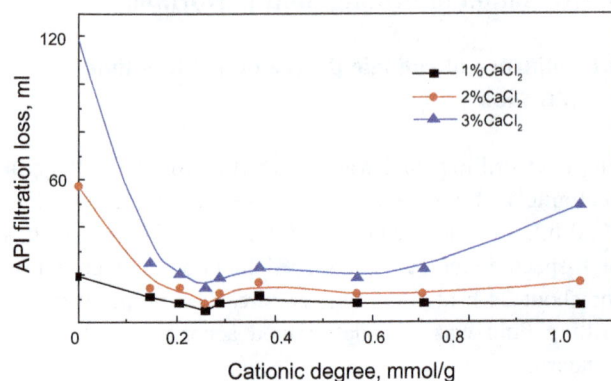

Fig. 3 Relationship between cationic degree and drilling fluid filtration loss at different concentrations of $CaCl_2$

a drilling fluid loss agent, and have good resistance to calcium chloride contamination.

Acknowledgments This work was supported by China National High Technology Research and Development Program (863 Program, 2013AA064803).

References

Ahmed MA, Ahmed AK, Mahmoud IA, et al. Investigation of some copolymers based on acrylic salts as circulation loss control agents. Egypt J Pet. 2013;22(4):481–91.

Cui YM, Zhu LJ. Study on the new technique for producing Na HPAN. Spec Petrochem. 1999;9:26–8.

Celik M, Qudrat ML, Akyuz E, et al. Synthesis and characterization of acrylic fibers-g-polyacrylamide. Fibers Polym. 2012;13(2):145–52.

Li G, Zhang SM. The synthesis and application of cationic polyelectrolyte with high density charge. China Environ Sci. 1999;2:145–8.

Ma XP, Zhao JB, Wang WX, et al. Acrylamide/trimethyl allyl ammonium chloride copolymer as heat and salts tolerant clay stabilizer: synthesis and properties. Oilfield Chem. 1995;12(3):197–200.

Tang SZ, Hu XQ, Ma XP, et al. Synthesis of monomer of *N,N*-dimethyl hydroxyethyl acrylamide and the preparation of its cationic copolymer with acrylamide. J Southwest Pet Inst. 1995;17(4):101–9.

Warren B, Van der Horst PM, Van T Zelfde TA. Quaternary nitrogen containing amphoteric water soluble polymers and their use in drilling fluids. 2000 WO: 0060023.

Wang ZH. The development overview on the domestic drilling fluids and drilling fluid additives. Sino-Glob Energy. 2013;18(10):34–41.

Wang B, Zhang Y. Application of small cationic polymer drilling fluid in Mao Ping 1 well deviated. Pet Drill Tech. 1995;23(1):27–9.

Xie BQ, Qiu ZS, Cao J. A novel hydrophobically modified polyacrylamide as a sealing agent in water-based drilling fluid. Pet Sci Technol. 2013;31(18):1866–72.

Yu PZ, Wang XY, Geng TM, et al. Study of cationic modified polyacrylonitrile waste fibers used in cationic polymer drilling fluid. Mod Chem Ind. 2004;24(10):31–4, 36.

Zhang JG, Deng H. Filtrate reducer with temperature and salt resistance used in MMH drilling fluid. Drill Fluid Complet Fluid. 1998;15(2):36–7.

Zlotnikov II, Khilo PA, Zubritskii MI. Polymer-based lubricating fluids for mechanical treatment of glass. J Frict Wear. 2006;27(6):63–5.

20

Pore-scale investigation of residual oil displacement in surfactant–polymer flooding using nuclear magnetic resonance experiments

Zhe-Yu Liu[1] · Yi-Qiang Li[1] · Ming-Hui Cui[1] · Fu-Yong Wang[1] · A. G. Prasiddhianti[1]

Abstract Research on the Gangxi III area in the Dagang Oilfield shows that there was still a significant amount of oil remaining in oil reservoirs after many years of polymer flooding. This is a potential target for enhanced oil recovery (EOR). Surfactant–polymer (SP) flooding is an effective chemical EOR method for mobilizing residual oil and improving displacement efficiency macroscopically, but the microscopic oil displacement efficiency in pores of different sizes is unclear. Nuclear magnetic resonance (NMR) is an efficient method for quantifying oil saturation in the rock matrix and analyzing pore structures. In this paper, the threshold values of different pore sizes were established from the relationship between mercury injection curves and NMR T_2 spectrums. The distribution and migration of residual oil in different flooding processes was evaluated by quantitatively analyzing the change of the relaxation time. The oil displaced from pores of different sizes after the water flood, polymer flood, and the SP flood was calculated, respectively. Experimental results indicate that (1) the residual oil in medium pores contributed the most to the incremental oil recovery for the SP flood, ranging from 40 % to 49 %, and small pores usually contributed <30 %; (2) the residual oil after the SP flood was mainly distributed in small and medium pores; the residual oil in medium pores accounted for 47.3 %–54.7 %, while that trapped in small pores was 25.7 %–42.5 %. The residual oil in small and medium pores was the main target for EOR after the SP flood in oilfields.

Keywords Nuclear magnetic resonance (NMR) · Surfactant–polymer (SP) flood · Residual oil distribution · Displacement mechanism · Core displacement test

1 Introduction

After over 30 years of water flooding and polymer flooding, the Dagang Oilfield is now a mature oil field with its water cut reaching the economic limit. However, as much as 70 % of the original oil in place (OOIP) may remain in the reservoir after secondary recovery processes (Sorbie 1991). A large portion of the residual oil is capillary trapped (Lake 1989). To tackle the residual oil saturation and revitalize this reservoir, a tertiary recovery is required. Surfactant–polymer (SP) flooding has been proved to be an efficient tertiary method for most major oilfields in China. To apply SP flooding in the Dagang Oilfield and optimize the process in heterogeneous reservoirs, it is necessary to predict the residual oil after the SP flood and identify the displacement efficiency in pores of different sizes.

Several experimental/numerical techniques have been proposed to measure or predict the residual oil distribution after displacement processes. For example, a widely used traditional method is to measure residual oil through analysis of cast thin sections of a reservoir core (Zao et al. 2009). This method damages reservoir cores while obtaining slices. Furthermore, fractured cores, unconsolidated sands, and mud cannot be cut into slices using this method. Another experimental technique is to use a micro-visualized model instead of a reservoir core to simulate a displacement process as well as the distribution of residual oil (Wang et al. 2010). However, this method does not take into account the influence of interstitial matter on the distribution of residual oil. X-ray computed tomography is

✉ Yi-Qiang Li
lyq89731007@163.com

[1] EOR Research Institute, China University of Petroleum, Beijing 102249, China

Edited by Yan-Hua Sun

often used to detect the rock matrix, but it is not sensitive to fluid changes (Vinegar 1986; de Argandona et al. 1999; Liu 2013). Numerical simulations require some assumptions in order to achieve mathematical completeness. In addition, reservoir parameters are uncertain and hard to determine. Therefore, there is usually a discrepancy between simulation results and actual conditions (Li et al. 2006). Nuclear magnetic resonance (NMR) is a quick, accurate, non-destructive, and widely used technology for core testing (Kleinberg and Vinegar 1996; Xie and Xiao 2007; Zhao et al. 2011). In NMR measurements, the received signals originate only from fluids in pores. To differentiate hydrocarbons from brine, brine is doped with paramagnetic ions to shield the signals from water, so that the signals only come from the oil. NMR T_2 relaxation time represents the fluid content in the pores of different sizes. The longer T_2 relaxation time corresponds to the larger pores, and vice versa (Williams et al. 1991; Gleeson et al. 1993; Cowan 1997). The residual oil distributions in pores of different sizes are quantified through the T_2 distribution analysis, and the accurate oil saturation can be calculated to investigate the oil movement in pores of different sizes. In our study, the NMR technique was used to evaluate the residual oil distribution after water flooding, polymer flooding, and SP flooding to study the "kickoff" mechanism of the residual oil.

2 NMR test principles

NMR is commonly used to image oil and water distributions in reservoir rocks by analyzing the relaxation time of reservoir fluids in the petroleum industry. NMR test signals come from hydrogen atoms, so more hydrogen atoms lead to stronger signals (Wang et al. 2001; Liu et al. 2004). However, it is difficult to distinguish signals from water and oil phases due to the presence of hydrogen atoms in both water and hydrocarbons. In order to separate oil-phase signals from mixed signals, the core slugs are doped in a paramagnetic solution (Guo and Gu 2005). Paramagnetic ions are able to diffuse into the core samples to shield NMR T_2 signals from the water phase.

Mn^{2+} ions, as paramagnetic ions in experiments, are able to penetrate into the water phase but not the oil phase. As a result, NMR T_2 signals from the water phase reduce below the dead time of NMR, while the signals from oil remain detectable without any loss (Kleinberg and Vinegar 1996; Gong et al. 2006). At the same time, the nonionic surfactant was chosen to have a large optimum salinity window and low susceptibility to divalent cations. Therefore, the addition of Mn^{2+} ions in the surfactant would not significantly affect the interfacial tension. Because adding Mn^{2+} ions makes the branched chains of polymer

molecules crinkle and decreases the viscosity of polymer solutions, different polymer concentrations were studied to avoid loss of mobility control at low polymer concentrations. The residual oil saturation in the core sample is closely related to NMR T_2 signals, and hence the change of the residual oil distribution after water flooding, polymer flooding, and SP flooding can be calculated and compared by analyzing NMR T_2 signals. The NMR T_2 relaxation distribution is an analytical method for analyzing the residual oil distribution. The relation between the NMR T_2 relaxation time and pore sizes is described by the following equation (Ausbrooks et al. 1999):

$$\frac{1}{T_2} = \rho_2 \frac{S}{V}, \qquad (1)$$

where T_2 is the NMR T_2 relaxation time, ρ_2 is the interfacial relaxivity determined by the mineral constituents and surficial properties of the pores, and $\frac{S}{V}$ is the specific pore surface area per volume which is inversely correlated with the pore diameter.

The total relaxation time is the sum of the relaxation time from the individual phases from different pore sizes:

$$S(t) = \Sigma A_i \exp(-t/T_{2i}), \qquad (2)$$

where $S(t)$ is a dimensionless parameter representing the total relaxation time; A_i is the ratio of the phase of which relaxation time is T_{2i} to the total, or the ratio of the pores which is represented by T_{2i} to the total pore volume.

According to this method, the NMR spectrums of cores after different displacement processes are shown in Fig. 1, taking Core 2 as an example. The relaxation time T_2 represents the different pore radii, while the signal amplitude represents the oil saturation. Based on the Eqs. (1) and (2) and combined with the mercury intrusion data, a relationship between the T_2 distribution and the distribution of different pore radii was established. According to the

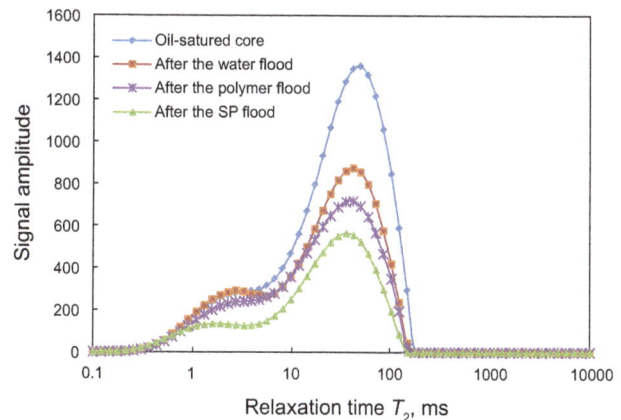

Fig. 1 NMR T_2 spectrums of Core 2 measured after different displacement processes

analysis, the small pores (less than 2 μm) were defined as the pores with T_2 less than 10 ms, the medium pores (2–10 μm) were defined as the pores with T_2 between 10 ms and 50 ms, and the large pores (more than 10 μm) were defined as the pores with T_2 longer than 50 ms.

3 Experimental method and procedures

3.1 Experimental materials and apparatus

Oil samples used were a mixture of crude oil from West Dagang III area and kerosene. It had a viscosity of 17 mPa s at 53 °C, similar to the crude oil at reservoir conditions. Brines were synthetic formation water prepared in the laboratory with a composition as listed in Table 1. Additional Mn^{2+} ions in the brines were the paramagnetic ions. Before use, brines were filtered by a 0.45 $μm^2$ filter membrane.

Partially hydrolyzed polyacrylamide (PHPA) had an average molecular weight of 25 million Daltons, a purity of over 88 %, and a degree of hydrolysis of 25 %. PHPA solutions of four concentrations were measured with a Brookfield DV II viscometer, and it was found that about 10 % of viscosity was lost after adding Mn^{2+} ions in PHPA solutions. DWS-3 nonionic surfactant (a mixed petroleum sulfonate surfactant) used in this study had an effective content of 40 %. No apparent change of the interfacial tension (IFT) was observed after adding Mn^{2+} ions to SP solutions. Experimental temperature was 53 °C (reservoir temperature). All these solutions were filtered by a 0.75 $μm^2$ sand core funnel before use. Natural cores were sampled from the West Dagang III Oilfield and shaped into cylinders with a diameter of 2.5 cm and a length of 6–8 cm. Some experimental parameters and displacement fluid characteristics are listed in Table 2.

A low-frequency NMR spectrometer, Reccore-04, developed by the Institute of Porous Fluid Mechanics at the Chinese Academy of Sciences was used to analyze the distribution of oil in cores after different displacement processes.

3.2 Surfactant–polymer solutions

Twelve different SP solutions were designed to investigate the effects of interfacial tension (IFT) reduction and mobility control on residual oil distribution. The viscosities of SP solutions were measured with a Brookfield DV II viscometer at a shear rate of 7.34 s^{-1}. The IFTs of SP solutions were measured with a TX500 spinning drop interface tensiometer at 5000 revolutions per minute (RPM) and 53 °C. The details of 12 SP solutions are listed in Table 2.

3.3 Experimental procedures

(1) The core sample was evacuated for 2 h to remove air and then fully saturated with brines. The core permeability to brine was evaluated.

(2) The core sample was oil flooded to residual water saturation (no further water production at the core outlet). Initial oil saturation distribution in the core was measured with an NMR spectrometer. After aging for 30 days, the core was water flooded until no further oil was produced at the core outlet and then the oil saturation distribution was tested again with the NMR spectrometer.

(3) The core was flooded with 3 pore volume (PV) of PHPA solutions with concentrations ranging from 1000 to 2500 mg/L (polymer displacement) and retested with the NMR spectrometer.

(4) The core was flooded with 3 PV of the SP solution (SP displacement), followed by water flooding until no further oil was produced. A final NMR test was then performed.

In core displacement tests, the injection rate was kept at 0.1 mL/min and the pressure during polymer displacement and SP displacement was controlled below 2.5 times of the water displacement pressure. A schematic of the core displacement system is shown in Fig. 2.

4 Results and discussion

4.1 Experimental results

Results of 12 core displacement tests are listed in Table 3. The overall oil recovery ranged from 52.6 % to 69.2 %. Oil recovery after the water flood was approximately 38.6 %–41.8 %, while an incremental oil recovery of 6.5 %–9.8 % over the water flood was achieved by the polymer flood and

Table 1 Ion composition of brine

Ion	$K^+ + Na^+$	Mg^{2+}	Ca^{2+}	HCO_3^-	CO_3^{2-}	Cl^-	Total salinity	Mn^{2+}	Cl^-	Total salinity with $MnCl_2$
Content, mg/L	2043	36	39	3126	135	1347	6726	1000	2000	9726

Table 2 Basic parameters of core samples and characteristics of SP displacement fluids

Core number	Porosity, %	Gas permeability, 10^{-3} μm^2	Oil saturation, %	Polymer concentration, mg/L	Surfactant concentration, %	Viscosity of the SP solution, mPa s	IFT, mN/m
1	29.6	636	73.8	1000	0.15	11.73	1.49×10^{-2}
2	30.0	629	72.8	1500	0.15	39.67	2.03×10^{-2}
3	29.6	638	72.9	2000	0.15	71.23	3.97×10^{-2}
4	28.8	624	72.5	2500	0.15	117.33	5.56×10^{-2}
5	29.2	635	72.6	1000	0.25	10.67	5.28×10^{-3}
6	29.4	632	72.8	1500	0.25	38.40	7.47×10^{-3}
7	29.9	626	73.7	2000	0.25	70.40	8.92×10^{-3}
8	30.0	630	73.0	2500	0.25	115.20	1.53×10^{-2}
9	29.1	628	73.9	1000	0.30	10.67	3.04×10^{-3}
10	29.6	637	73.9	1500	0.30	38.40	5.53×10^{-3}
11	30.0	636	73.9	2000	0.30	70.40	7.80×10^{-3}
12	29.0	627	72.6	2500	0.30	115.20	1.08×10^{-2}

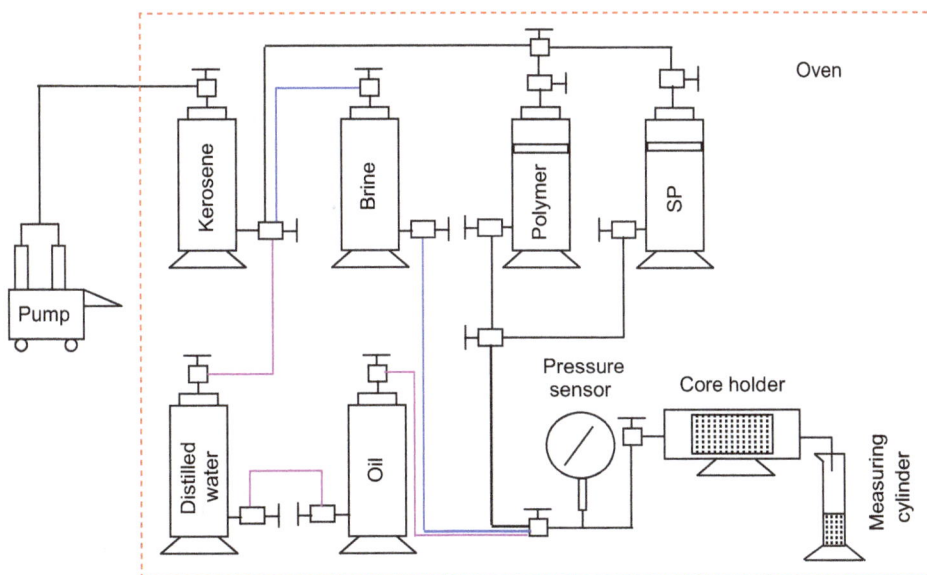

Fig. 2 Schematic illustration of the core displacement system

4.5 %–21.5 % over the polymer flood achieved by the subsequent SP flood. As shown in Table 3, with a constant surfactant concentration in the case of SP flood, the incremental oil recovery increased when the polymer concentration changed from 1000 mg/L to 2500 mg/L. The ultimate incremental oil recovery had a positive correlation with the polymer concentration.

Xia et al. (2006) proposed that a reduction in residual oil saturation by polymer flooding might be owing to the viscoelasticity of a polymer solution. When the polymer viscoelasticity increased with an increase in its concentration, the interaction between the polymer solution and the residual oil increased, and more residual oil would be displaced from reservoir pores. However, after the polymer concentration exceeded a limit, the inaccessible pore volume would increase and the incremental oil recovery owing to additional increase in the polymer concentration would be insignificant (Guo et al. 2014). The sweep efficiency by the polymer flood may even decrease due to incompatibility between the polymer molecules and the pore size.

Figure 3 indicates that for a constant polymer concentration, an incremental oil recovery of 7.5 %–10.5 % was achieved after the SP flood when the surfactant concentration changed from 0.15 % to 0.25 % (curves in Fig. 3 are steep when the surfactant concentration is low). This

Table 3 Data on 12 core displacement tests

Core number	Oil recovery by the water flood, %	Incremental oil recovery by the polymer flood, %	Incremental oil recovery by the SP flood, %	Overall oil recovery, %
1	40.4	7.7	4.5	52.6
2	39.8	8.0	5.8	53.6
3	41.0	6.5	10.1	57.6
4	41.8	7.0	10.4	59.2
5	39.1	8.1	12.0	59.1
6	40.6	9.1	14.3	64.0
7	39.2	8.0	18.9	66.0
8	39.4	8.1	20.9	68.4
9	41.5	8.8	13.0	63.2
10	39.1	9.8	15.7	64.6
11	41.7	6.8	19.2	67.6
12	38.6	9.1	21.5	69.2

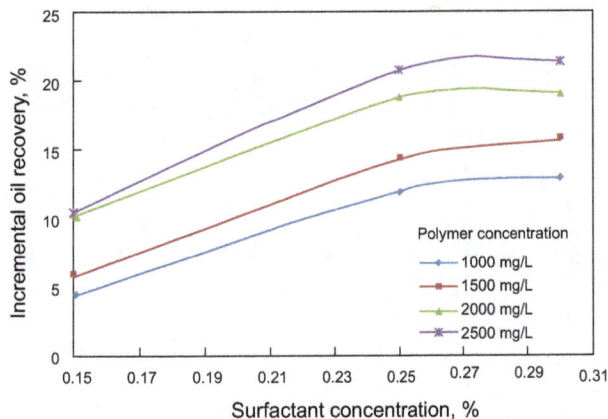

Fig. 3 Incremental oil recovery as a function of surfactant concentration for SP displacement processes of different polymer concentrations

increment was due to an increase in the surfactant concentration. However, when the surfactant concentration changed from 0.25 % to 0.30 %, the incremental oil recovery increased slowly from 0.3 % to 1.4 % (curves in Fig. 3 are relatively flat when the surfactant concentration is high).

From Tables 2 and 3, we concluded that the residual oil was obviously removed by the displacement solution when the IFT was lowered to 10^{-3} mN/m. The overall oil recovery had a positive correlation with the surfactant concentration. The addition of surfactant would reduce the interfacial tension between crude oil and water, increase the capillary number, and thus decrease the adhesive force between oil and rock surfaces (Shi et al. 2012). On the other hand, when the injected surfactant solution contacted crude oil, an emulsion was formed, in which oil was the

dispersed phase. Oil was removed, as the emulsion was displaced through the formation, rather than tending to adhere to pore surfaces, resulting in an incremental oil recovery. In summary, the incremental oil recovery was attributed to the synergism between polymer and surfactant (Kendall and Martin 2007; Lai et al. 2014).

4.2 Quantitative calculations of oil recovery from different pores

With the correlation between pore throat sizes and NMR T_2 time distribution (Li et al. 2008), recovery factors by the water flood, polymer flood, and the SP flood from pores of different sizes were calculated and are listed in Table 4. The recovery factor is defined as the ratio of the volume of oil displaced from pores of a specific size to the total volume of oil trapped in the pores of the same size.

Table 4 shows that the oil was first mobilized from pores with relatively large pore sizes by the water flood, rather than that trapped in small or medium pores. During polymer flooding, the mobility of the displacing phase was controlled by the polymer. The residual oil in small pores could be "kicked off" due to the pressure increase. However, with an increase in the polymer concentration, the inaccessible pore volume increased and the polymer could not enter into small pores. Consequently, when the polymer concentration increased, the oil recovery from small pores reduced. During SP flooding, the surfactant was used to reduce IFT and the polymer could control mobility of the displacing fluid. As a result, both the displacement efficiency and the sweep efficiency were improved for naturally heterogeneous cores.

Figure 4 shows that the recovery factor from small pores after the SP flood decreased with an increase in the

Table 4 Recovery factors of oil from pores of different sizes after the water flood, polymer flood, and the SP flood, respectively

Core number	Polymer concentration, mg/L	Surfactant concentration, %	Recovery factor by the water flood, %			Recovery factor by the polymer flood, %			Recovery factor by the SP flood, %		
			Small pores	Medium pores	Large pores	Small pores	Medium pores	Large pores	Small pores	Medium pores	Large pores
1	1000	0.15	8.4	41.5	59.4	21.0	5.8	5.1	16.7	3.1	4.4
2	1500		8.1	40.8	59.8	20.0	6.3	6.1	15.7	4.8	5.6
3	2000		8.3	41.1	60.6	14.1	6.5	9.0	13.8	9.9	10.1
4	2500		8.5	41.5	61.1	12.3	7.4	9.6	11.5	10.7	12.2
5	1000	0.25	8.0	40.9	60.2	20.5	6.0	5.4	32.9	9.7	10.3
6	1500		8.6	42.1	61.2	20.1	6.8	6.2	30.5	12.0	14.1
7	2000		8.3	40.7	60.2	13.2	7.1	8.9	24.7	18.8	17.1
8	2500		8.2	40.8	61.2	12.0	7.9	9.9	19.1	22.0	18.2
9	1000	0.30	8.1	41.1	62.0	21.0	6.2	5.4	33.8	11.2	11.8
10	1500		8.2	40.9	60.2	18.2	6.9	6.0	30.6	12.7	14.6
11	2000		8.1	42.0	62.8	13.9	7.2	8.4	27.2	19.9	17.9
12	2500		7.9	41.8	60.8	11.9	8.0	9.8	23.5	22.9	18.5

polymer concentration when the surfactant concentration was fixed. This phenomenon was due to the incompatibility between the polymer molecular size and the pore size (Lu et al. 2009; Li et al. 2014; Yin et al. 2014). However, when the polymer concentration was fixed in the SP flood, the IFT reduced and the capillary number increased with an increase in the surfactant concentration, so the residual oil saturation reduced and the recovery factor from small pores enhanced.

Figure 5 shows that the recovery factor from medium pores after the SP flood increased significantly with an increase in the polymer concentration when the surfactant concentration was fixed. This improvement of oil recovery was due to the increased viscosity or viscoelasticity of the SP solution.

Figure 6 shows that the recovery factor from large pores increased gradually with an increase in the polymer concentration when the surfactant concentration was fixed

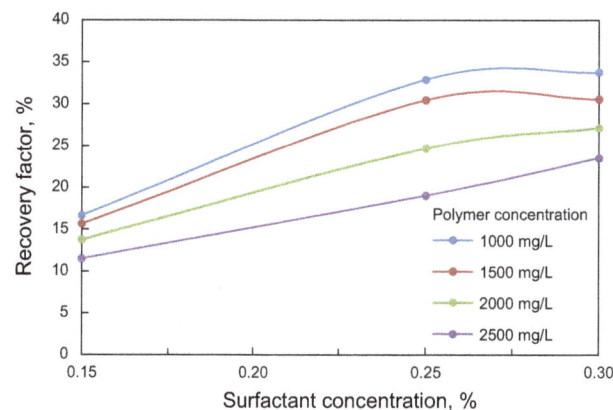

Fig. 4 Recovery factor of oil from small pores after SP floods

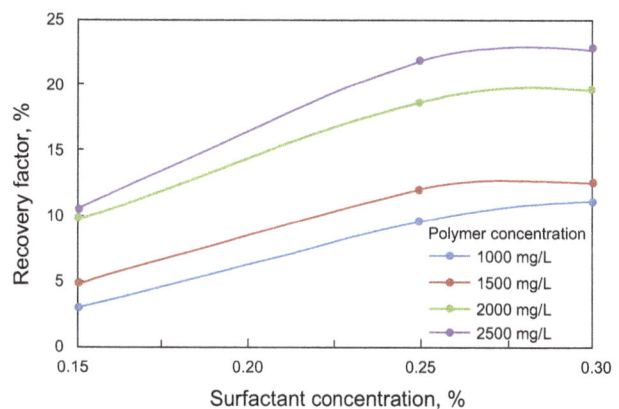

Fig. 5 Recovery factor of oil from medium pores after SP floods

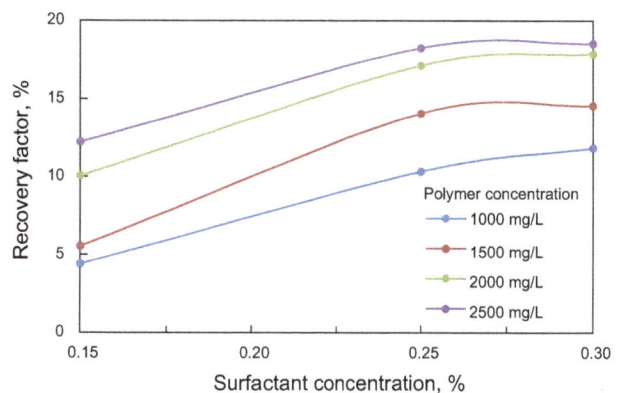

Fig. 6 Recovery factor of oil from large pores after SP floods

during SP flooding. However, the increasing trend became less steep after the polymer concentration increased to 2000 mg/L.

We concluded from Figs. 5 and 6 that if the polymer concentration was fixed the oil recovery from medium or large pores was enhanced significantly when the surfactant concentration increased. After the surfactant concentration increased to 0.25 %, the increasing trend became less steep.

4.3 Contribution of different pores to oil recovery during SP flooding

As shown in Table 5, with the surfactant concentration fixed at 0.15 %, when the polymer concentration increased the large pore contribution did not change significantly, and the medium pore contribution increased while the small pore contribution decreased. The reason was that there was little residual oil in large pores. Therefore, if the IFT could not be reduced, the residual oil would be very difficult to displace. In this case, the flow resistance increased as polymer molecules entered large pores, thus the SP solution was diverted into medium pores, which led to an increase in the sweep efficiency in medium pores (Dang et al. 2011).

When the surfactant concentration was fixed at 0.25 % or 0.30 %, the contribution from medium and large pores increased with an increase in the polymer concentration, especially medium pores, but the small pore contribution decreased. The reason was that the hydrodynamic size of the polymer was larger than the size of small pores. Therefore, polymer molecules could not penetrate into small pores (Pye 1964; Gogarty 1967; Gogarty et al. 1970; Lu and Gao 1996). The synergy effect of surfactant and polymer also influenced IFT and emulsification. The flow resistance was reduced due to the reduction in IFT achieved by the SP system. On the other hand, the

Table 6 Distribution of residual oil in different pores after the SP flood

Core number	Polymer concentration, mg/L	Surfactant concentration, %	Residual oil in different pores, %[a]		
			Small pores	Medium pores	Large pores
1	1000	0.15	12.8	26.0	9.3
2	1500		12.5	24.8	9.2
3	2000		12.9	21.7	7.9
4	2500		13.8	19.5	7.5
5	1000	0.25	10.3	19.8	10.5
6	1500		10.7	18.3	7.1
7	2000		11.7	17.4	4.8
8	2500		13.4	14.9	3.2
9	1000	0.30	10.9	18.8	7.0
10	1500		10.6	18.2	6.5
11	2000		11.3	16.8	4.2
12	2500		13.0	14.7	3.1

[a] The ratio of the volume of residual oil in the specific pores after the SP flood to the total volume of oil in all pores before the water flood

stable emulsion slug increased the pressure gradient in the pores. Some residual oil was carried away with the emulsion. In summary, the medium pore contribution after the SP flood was the highest, ranging from 40 % to 49 %. Small and large pore contributions were relatively lower.

4.4 Residual oil in pores

As shown in Table 6, most of residual oil was found in medium pores, followed by small pores and large pores. When the polymer concentration was fixed, the recovery from all pores of different sizes increased significantly. When the surfactant concentration was fixed, residual oil in the large and medium pores was reduced with increasing polymer concentration. Meanwhile, the residual oil in small pores increased slightly. Therefore, the IFT should be reduced to ultralow (10^{-3} mN/m) level to improve displacement efficiency in small pores.

5 Conclusions

(1) For an SP flood with a fixed surfactant concentration, when the polymer concentration increased, the oil recovery from small pores was reduced due to the incompatibility between the polymer molecule size and the small pore size. Meanwhile, the oil recovery from the medium and large pores increased due to increased viscosity or viscoelasticity of the SP solution.

Table 5 Recovery contribution of different pores during SP flooding

Core number	Polymer concentration, mg/L	Surfactant concentration, %	Contribution to oil recovery in the SP flood, %		
			Small pores	Medium pores	Large pores
1	1000	0.15	58.4	18.8	24.9
2	1500		53.3	23.0	23.6
3	2000		36.8	37.8	25.4
4	2500		26.7	49.6	23.7
5	1000	0.25	36.3	39.6	24.1
6	1500		32.4	40.7	26.9
7	2000		27.0	42.8	30.2
8	2500		19.7	49.6	30.6
9	1000	0.30	35.9	40.3	23.7
10	1500		34.8	42.8	29.9
11	2000		29.5	43.2	29.4
12	2500		20.6	49.6	29.9

(2) For an SP flood with a fixed polymer concentration, when the surfactant concentration increased, the oil recovery from all pore sizes increased significantly. When the surfactant concentration reached 0.25 %, the incremental oil recovery became negligible. After the IFT was reduced to an ultralow level of about 10^{-3} mN/m, the oil recovery could not be improved significantly by increasing the surfactant concentration.

(3) After the polymer and surfactant concentrations increased to critical values, additional increases in their concentrations led to a slight improvement in the overall oil recovery. For economic reasons, an SP system of 2000 mg/L polymer and 0.25 % surfactant was recommended.

(4) Contribution from medium pores was the highest among all the three groups of pores, accounting for about 40 %–49 %, during SP flooding. Although the SP system mainly displaced the residual oil in medium pores, its effect on the residual oil in small and large pores cannot be neglected. After the SP flood, the residual oil was mainly trapped in small and medium pores, so the oil in small and medium pores was the main target for enhanced oil recovery after the SP flood.

Acknowledgments The authors would like to thank Scientific Research Foundation of China University of Petroleum, Beijing (No. 2462013YJRC033) and (No. 01JB0177) for funding this research.

References

Ausbrooks R, Hurley NF, May A, et al. Pore-size distribution in vuggy carbonates from core images, NMR, and capillary pressure. In: SPE annual technical conference and exhibition, 3–6 October, Houston, Texas; 1999. doi:10.2118/56506-MS.

Cowan B. Nuclear magnetic resonance and relaxation. New York: Cambridge University Press; 1997.

Dang CTQ, Chen ZJ, Nguyen NTB, et al. Development of isotherm polymer/surfactant adsorption models in chemical flooding. In: SPE Asia Pacific oil and gas conference and exhibition, 20–2 September, Jakarta, Indonesia; 2011. doi:10.2118/147872-MS.

de Argandona VGR, Rey AR, Celorio C, et al. Characterization by computed X-ray tomography of the evolution of the pore structure of a dolomite rock during freeze-thaw cyclic tests. Phys Chem Earth Part A. 1999;24(7):633–7. doi:10.1016/S1464-1895(99)00092-7.

Gleeson JW, Woessner DE, Jordan CF Jr. NMR imaging of pore structures in limestones. SPE Formation Evalu. 1993;8(2):123–7. doi:10.2118/20493-PA.

Gogarty WB. Mobility control with polymer solutions. SPE J. 1967;7(2):161–73. doi:10.2118/1566B-PA.

Gogarty WB, Meabon HP, Milton HW Jr. Mobility control design for miscible-type waterfloods using micellar solutions. J Pet Technol. 1970;22(2):141–7. doi:10.2118/1847-E-PA.

Gong G, Sun B, Liu M, et al. NMR relaxation of the fluid in rock porous media. Chin J Magn Reson. 2006;23(3):380–94 (in Chinese).

Guo A, Geng Y, Zhao L, et al. Preparation of cationic polyacrylamide microsphere emulsion and its performance for permeability reduction. Pet Sci. 2014;11(3):408–16. doi:10.1007/s12182-014-0355-0.

Guo G, Gu C. Oil saturation in rock cuttings measured by nuclear magnetic resonance. Chin J Magn Reson. 2005;22(1):67–72 (in Chinese).

Kendall TA, Martin ST. Water-induced reconstruction that affects mobile ions on the surface of calcite. J Phys Chem A. 2007;111(3):505–14. doi:10.1021/jp0647129.

Kleinberg R, Vinegar HJ. NMR properties of reservoir fluids. Log Anal. 1996;37(6):20–32.

Lai N, Zhang X, Ye Z, et al. Laboratory study of an anti-temperature and salt-resistance surfactant-polymer binary combinational flooding as EOR chemical. J Appl Polym Sci. 2014. doi:10.1002/app.39984.

Lake LW. Enhanced oil recovery. Englewood Cliffs: Prentice Hall; 1989.

Li H, Liu Q, Wen C, et al. Residual oil distribution and potential tapping study. Spec Oil Gas Reserv. 2006;13(3):8–11 (in Chinese).

Li H, Zhu J, Guo H. Methods for calculating pore radius distribution in rock from NMR T_2 spectra. Chin J Magn Reson. 2008;25(1):67–72 (in Chinese).

Li Y, Gao J, Yin D, et al. Study of the matching relationship between polymer hydrodynamic characteristic size and pore throat radius of target blocks based on the micro porous membrane filtration method. J Chem. 2014. doi:10.1155/2014/569126.

Liu H. Study of meso-structure and damage mechanical characteristics of frozen rock based on CT image processing. Xi'an: Xi'an University of Science and Technology; 2013 (in Chinese).

Liu Y, Liu Y, Sui X. Characteristics of pore structure of transitional zone of Saertu central area in Daqing Oilfield. J Daqing Pet Inst. 2004;28(1):109–11 (in Chinese).

Lu G, Zhang X, Shao C, et al. Molecular dynamics simulation of adsorption of an oil-water-surfactant mixture on calcite surface. Pet Sci. 2009;6(1):76–81. doi:10.1007/s12182-009-0014-z.

Lu X, Gao Z. Pore throat radius to coil gyration radius ratio as characteristic of adaptively of polymer molecular mass to core permeability. Oilfield Chem. 1996;13(1):72–5 (in Chinese).

Pye DJ. Improved secondary recovery by control of the water mobility. J Pet Technol. 1964;16(8):911–6. doi:10.2118/845-PA.

Shi L, Chen L, Ye Z, et al. Effect of polymer solution structure on displacement efficiency. Pet Sci. 2012;9(2):230–5. doi:10.1007/s12182-012-0203-z.

Sorbie KS. Polymer-improved oil recovery. Baca Raton: CRC Press; 1991.

Vinegar HJ. X-ray CT and NMR imaging of rocks. J Pet Technol. 1986;38(3):257–9. doi:10.2118/15277-PA.

Wang W, Guo H, Ye C. Experimental studies of NMR properties of continental sedimentary rocks. Chin J Magn Reson. 2001;18(2):113–21 (in Chinese).

Wang Y, Zhao F, Bai B, et al. Optimized surfactant IFT and polymer viscosity for surfactant-polymer flooding in heterogeneous formations. In: SPE improved oil recovery symposium, 24–8 April, Tulsa, Oklahoma; 2010. doi:10.2118/127391-MS.

Williams JLA, Taylor DG, Maddinelli G, et al. Visualisation of fluid displacement in rock cores by NMR imaging. Magn Reson

Imaging. 1991;9(5):767–73. doi:10.1016/0730-725X(91)90374-U.

Xia H, Wang D, Wang G, et al. Elastic behavior of polymer solution to residual oil at dead-ends. Acta Petrolei Sinica. 2006;27(02):72–6 (in Chinese).

Xie R, Xiao L. Dispersion properties of NMR relaxation for crude oil. Pet Sci. 2007;4(2):35–8. doi:10.1007/BF03187439.

Yin D, Li Y, Chen B, et al. Study of compatibility of polymer hydrodynamic size and pore throat size for Honggang reservoir. Int J Polym Sci. 2014. doi:10.1155/2014/729426.

Zao M, Guo Z, Qing H, et al. Identification of thin rock cast section and application of microscopic image analysis technology. West-China Explor Eng. 2009;3:66–7 (in Chinese).

Zhao Y, Song Y, Liu Y, et al. Visualization of CO_2 and oil immiscible and miscible flow processes in porous media using NMR micro-imaging. Pet Sci. 2011;8(2):183–93. doi:10.1007/s12182-011-0133-1.

PERMISSIONS

All chapters in this book were first published in PS, by Springer; hereby published with permission under the Creative Commons Attribution License or equivalent. Every chapter published in this book has been scrutinized by our experts. Their significance has been extensively debated. The topics covered herein carry significant findings which will fuel the growth of the discipline. They may even be implemented as practical applications or may be referred to as a beginning point for another development.

The contributors of this book come from diverse backgrounds, making this book a truly international effort. This book will bring forth new frontiers with its revolutionizing research information and detailed analysis of the nascent developments around the world.

We would like to thank all the contributing authors for lending their expertise to make the book truly unique. They have played a crucial role in the development of this book. Without their invaluable contributions this book wouldn't have been possible. They have made vital efforts to compile up to date information on the varied aspects of this subject to make this book a valuable addition to the collection of many professionals and students.

This book was conceptualized with the vision of imparting up-to-date information and advanced data in this field. To ensure the same, a matchless editorial board was set up. Every individual on the board went through rigorous rounds of assessment to prove their worth. After which they invested a large part of their time researching and compiling the most relevant data for our readers.

The editorial board has been involved in producing this book since its inception. They have spent rigorous hours researching and exploring the diverse topics which have resulted in the successful publishing of this book. They have passed on their knowledge of decades through this book. To expedite this challenging task, the publisher supported the team at every step. A small team of assistant editors was also appointed to further simplify the editing procedure and attain best results for the readers.

Apart from the editorial board, the designing team has also invested a significant amount of their time in understanding the subject and creating the most relevant covers. They scrutinized every image to scout for the most suitable representation of the subject and create an appropriate cover for the book.

The publishing team has been an ardent support to the editorial, designing and production team. Their endless efforts to recruit the best for this project, has resulted in the accomplishment of this book. They are a veteran in the field of academics and their pool of knowledge is as vast as their experience in printing. Their expertise and guidance has proved useful at every step. Their uncompromising quality standards have made this book an exceptional effort. Their encouragement from time to time has been an inspiration for everyone.

The publisher and the editorial board hope that this book will prove to be a valuable piece of knowledge for researchers, students, practitioners and scholars across the globe.

LIST OF CONTRIBUTORS

Qian-Qian Song, Qing-Zhe Jiang and Zhao-Zheng Song
State Key Laboratory of Heavy Oil Processing, China University of Petroleum, Beijing 102249, China

Taraneh Jafari Behbahani
Research Institute of Petroleum Industry (RIPI), P.O.Box 14665-1998, Tehran, Iran

Hong-Qi Liu and You-Ming Deng
State Key Laboratory of Oil and Gas Reservoir Geology and Exploitation, Southwest Petroleum University, Chengdu 610500, Sichuan, China

Yan Jun
Centrica Energy (E&P) Upstream Kings Close, 62 Huntly Street, Aberdeen AB 101 RS, UK

Zheng-Xiang Lü and Xiang Yang
State Key Laboratory of Oil and Gas Reservoir Geology & Exploitation, Chengdu University of Technology, Chengdu 610059, Sichuan, China

Su-Juan Ye
Exploration and Production Research Institute, Sinopec Southwest, Chengdu 610041, Sichuan, China

Rong Li
Chengdu Institute of Geology and Mineral Resources, Chengdu 610081, Sichuan, China

Yuan-Hua Qing
State Key Laboratory of Oil and Gas Reservoir Geology & Exploitation, Chengdu University of Technology, Chengdu 610059, Sichuan, China
PetroChina Tarim Oilfield Company, Korla 841000, Xinjiang, China

Shi-Cheng Zhang, Xin Lei, Yu-Shi Zhou and Guo-Qing Xu
College of Petroleum Engineering, University of Petroleum, Beijing 102249, China

Wen-Tao Li
School of Energy Resources, China University of Geosciences, Beijing 100083, China
Geoscience Research Institute, Shengli Oilfield Company SINOPEC, Dongying 257015, Shandong, China

Yang Gao and Chun-Yan Geng
Geoscience Research Institute, Shengli Oilfield Company SINOPEC, Dongying 257015, Shandong, China

Mahmoud O. Elsharafi
McCoy School of Engineering, Midwestern State University, Wichita Falls, TX 76308, USA

Baojun Bai
Department of Geological Sciences & Engineering, Missouri University of Science and Technology, Rolla, MO 65409, USA

De-Li Gao and Wen-Jun Huang
MOE Key Laboratory of Petroleum Engineering, China University of Petroleum, Beijing 102249, China

Xiao-Hua Che, Wen-Xiao Qiao, Xiao-Dong Ju, Jun-Qiang Lu, Jin-Ping Wu and Ming Cai
State Key Laboratory of Petroleum Resources and Prospecting, China University of Petroleum, Beijing 102249, China
Key Laboratory of Earth Prospecting and Information Technology, China University of Petroleum, Beijing 102249, China

Roman Pogreb
Physics Faculty, Ariel University, 40700 Ariel, Israel

Edward Bormashenko
Physics Faculty, Ariel University, 40700 Ariel, Israel
Department of Chemical Engineering and Biotechnology, Ariel University, 40700 Ariel, Israel

Hadas Aharoni
Department of Chemical Engineering and Biotechnology, Ariel University, 40700 Ariel, Israel

Revital Balter
Department of Chemical Engineering and Biotechnology, Ariel University, 40700 Ariel, Israel
Department of Chemistry, Bar-llan University, 52900 Ramat Gan, Israel

Doron Aurbach
Department of Chemistry, Bar-llan University, 52900 Ramat Gan, Israel

Vladimir Strelnikov
Institute of Technical Chemistry of Ural Division of Russian Academy of Science, Perm, Russia

Ping-Sheng Wei and Ming-Jun Su
Petrochina Research Institute of Petroleum Exploration & Development-Northwest, Lanzhou 730020, Gansu, China

Pu Zhang, Sheng Wang, Kai-Bo Zhou and Li Kong
School of Automation, Huazhong University of Science and Technology, Wuhan 430074, Hubei, China

Hua-Xiu Zeng
China Petroleum Logging Co., Ltd, Xi'an 710077, Shaanxi, China

Majid Goodarzi
Department of Civil Engineering and Geoscience, University of Newcastle, Newcastle upon Tyne NE1 7RU, UK

Soheil Mohammadi
School of Civil Engineering, University of Tehran, Tehran 11365-4563, Iran

Ahmad Jafari
School of Mining Engineering, University of Tehran, Tehran 515-14395, Iran

Jian Hou
State Key Laboratory of Heavy Oil Processing, China University of Petroleum, Qingdao 266580, Shandong, China

Jian Hou and Kang Zhou
School of Petroleum Engineering, China University of Petroleum, Qingdao 266580, Shandong, China

Xian-Song Zhang and Xiao-Dong Kang
CNOOC Research Institute, Beijing 100027, China

Hai Xie
ZTE Corporation, Shenzhen 518055, Guangdong, China

Ming-De Yang
Chemical Institute of Chemical Industry, Gannan Normal University, Jiangxi 341000, China

De-Run Hua
Chemical Institute of Chemical Industry, Gannan Normal University, Jiangxi 341000, China
Institutes of Nuclear and New Energy Technology, Tsinghua University, Beijing 100084, China

Yun-Feng Liu, Yu Chen, Xin-Ning Lu and Jian Li
Institutes of Nuclear and New Energy Technology, Tsinghua University, Beijing 100084, China

Yu-Long Wu
Institutes of Nuclear and New Energy Technology, Tsinghua University, Beijing 100084, China
Beijing Engineering Research Center for Biofuels, Tsinghua University, Beijing 100084, China

Zhi-Guo Mao, Ru-Kai Zhu, Jing-Hong Wang, Ling Su and Shao-Min Zhang
Research Institute of Petroleum Exploration & Development, Petro China, Beijing 100083, China
State Key Laboratory of Enhanced Oil Recovery, Beijing 100083, China

Jing-Lan Luo
Northwest University, Xi'an 710069, Shaanxi, China
State Key Laboratory of Continental Dynamics, Xi'an 710069, Shaanxi, China

Zhan-Hai Du
International Iraq FZE, PetroChina, Beijing 100034, China

Shao-Min Zhang
China University of Petroleum, Qingdao 266580, Shandong, China

Lei-Ting Shi, Jian Zhang, Xin-Sheng Xue and Wei Zhou
State Key Laboratory of Offshore Oil Exploitation, Beijing 100027, China

Lei-Ting Shi, Shi-Jie Zhu and Zhong-Bin Ye
State Key Laboratory of Oil and Gas Reservoir Geology and Exploitation, Southwest Petroleum University, Chengdu 610500, Sichuan, China

Song-Xia Wang
Research Institute of Exploration and Development, PetroChina Southwest Oil & Gas Field Company, Chengdu 610041, Sichuan, China

Cun-Ge Liu
State Key Laboratory of Oil and Gas Reservoir Geology and Exploitation, Chengdu University of Technology, Chengdu 610059, Sichuan, China
Northwest Oilfield Company, China Petroleum & Chemical Corporation, Urumqi 830011, Xinjiang, China

Li-Xin Qi, Yong-Li Liu, Ming-Xia Luo, Xiao-Ming Shao and Peng Luo
Northwest Oilfield Company, China Petroleum & Chemical Corporation, Urumqi 830011, Xinjiang, China

Zhi-Li Zhang
Institute of Petroleum Exploration and Development, China Petroleum & Chemical Corporation, Beijing 100083, China

Pei-Zhi Yu
School of Engineering and Technology, China University of Geosciences (Beijing), Beijing 100083, China

Zhe-Yu Liu, Yi-Qiang Li, Ming-Hui Cui, Fu-Yong Wang and A. G. Prasiddhianti
EOR Research Institute, China University of Petroleum, Beijing 102249, China

Index

www.ingramcontent.com/pod-product-compliance
Lightning Source LLC
Chambersburg PA
CBHW080625200326
41458CB00013B/4513